T0319394

RADIO RESOURCE MANAGEMENT IN MULTI-TIER CELLULAR WIRELESS NETWORKS

Wiley Series on
Adaptive and Cognitive Dynamic Systems

Editor: Simon Haykin

A complete list of titles in this series appears at the end of this volume.

RADIO RESOURCE MANAGEMENT IN MULTI-TIER CELLULAR WIRELESS NETWORKS

EKRAM HOSSAIN

University of Manitoba
Canada

LONG BAO LE

INRS-EMT
Quebec, Canada

DUSIT NIYATO

Nanyang Technological University
Singapore

Published by John Wiley & Sons, Inc., Hoboken, New Jersey.
Published simultaneously in Canada.

For general information on our other products and services or for technical support, please contact our
Customer Care Department within the United States at (800) 762-2974, outside the United States at
(317) 572-3993 or fax (317) 572-4002.

Wiley also publishes its books in a variety of electronic formats. Some content that appears in print may
not be available in electronic formats. For more information about Wiley products, visit our web site at
www.wiley.com.

Library of Congress Cataloging-in-Publication Data:

Hossain, Ekram, 1971–
 Radio resource management in multi-tier cellular wireless networks / Ekram Hossain,
Long Bao Le, and Dusit Niyato. – 1st edition.
  pages cm
 Includes index.
 ISBN 978-1-118-50267-9 (hardback)
 1. Wireless communication systems. 2. Femtocells. 3. Radio resource management
(Wireless communications)   I. Niyato, Dusit.   II. Le, Bao Long, 1976– III. Title.
 TK5103.2.H666 2013
 621.3845′6–dc23

                                                                                    2013016956

Printed in the United States of America.

10 9 8 7 6 5 4 3 2 1

For our families

CONTENTS IN BRIEF

CONTENTS

PREFACE

The phenomenal increase in mobile data traffic and the high data rate and improved quality-of-service (QoS) user requirements has created a huge demand for network capacity in the cellular wireless networks. Multi-tier cellular wireless networks, consisting of macrocells overlaid with "small cells," will provide a fast, flexible, cost-efficient solution to satisfy this capacity demand.

"Small cell" is an umbrella term for low-power radio access nodes that operate in both licensed and unlicensed spectra and have a range of ten to several hundred meters. These contrast with a typical mobile macrocell, which might have a range of up to several kilometers or even higher. The term "small cell" encompasses femtocells, picocells, microcells, and metrocells. The multi-tier cellular wireless networks, including macrocells and small cells of all types (which are also referred to as heterogeneous networks (HetNets) or small cell networks (SCNs)), are expected to provide improved spectrum efficiency (bps/Hz/km^2), capacity, and coverage in future wireless networks.

Small cells can support wireless applications for homes and enterprises as well as metropolitan and rural public spaces. Small cell technology is applicable to the entire range of licensed spectrum mobile technologies, such as those standardized by the 3GPP, the 3GPP2, and the WiMAX forum. When compared with unlicensed small cells (e.g., Wi-Fi), small cells operating in the licensed band (i.e., licensed small cells) provide support for legacy handsets, operator managed QoS, seamless continuity with the macro networks through better support for mobility/handoff, and improved security.

Deployment of small cells poses many challenges, among which the radio resource management (i.e., interference management, admission control, load balancing) is the most significant. The aim of this book is to provide an in-depth overview of the radio resource management problem in multi-tier networks considering both code-division multiple access (CDMA)-based (e.g., 3G) and orthogonal frequency-division multiple access (OFDMA)-based (e.g., LTE, WiMAX) small cells, and the state-of-the-art research on this problem.

The book consists of ten chapters. In Chapter 1, after a brief overview of the multi-tier cellular networks, LTE-Advanced networks, LTE, and 3G small cells (femtocells, in particular), we outline the major challenges in the successful deployment of small cells in next generation cellular wireless systems. In particular, the challenges related to resource allocation, co-tier and cross-tier interference management and admission control, mobility and handoff management, auto-configuration, timing and synchronization, and security are discussed.

In Chapter 2, after discussing the design issues for resource allocation in multi-tier networks, we provide a comprehensive overview of state-of-the-art techniques for resource allocation and interference management in small cell networks. In particular, several concepts for resource allocation and interference management in two-tier macrocell-femtocell networks, including the femto-aware spectrum arrangement approach, the graph-based clustering approach, the adaptive power control approach, the transmit beamforming approach, the collaborative frequency scheduling approach, the cognitive radio-based approach, the game theoretic approach, and the fractional frequency reuse (FFR)-based approach are discussed. Several important research directions are also outlined.

Since the adoption of OFDMA as the radio transmission technology for LTE/LTE-Advanced networks, radio resource allocation in the OFDMA-based cellular networks has become a significant research topic. In Chapter 3, we provide a review of the resource allocation methods for OFDMA-based single-tier cellular wireless networks. Then, a resource allocation framework for uplink transmission in a two-tier OFDMA-based macrocell–femtocell network is discussed which provides max–min fairness to the femtocell users and robust SINR (signal-to-interference-plus noise ratio) protection to macrocell users. The complexity of solving the problem for optimal solution is discussed. Subsequently, a suboptimal and distributed solution is proposed. To this end, several open issues related to adaptive radio resource allocation in OFDMA-based multi-tier networks are discussed.

Cooperation of small cells through clustering (or grouping) is an effective technique to mitigate both the cross-tier and co-tier interferences in OFDMA-based two-tier networks, especially in dense deployment scenarios. When the small cells in a cluster cooperate, the co-tier interference among these small cells in the same cluster is completely eliminated. In Chapter 4, we study the problem of radio resource allocation (i.e., subchannel and power allocation) in clustered femtocells in a two-tier macrocell–femtocell network. The problem of joint subchannel and power allocation in a clustered femtocell network is formulated as an optimization problem (more specifically, as a mixed-integer non-linear program) under constraints on both cross-tier and co-tier interferences, as well as constraints on data rates. To solve the problem suboptimally, different approaches are proposed, which offer close to optimal performances, but incur much lower computational complexity. The effects of different cluster configurations (in terms of cluster size) are evaluated.

In Chapter 5, we address the FFR-based interference management method in OFDMA-based two-tier networks. The concept of FFR is to partition a macrocell service area into regions and assign different frequency sub-bands to each region. When operating on a large timescale, this is referred to as a static FFR scheme. A static FFR scheme for interference management through spatial channel allocation requires minimal cooperation among the base stations and has a simple operational mechanism. We discuss four FFR schemes for OFDMA-based two-tier macrocell-femtocell networks and compare their performances in terms of outage probability of users, average network sum-rate, and spectral efficiency.

An important network functionality, closely related to radio resource allocation, is the call admission control (CAC), which is responsible for admission or rejection of an incoming call request from a user. The decision of admission or rejection is made based on the current network load and the QoS requirements of the existing users and the potential incoming user. An efficient CAC scheme will be required for multi-tier networks to achieve high spectrum utilization while satisfying the QoS requirements of the users in all network tiers. In Chapter 6, we present a CAC method for a two-tier macrocell–femtocell network, which uses a sector-based FFR for spatial channel allocation for macrocells and femtocells. For this sector-based FFR, the system parameters (i.e., spatial channel allocation parameters) are optimized to maximize the total network throughput, subject to a minimum rate requirement for every user. The CAC method can be executed in a decentralized manner at each macro base station (i.e., MeNB) or femto access point (i.e., HeNB) and the CAC policy determines whether to admit or reject the arriving calls in the MeNB and HeNBs. All types of calls in the network (e.g., new calls to MeNB, new calls to HeNB, inter-sector macro–macro handoff calls, inter-sector femto–femto handoff calls, intra-sector macro–macro handoff calls, intra-sector femto–femto handoff calls, inter-sector macro–femto and femto–macro handoff calls) are considered, as is the random mobility of the users. The CAC problem is formulated as a semi-Markov decision process (SMDP), and a value iteration algorithm is used to obtain the admission control policy.

Game theory, which provides a rich set of mathematical tools to analyze interactions among independent rational entities, can be used to model and analyze radio resource management problems in multi-tier cellular networks. In Chapter 7, we discuss the applications of game theory for radio resource management in two-tier macrocell–femtocell networks. A basic introduction to the different game models is provided, and subsequently, several examples of game formulations for resource/interference management (subchannel and power allocation), spectrum sharing, and pricing in two-tier cellular networks are discussed. Several potential research directions are outlined.

In Chapter 8, we focus on the radio resource allocation problem in CDMA-based multi-tier networks. We provide a review of the existing literature on resource allocation and QoS support in single-tier CDMA networks. Then we demonstrate how game theory and optimization theory can be used to develop distributed resource allocation algorithms for CDMA-based multi-tier cellular networks, considering the tradeoff between efficiency and signaling complexity. We present example resource allocation algorithms for such networks, which provide robust QoS protection for macro users and converge to desirable operating points. Several open issues are outlined.

To reduce the capital expenditure (CAPEX) and operation expenditure (OPEX) of the deployment of small cells in multi-tier cellular networks, the small cells are required to have self-organizing network (SON) functionalities. Self-organizing small cells are expected to have self-configuration, self-optimization, and self-healing properties. While the self-configuration functionality is related to the preoperational stage of the network (e.g., installation and initialization), the self–optimization and self-healing functionalities are required at the operational stage. Different techniques can be adopted (e.g., from control theory, game theory) to implement the SON functionalities in small cells. In Chapter 9, we focus on self-organizing SCNs. We discuss the motivations behind self-organization and the use cases of self-organizing SCNs. We also classify these networks based on the timescale of self-organization and the deployment phase. We review selected works in the literature on self-configuration, self-optimization, and self-healing of two-tier macrocell–femtocell networks. Also, we outline several open research challenges.

The concept of cognitive radio for dynamic spectrum access in wireless networks can be exploited in designing self-organizing SCNs. For example, cognitive small cell base stations (SBSs) should be able to monitor the radio environment (i.e., spectrum usage) and opportunistically access the radio spectrum so that major interference sources can be avoided. In Chapter 10, we focus on cognitive small cells. We discuss the approaches for traffic offloading to small cells, which affects the resource allocation performance in the network. Then we discuss two spectrum access techniques for the cognitive small cells and compare their performances, and outline future research directions.

Multi-tier wireless networking has emerged as a new frontier in cellular radio technology. This is a fertile area of research and offers significant challenges to "wireless" researchers. We would be pleased to learn that researchers find this book useful in their pursuit of progress in this area.

Last but not least, for this book project, we acknowledge the research support from the Natural Sciences and Engineering Research Council of Canada through the Strategic Project Grant (STPGP 430285).

Ekram Hossain
Long Bao Le
Dusit Niyato

CHAPTER 1

OVERVIEW OF MULTI-TIER CELLULAR WIRELESS NETWORKS

1.1 INTRODUCTION

The demand of wireless services (e.g., data, voice, multimedia, e-Health, online gaming, etc.) through the cellular networks is ever-increasing. A recent statistics in [1] shows that, in March 2009, there were approximately 4.8 billion of mobile subscribers all over the world and this number is expected to reach 5.9 billion by 2013. The Global Mobile Data Traffic Forecast Report presented by Cisco predicts 2.4 exabytes (1 exabyte $= 10^{18}$ bytes) mobile data traffic per month for the year 2013 [2]. It has been indicated in [3] that the global mobile data traffic has tripled each year since 2008 which is projected to increase up to 26-fold between 2010 and 2015. This results up to 6.9 exabytes of mobile data traffic per month for the year 2015 [2].

Satisfying the posed capacity demand has become very challenging, and the challenge has been even more acute with the introduction of the machine-to-machine (M2M) communications and the Internet of Things (IoT). With the introduction of M2M, the network usage goes beyond the conventional voice and data usage and needs innovative solutions to handle the growing amount of traffic and user population. It is expected that by 2020 there will be more than 50 billion connected devices, which is almost 10 times the number of currently existing connected devices [4]. Note that at least 60% of the Internet traffic will be transferred via wireless access [5].

Figure 1.1 shows the three evolution phases of the user population defined by the industry, namely, the connected consumer electronics phase, the connected industry phase, and the connected everything phase [4]. In the connected consumer electronics phase, the majority of the connected devices are smart phones, tablets, computers, IP TVs, and phones. In the connected industrial phase, sensor networks, industry and buildings automation, surveillance, and e-Health applications contribute significantly to the population of wireless devices. Finally, in the connected everything phase or the IoT phase, every machine we know will have a ubiquitous Internet connectivity to be remotely operated and/or to periodically report its status.

Radio Resource Management in Multi-Tier Cellular Wireless Networks, First Edition.
Ekram Hossain, Long Bao Le, and Dusit Niyato
© 2014 John Wiley & Sons, Inc. Published 2014 by John Wiley & Sons, Inc.

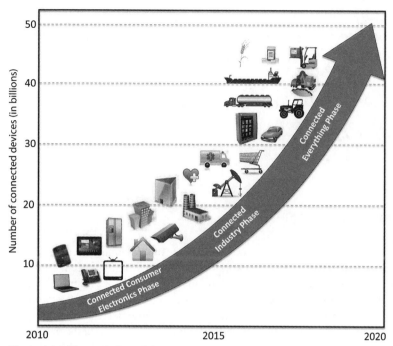

Figure 1.1   The evolution of the population of wireless devices.

One of the major challenges for next generation cellular wireless communication networks is therefore to accommodate the exponentially growing mobile data traffic by improving the capacity (e.g., spectral efficiency per unit area) of the networks. Also, the coverage (both indoor and outdoor) of the presently available cellular systems needs to be improved and high data rate services with enhanced quality-of-service (QoS) need to be provided to the subscribers. The current cellular standards and technologies such as the High Speed Packet Access (HSPA), Long-Term Evolution (LTE), LTE-Advanced (LTE-A), and Worldwide Interoperability for Microwave Access (WiMAX) systems are evolving toward meeting these requirements.

The conventional cellular systems use a macrocell-based planned homogeneous network architecture, where a network of macrocell base stations (referred to as Macrocell evolved Node B or MeNBs) provides coverage to user equipments (UEs) in each cell. In such a homogeneous network, the MeNBs have similar transmission power levels, antenna patterns, access schemes, modulation technique, receiver noise floors, and backhaul connectivity to offer similar QoS to the UEs across all cells [6, 7]. However, such a deployment especially degrades the coverage and capacity of the cell-edge users.

One of the approaches to solving this problem is to use the concept of cell splitting. However, this approach may not be economically feasible since it involves deploying more MeNBs within the network, and site acquisition for MeNBs in dense urban areas becomes a difficult proposition for the operators [6]. Therefore, the evolving LTE-Advanced systems are adopting a more flexible and scalable deployment

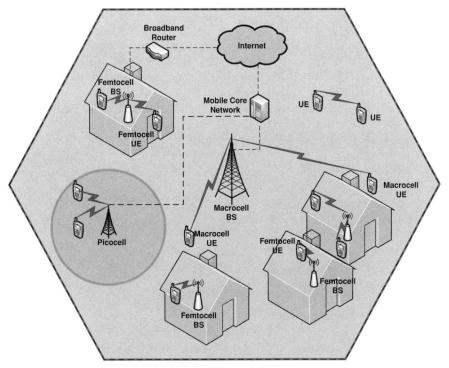

Figure 1.2   A heterogeneous cellular wireless network.

approach using a hierarchical cell deployment model where *small cells* are overlaid on the macrocells. The resulting network architecture is also referred to as a heterogeneous network (HetNet) architecture.[1] Such a deployment approach is beneficial to both the operators and end users. This approach is expected to not only increase the coverage and capacity of the cell but also improve the broadband user experience within the cell in a ubiquitous and cost-effective manner [6].

1.2   SMALL CELLS: FEMTOCELLS, PICOCELLS, AND MICROCELLS

Small cells can support wireless applications for homes and enterprises as well as metropolitan and rural public spaces. Different types of small cells include femtocells, picocells, and microcells. Due to the smaller coverage area, the same licensed frequency band can be efficiently reused multiple times within the small cells in a HetNet (Figure 1.2), thus improving the spectral efficiency per unit area (and hence the capacity) of the network. In a HetNet, small cells are envisioned as traffic off-loading spots in the Radio Access Network (RAN) to decrease the congestion in macrocells,

[1]Throughout this book, the terms "HetNets," "multi-tier networks," and "small cell networks" are used interchangeably.

and enhance the users' QoS experience [5]. The small cells in the licensed bands can be used in the cellular networks standardized by 3GPP, 3GPP2, and the WiMAX forum. When compared to unlicensed small cells (e.g., Wi-Fi), the small cells operating in the licensed band (i.e., licensed small cells) provide support for legacy handsets, operator-managed QoS, seamless continuity with the macro networks through better support for mobility/handoff, and improved security.

A *femtocell* is a small area covered by a small base station, called the femtocell access point (FAP), intended for residential indoor applications, which is installed and managed by the customers. The FAP is characterized with its limited transmission power (10~100 mW), small coverage range (10 ~ 30 m), IP backhauling, and low deployment cost. Femtocells operating in the licensed spectrum owned by the mobile operator providing Fixed Mobile Convergence (FMC) service (i.e., seamless transition for the user between wired and wireless communication devices) by connecting to the cellular network via broadband communication links (e.g., digital subscriber line [DSL]) [8].

One of the main advantages of femtocell deployment is the improvement of indoor coverage where macrocell base station or MeNB signal is weak. Femtocells provide high data rate and improved QoS to the subscribers or UEs. It also lengthens the battery life of the mobile phones since the mobile phones do not need to communicate with a distant macrocell base station. By off-loading traffic to the femtocells, the macrocell load can be reduced and hence more resources can be made available to each macro user. Deployment of femtocells can improve the utilization of radio frequency spectrum significantly. Femtocells can easily be deployed by the end users in indoor environments on a "plug-and-play" basis. It saves the backhaul cost for the mobile operators since femtocell traffic is carried over wired residential broadband connections and reduces the traffic intensity at the macrocell network. Femtocell technology has the potential to offer new services to the mobile phone users. Finally, femtocells can also be considered as an option toward the convergence of landline and mobile services. A recent study conducted by a market research company Informa Telecoms & Media estimates that by 2014, 114 million mobile users will be accessing mobile networks through femtocells [9]. This signifies that in the upcoming years femtocells could be an integral part of the next generation wireless communication systems.

In recent years, different types of femtocells have been designed and developed based on various air-interface technologies, services, standards, and access control strategies. Different operators such as Sprint Nextel, Verizon, and AT&T in the United States, Vodafone in Europe, NTT DoCoMo, Softbank mobile, and China Unicom in Asia have already successfully deployed their femtocell systems. Due to the flexibility in spectrum allocation, LTE-Advanced femtocells will use orthogonal frequency-division multiple access (OFDMA) as the air-interface technology. This is one of the most innovative technologies that will shape the future generations of the cellular wireless systems. In this standard, the FAPs are referred to as Home evolved Node Bs (HeNBs). The FAPs can use different access modes [10] as will be discussed in Section 1.6.2.

The term *picocell* is typically used to describe low power compact base stations (BSs) used in enterprise or public indoor areas and sometimes in outdoor areas as well. Picocells are usually deployed to eliminate coverage holes in a homogeneous system and to improve the capacity of the network. The coverage area of picocells usually varies between 40 and 75 m [7]. The picocells consist of omnidirectional antennas with about 5-dBi antenna gain providing better indoor coverage to the UEs in the public places such as airports and shopping malls [7].

The term *microcell* is used to describe an outdoor short-range BS aimed at enhancing the coverage for both indoor and outdoor users where macro coverage is insufficient. The term *metrocell* has recently been used to describe small cell technologies designed for high capacity metropolitan areas and can include technologies such as femtocells, picocells, and microcells. The evolving HetNets including macrocells and small cells of all types (and in some cases Wi-Fi access points operating in the unlicensed bands as well) with handoff capabilities among them are envisioned to provide improved spectrum efficiency (bps/Hz/km^2), capacity, and coverage in future wireless networks.

A comparison among the different types of small cell specifications is provided in Table 1.1.

Among all the small cells, femtocells or HeNBs, are of great interest and importance to the research community and mobile operators. A study by ABI research shows that in the future, more than 50% of voice calls and more than 70% of mobile data traffic are expected to originate from indoor UEs [11]. Another survey shows that 30% of business and 45% of household users experience poor indoor coverage [12]. From now on, our discussions will focus on femtocells; however, the concepts and techniques to be discussed throughout this book can apply to other types of small cells.

TABLE 1.1   Small cell specifications

Attribute	MeNB	Picocell	HeNB	Wi-Fi
Coverage	Wide area	Hot spot	Hot spot	Hot spot
Type of coverage	Outdoor	Outdoor, indoor	Indoor	Indoor
Density	Low	High	High	High
BS installation	Operator	Operator	Subscriber	Customer
Site acquisition	Operator	Operator	Subscriber	Customer
Transmission range	300–2000 m	40–100 m	10–30 m	100–200 m
Transmission power	40 W (approx.)	200 mW–2 W	10–100 mW	100–200 mW
Band license	Licensed	Licensed	Licensed	Unlicensed
System bandwidth	5, 10, 15, 20 MHz (up to 100 MHz)	5, 10, 15, 20 MHz (up to 100 MHz)	5, 10, 15, 20 MHz (up to 100 MHz)	5, 10, 20 MHz
Transmission rate	up to 1 Gbps	up to 300 Mbps	100 Mbps–1 Gbps	up to 600 Mbps
Cost (approx.)	$60,000/yr	$10,000/yr	$200/yr	$100–200/yr
Power consumption	High	Moderate	Low	Low
Backhaul	S1 interface	X2 interface	IP	IP
Mobility	Seamless	Nomadic	Nomadic	Nomadic
QoS	High	High	High	Best-effort

1.3 HISTORICAL PERSPECTIVE

The concept of self-optimizing home BSs is not completely new [13]. In March 1999, Alcatel first announced its plan to launch a GSM home BS in 2000 which would be compatible with existing standard GSM phones. Although demonstration of these home BSs, which were basically dual mode DECT/GSM units, was successful, they were not commercially viable due to the high cost of 2G chipsets. In 2002, Motorola engineers in Swindon, UK, claimed to have built the first complete 3G home BSs. In 2004, two UK-based startup companies, namely, Ubiquisys and 3Way networks (now part of Airvana), were using 3G chipsets developed by the chipset design company picoChip to develop their own 3G cellular home BSs. Around 2005, the term femtocell was adopted for a standalone, self-configuring home BS. Rupert Baines, VP marketing of picoChip and Will Franks, CTO of Ubiquisys are both said to have coined the term during this period, although picoChip registered the website URL www.femtocell.com first (in April 2006). The femtocell products were demonstrated by several vendors in February 2007. The "Femto Forum" (www.femtoforum.org) was formed in 2007 and grew to represent industry players and to advocate this technology. The Femto Forum is active in standardization, regulation, and inter-operability issues as well as marketing and promotion of femtocell solutions.

In late 2007, Sprint Nextel started deployment of its 2G femtocell system in the United States which it developed by teaming up with Samsung Electronics. This system worked with all Sprint handsets. Sprint launched its commercial femtocell service in August 2008. In the United Kingdom, O2 is one of the leaders in femtocell technology who developed 3G femtocells by partnering with NEC. Vodafone is another player in this technology which has started deployment and testing of femtocells in Spain. In Asia, Softbank Japan launched their 3G femtocell systems in January 2009.

Recently, research on HetNets and small cell technology in general has attracted significant interest in both academia and industry [5] and standardization efforts are ongoing since the 3rd Generation Partnership Project (3GPP) Release-8 (http://www.3gpp.org/Release-8). In recent years, different types of femtocells have been designed and developed based on various air-interface technologies, services, standards, and access control strategies. For example, 3G femtocells use Wideband Code-Division Multiple Access (WCDMA)-based air interface of Universal Mobile Telecommunication System (UMTS), which is also known as UMTS Terrestrial Radio Access (UTRA). The 3GPP refers to these 3G femtocells as Home Node Bs (HNBs). On the other hand, WiMAX and LTE femtocells use OFDMA. The LTE femtocells are referred to as HeNBs.

1.4 OVERVIEW OF LTE NETWORKS

The LTE standard was first published in March 2009 as part of the 3GPP Release-8 specifications with the aim of providing higher data rates and improved QoS. LTE corresponds to a packet-switched optimized network that encompasses

TABLE 1.2   Major specifications of the LTE standard [15–18]

Specifications	LTE					
Standard	3GPP Release 8					
Frequency bands	700 MHz, 1.5 GHz, 1.7/2.1 GHz, 2.6 GHz					
Access scheme – Uplink	Single Carrier Frequency Division					
	Multiple Access (SCFDMA)					
Access scheme – Downlink	Multiple Access (OFDMA)					
	Orthogonal Frequency Division					
Channel bandwidth (MHz)	1.4	3	5	10	15	20
Number of sub-channels	6	15	25	50	75	100
Number of sub-carriers (1 sub-channel consists of 12 sub-carriers)	72	180	300	600	900	1200
IDFT/DFT size	128	256	512	1024	1536	2048
Data modulation	QPSK, 16 QAM, 64 QAM					
Duplexing	Frequency-Division Duplexing (FDD)					
	Time-Division Duplexing (TDD)					
Frame size	1-ms sub-frames					
Sub-carrier spacing	15 KHz					
Channel coding	Convolutional and Turbo Coding					
	Rate: 78/1024–948/1024					
Cyclic prefix length – Short	4.7 µs					
Cyclic prefix length – Long	16.7 µs					
Peak uplink data rate	75 Mbps					
	(Channel Bandwidth: 10 MHz)					
Peak downlink data rate	150 Mbps					
	(2 × 2 MIMO, Channel Bandwidth: 20 MHz)					
User-plane latency	5–15 ms					

high capacity, high spectral efficiency due to robustness of the air-interface technologies, co-existence and inter-networking with 3GPP and non-3GPP systems, low latency of the network (i.e., short call setup time, short handover latency, etc.), and supports for Self-Organizing Network (SON) operation [6, 14]. Some of the major specifications of LTE standard are listed in Table 1.2 [15–18].

A generic LTE cell architecture is shown in Figure 1.3. The LTE networks correspond to a much simpler RAN in comparison to its predecessor, the UMTS Terrestrial Access Network (UTRAN) [19]. LTE is designed to support packet-switched services based on Evolved Packet System (EPS). EPS aims to provide a uniform user experience to the mobile users or UEs anywhere inside a cell through establishing an uninterrupted Internet Protocol (IP) connectivity between the UE and Packet Data Network (PDN). The network core component of EPS is referred to as Evolved Packet Core (EPC) or System Architecture Evaluation (SAE) [20]. The major components of LTE at the EPC are the Mobility Management Entity (MME)

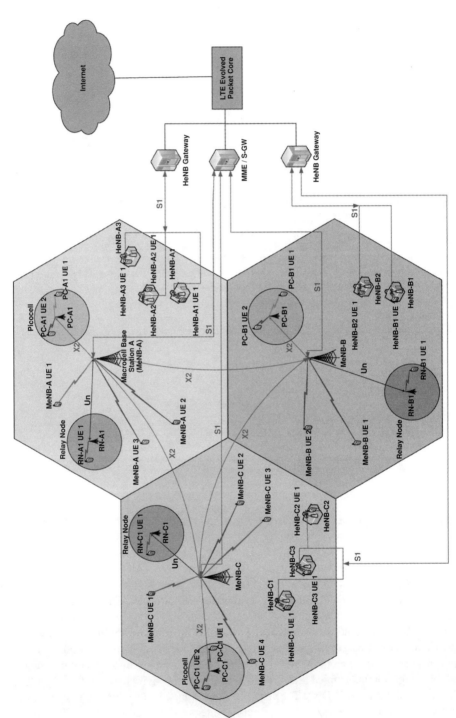

Figure 1.3   A generic LTE cell architecture [3].

and Serving GateWay (S-GW), and at the RAN are evolved Node-Bs (referred to as eNBs), that is, radio BSs.

1.4.1 The Core Network

The Core Network (CN) or the EPC of LTE comprises some logical nodes (e.g., MME and S-GW) which are responsible for the overall control of the UE and the establishment of the EPS bearer with desired QoS [20]. A brief description of the components of the CN is provided below.

Mobility Management Entity: The MME is a logical node that processes the control signaling in order to establish connection and security between the core network and the UE. Some of the major functionalities supported by the MME are as follows: (i) EPS bearer management, which includes the establishment, control, maintenance, and release of the bearers (an EPS bearer is an IP packet flow with a defined QoS [20]), (ii) Non-Access Stratum (NAS) security,[2] (iii) mobility anchoring for UEs in the idle state, and (iv) inter-working with other 3GPP or non-3GPP networks which involves handing over the voice calls.

Serving GateWay: The S-GW works as a mobility anchor for the UEs and transfers the IP packets when UEs move between eNBs. The S-GW also gathers call charging information (e.g., the volume of data sent and/or received from the UE) and serves as the mobility anchor to enable handover of voice calls to other 3GPP or non-3GPP networks. The basic functionalities of eNBs include scheduling, radio access control, inter-cell radio resource allocation, mobility management, interference management, and call admission control (CAC). The detailed functionalities of the network components of LTE are elaborated in [21].

1.4.2 The Access Network

The access network of LTE is referred to as E-UTRAN and consists of eNBs connected via different interfaces. The architecture of E-UTRAN is called a *flat architecture* since there is no centralized controller for data traffic in E-UTRAN. The eNBs are inter-connected with each other by an interface called X2 and with the core network via an interface called S1. More specifically, the eNBs are connected to the MME and S-GW through S1-MME (known as S1 control-plane) and S1-U interfaces (known as S1 user-plane), respectively. The S1-U interface carries data traffic between the serving gateway and eNB using General Packet Radio Service (GPRS) tunneling protocol which is referred to as GTP. On the other hand, the signaling information between eNB and the MME is carried via S1-MME interface which uses the Stream Control Transmission Protocol (SCTP). The network management functionalities of EPC, for example, radio access bearer management, load balancing between MMEs, paging, and instantaneous intra-LTE and/or inter-3GPP handovers

---

[2]NAS are the protocols running between the UE and the core network [22].

are performed via S1 interface [19]. Similar to the S1 interface, the X2 interface is split into two parts: X2 user-plane (X2-U) and X2 control-plane (X2-C). Based on the User Datagram Protocol (UDP), the X2-U interface carries user data traffic between the inter-connected eNBs. The X2-C interface is used for error handling functionalities and intra-LTE mobility between the serving eNB and the target eNB. The specific functions supported by each components and interfaces are provided in [21, 23].

1.4.3 The Air Interface

LTE uses OFDMA for the downlink (DL) communication. However, for the uplink (UL) communication, it uses Single Carrier Frequency-Division Multiple Access (SCFDMA), which is a cost-effective and power efficient (saves the battery life of mobile terminal) transmission scheme. LTE supports both TDD and FDD duplexing modes. The frame duration in LTE is 10 ms. In case of FDD, the entire frame is used for UL or DL transmissions. In the case of TDD, the frame is divided for UL and DL communication. Each frame consists of 10 sub-frames and the transmission duration of each sub-frame is 1 ms. Each sub-frame is divided into 2 time slots. Each time slot has a duration of 0.5 ms. There are two types of cyclic prefix (CP) used in LTE, that is, a short CP of 4.7 μs for short cell coverage and a long CP of 16.7 μs for large cell coverage. The time slot will consist of seven and six OFDM symbols when short and long CPs are used, respectively. The resource block (RB) in LTE refers to a time slot spanned with 12 sub-carriers where the bandwidth of each sub-carrier is 15 KHz. LTE supports different types of modulation technique such as, QPSK, 16-QAM, and 64-QAM. The peak data rate in UL/DL transmission mode depends on the modulation used between eNB and UE. For example, the peak data rate of LTE in the DL transmission is approximately 150 Mbps (assuming 20 MHz channel bandwidth, short CP, 2×2 multiple-input and multiple-output (MIMO)-based system with 64-QAM). This can be obtained as follows.

- The number of resource elements per sub-frame is calculated first. Since the channel bandwidth is 20 MHz and the short CP is used, the number of resource elements per sub-frame is: 12 (sub-carriers) × 7 (OFDM symbols) × 100 (RBs) × 2 (time slots) = 16,800.
- Each resource element is carried by a modulation symbol. Since 64-QAM is used, one modulation symbol will consist of 6 bits. The total number of bits in a sub-frame is: 16,800 (modulation symbols) × 6 (bits/modulation symbol) = 100,800 bits.
- The duration of each sub-frame is 1 ms. The data rate is: 100,800 bits/1 ms = 100.8 Mbps. With 2×2 MIMO, the peak data rate: 100.8 (Mbps) × 2 = 201.6 Mbps.
- Considering 25% overhead, the peak date rate is approximately 150 Mbps (= 201.6 Mbps × 0.75).

1.4.4   Radio Base Stations in LTE

Although 3GPP-LTE employs a flat architecture, the radio BSs or cells in LTE can be classified in terms of their transmission powers, antenna heights, the type of an access mechanism provided to UEs, the air-interface and the backhaul connection to other cells, or the core network. In this regard, the radio BSs in LTE can be classified as follows.

- Macrocell base stations or Macrocell e-Node Bs (MeNBs) have a large coverage area (e.g., cell radius of 500 m–1 km) with high transmission power ($\sim$46 dBm or 20 W) and provide service to all the UEs in its coverage area [3].

- Picocells are usually deployed to eliminate coverage holes in a homogeneous cellular network and improve the capacity of the network. The coverage area of picocells usually varies between 40 and 75 m [7]. The picocells consist of omni-directional low transmission powered (in comparison to MeNB) antennas with about 5-dBi antenna gain providing significant indoor coverage to the UEs in the public places such as airports and shopping malls [7].

- Similar to picocells, relay nodes are also used to improve coverage in new areas (e.g., events, exhibitions, etc.). However, relay nodes transfer their data traffic via wireless link to a Donor eNodeB (e.g., MeNB). There are two types of relay nodes: inband relays and out of band relays. In the case of inband relay nodes, the same frequency spectrum is used for relay link (relay node to UE) and backhaul link (relay node to donor eNB). On the other hand, in case of out of band relaying, the relay nodes use different frequency spectrum for relay link and backhaul link.

- Femtocells or Home e-Node Bs (referred to as HeNBs) are short-range (10 $\sim$ 30 m), low power (10 $\sim$ 100 mW), and cost-effective (\$100 – \$200) home BSs deployed by the mobile subscribers.

1.4.5   Mobility Management

In LTE, the mobility management functionalities are divided into three categories: (i) intra-LTE mobility, (ii) inter-3GPP mobility, and (iii) inter-radio access technologies (RAT) mobility. The intra-LTE mobility corresponds to the UE mobility within the LTE system, whereas mobility to other 3GPP systems (e.g., UMTS) is referred to as Inter-3GPP mobility. Mobility of UEs between the LTE system and other non-3GPP (e.g., Global System for Mobile communications or GSM) is referred to as inter-RAT mobility. Intra-LTE mobility is handled via S1 or X2 interface. The mobility through X2 interface occurs when a UE moves from one MeNB to another MeNB within the same RAN connected to the same MME. On the other hand, when the serving MeNB and the target MeNB are not connected through X2 interface, then the mobility of UE between the MeNBs takes place over the S1 interface. In addition, when the UE moves from one MeNB to another MeNB belonging to different RAN attached to different MME, then the mobility is handled via the S1 interface. Figure 1.4 illustrates

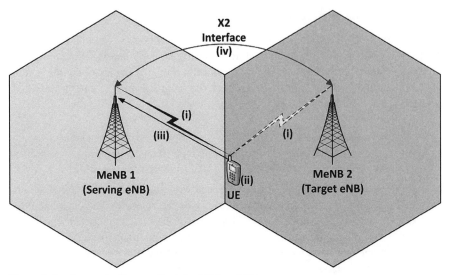

Figure 1.4   A generic diagram for intra-LTE mobility.

a generic intra-LTE mobility via X2 interface. In this handover scenario, the basic mechanism comprises the following steps.

1. The UE measures the DL signal strength.
2. The UE processes the measurement results.
3. The UE sends the measurement report to the serving MeNB (i.e., MeNB 1).
4. The serving MeNB makes the handover decision via X2 interface.

The mobility over the X2 interface comprises three phases [18]: (i) handover preparation phase, (ii) handover execution phase, and (iii) handover completion phase. A generic handover call flow for inter-MeNB mobility is shown in Figure 1.5.

Handover preparation phase: The serving MeNB makes the handover decision based on the UE measurement report. Once the handover decision is made by the serving MeNB, it sends a handover request message to the target MeNB. Based on the handover request message, the target MeNB allocates the resources to the UE. Once the preparation for admission control of the UE is completed, a handover request acknowledgment (ACK) message is sent back to the serving MeNB.

Handover execution phase: As the serving MeNB receives the handover request ACK, it sends a handover command to the UE. At the same time, the serving MeNB transfers the buffered data packets of UE to the target MeNB. Upon receiving the handover command from the serving MeNB, the UE synchronizes with the target MeNB.

Handover completion phase: The UE sends a handover confirmation message to the target MeNB when the handover procedure is completed. Upon receiving this confirmation message, the target MeNB sends a path switch request message to the MME/S-GW. Then, the S-GW switches the GTP from the serving MeNB to the target

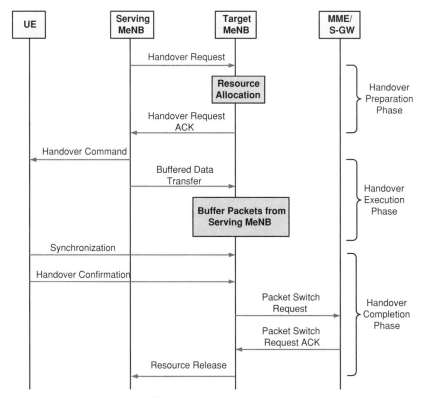

Figure 1.5   Inter-MeNB mobility management via X2 interface.

MeNB. As the data path in the user-plane for the UE is switched, the target MeNB sends a message to the serving MeNB to release the resources that were used by the UE.

The mobility via the S1 interface is similar to the mobility over the X2 interface and the details are elaborated in [18–20].

1.5   OVERVIEW OF LTE-ADVANCED NETWORKS

3GPP has been working on various aspects to improve LTE performance in the framework of LTE-Advanced [18, 24]. The summary of the LTE-Advanced target requirements is listed in Table 1.3. Some of the technologies that are being considered in LTE-Advanced include the following [18].

- In LTE-A multi-tier networks, orthogonal frequency-division multiplexing (OFDM) is used for DL and single-carrier FDM (SC-FDM) waveform is used for UL communications over 20 MHz bandwidth. The OFDM (SC-FDM) symbols are grouped into sub-frames of 1-ms duration. Each sub-frame is composed of two 0.5-ms slots. The minimum scheduling unit for the DL and

TABLE 1.3   Major specifications of the LTE-Advanced standard [18, 24]

Standard	Target Requirements of LTE-Advanced
Peak data rate	Uplink : 500 Mbps
	Downlink : 1 Gbps
	(Assuming low mobility and 100-MHz channel bandwidth)
Peak spectral efficiency	Uplink : 15 b/s/Hz (up to 4×4 MIMO)
	Downlink : 30 b/s/Hz (up to 8×8 MIMO)
Average downlink cell	2.4 b/s/Hz (up to 2×2 MIMO)
spectral efficiency	2.6 b/s/Hz (up to 4×2 MIMO)
	3.7 b/s/Hz (up to 4×4 MIMO)
Average downlink	0.07 b/s/Hz (up to 2×2 MIMO)
cell-edge spectral efficiency	0.09 b/s/Hz (up to 4×2 MIMO)
	0.12 b/s/Hz (up to 4×4 MIMO)
Mobility	Considered up to 500 km/h
Duplexing	FDD and TDD
User-plane latency	Less than 10 ms

UL of LTE is referred to as an RB. One RB consists of 12 sub-carriers in the frequency domain (180 kHz) and one sub-frame in the time domain (1 ms). The sub-frames are further grouped in 10-ms radio frames (Figure 1.6 [25]). A reference or pilot signal, referred to as a common reference signal (CRS), is used for mobility measurements as well as for demodulation of the DL control and data channels. The CRS transmissions are distributed in time and frequency.

- Higher order MIMO-based system (up to 8×8 MIMO) along with beamforming technique: Multiple antennas at the macrocell base stations can transmit the same signal with appropriate weight for each antenna element in such a way that the transmitted beam focuses in the direction of the receiver to improve the received signal-to-interference-plus-noise ratio (SINR) at the mobile station (MS) or UE [18]. The beamforming technique provides improvement in macrocell coverage and network capacity, and reduces the power consumption at the macrocell base stations.

- Increase in spectrum-efficiency and network throughput by using carrier aggregation (CA) mechanism: The available spectrum in LTE is divided into component carriers (CCs) with bandwidth of 1.4, 3, 5, 10, 15, and 20 MHz. The CA technique involves aggregation of the CCs at the macrocell base stations to allow higher data rates for the UEs. The current LTE-Advanced standard allows up to five 20-MHz CCs to be aggregated resulting in a maximum aggregated channel bandwidth of 100 MHz.

- Network MIMO to improve the overall system performance: Network MIMO corresponds to macrocell diversity and efficient coordination among macrocell base stations. In the DL transmission mode, multiple macrocell base stations transmit to a UE and in the UL transmission mode, the transmitted signal from the UE is received by one or more macrocell base stations. The network MIMO

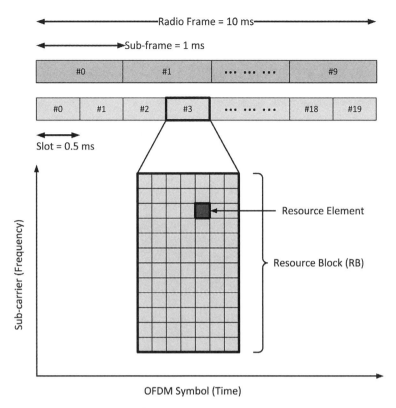

Figure 1.6   LTE-advanced frame structure.

allows the cooperating macrocell base stations to use the spectrum efficiently so that the interference from the adjacent macrocells is minimal.

- Implementation of multi-tiered network, that is, macrocell base stations over-laid with small cells, for example, femtocells, picocells, and microcells, to improve network coverage and increase the network capacity.

- Efficient inter-cell interference coordination (ICIC) and inter-cell interference (ICI) cancellation techniques: ICIC involves techniques such as spectrum split-ting, power control, etc. Inter-cell interference cancellation techniques involve decoding or demodulation of the desired information, which is further used along with the channel estimates to eliminate (or reduce) the interference from the received signal [26]. Successive interference cancellation (SIC) and parallel interference cancellation (PIC) are the two techniques that are extensively used in wireless communication systems.

- The eNB serving the RN denoted as donor eNB (DeNB): The same eNB can be the DeNB for one RN and the regular serving node for UE. The X2 interface, defined as a direct eNB-to-eNB interface, allows for ICIC. The S1 and S11 interfaces support transfer of user and data traffic between the corresponding nodes. The Un interface refers to an air interface between the DeNB and RN. Un is based on a modified interface between the eNB and UE in order to allow half duplex operation of the RN.

- Backward compatibility to ensure the reuse of the LTE architecture to coexist with other 3GPP and non-3GPP systems: The LTE devices should comply with the standards of LTE-Advanced system.

1.6   LTE FEMTOCELLS (HeNBs)

LTE or 3GPP Release-8 onward and LTE-Advanced (or Release 10 onward) femtocells use OFDMA as the air-interface technology. In LTE femtocells, resource allocation can be done in a more flexible manner due to dynamic allocation of time and frequency slots; however, large amount of coordination may be necessary. Also, semi-static allocation of frequency channels can be performed for interior, edge, and femtocell users, along with power control.

The major motivations behind the deployment of HeNBs are as follows.

- Unlike Wi-Fi, HeNBs operate in licensed spectrum offered by the mobile operator to extend the indoor coverage, improve QoS provisioning to the UEs, and off-load the indoor mobile data traffic.

- HeNBs offer increased spectrum utilization. Since the HeNBs are short-ranged BSs, the same licensed frequency band can be efficiently reused multiple times within the service area of the MeNB.

- HeNBs reduce the MeNB load which results in more resources (i.e., frequency and power) for macrocell UEs (referred to as MUEs). At the same time, the expected battery life of mobile phones are extended since the UEs do not require to communicate with the distant MeNBs.

- HeNBs are deployed by the mobile users on "plug-and-play" basis and they are self-organizing. Such deployment saves the additional macrocell base station installation cost for mobile operators. In addition to that, HeNB corresponds to low cost home BS that is capable of providing fixed mobile coverage and high rate to the UEs.

1.6.1   Access Network for HeNBs

The Home eNodeB Subsystem (HeNS) (Figure 1.3) consists of a Home eNodeB (HeNB) and optionally a Home eNodeB Gateway (HeNB-GW). An HeNB connects to the EPC through the S1-U and S1-MME interfaces. To support large number of HeNBs, an HeNB gateway is employed between the HeNBs and EPC. The HeNB gateway appears as an HeNB to the MME. S1 interface is used between the HeNBs and HeNB gateway as well as between the HeNB gateway and EPC. The basic functionalities of HeNB are the same as the eNBs. In addition, an HeNB performs access control or CAC for UEs based on the access modes incorporated with the HeNB. More detailed architectural overview of HeNBs can be found in [27].

1.6.2   Access Modes of HeNBs

In general, femtocells are designed to operate in one of three different access modes, that is, closed access mode, open access mode, and hybrid access mode [10]. In closed access mode, a set of registered UEs belonging to Closed Subscriber Group (CGS) are allowed to access a femtocell. This type of femtocell access control strategy is usually applicable to residential deployment scenarios. However, in public places such as airports and shopping malls, open access mode of femtocells can also be used where any UE can access the femtocell and benefit from its services. This access mode is usually used to improve indoor coverage and minimize the coverage holes in macrocell footprint. In hybrid access mode, any UE may access the femtocell but preference would be given to those UEs which subscribe to the femtocell. In small business or enterprise deployment scenarios hybrid access mode of femtocells may be used [10].

1.6.3   Mobility Management

The mobility management functions in a two-tier femtocell network can be categorized into three groups (Figure 1.7): (i) hand-in or inbound mobility (i.e., mobility from MeNB to HeNB), (ii) hand-out or outbound mobility (i.e., mobility from HeNB to MeNB), and (iii) inter-HeNB mobility (i.e., mobility between HeNBs). The mobility management in femtocell networks is handled over the S1 interface.

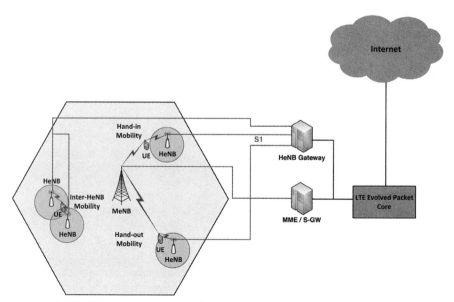

Figure 1.7   Mobility management in a femtocell network.

The handover procedure may be triggered based on different attributes such as received signal strength indicator (RSSI), QoS, UE velocity, etc. The mobility over the S1 interface consists of three phases: (i) handover preparation phase (measurement control/report, admission decision, resource allocation), (ii) handover execution phase (synchronization with target BS), and (iii) handover completion phase (path switch and resource release). The message/signaling involved in the mobility management for femtocell network is elaborated in [21].

1.7   3G FEMTOCELLS

A generic network architecture for cdma2000 femtocell enabled systems is shown in Figure 1.8. A brief description of the major network entities and the air interfaces is provided below.

1.7.1   Network Entities

Femtocell Access Point (FAP): The femtocell access point is a short-ranged, low power wireless access point that operates in licensed frequency to connect a MS or HRPD AT (known as High Rate Packet Data Access Terminal, a device to provide data connectivity to a user) to the Internet. The FAP provides access control for the MSs and supports cdma2000 1x packet data and HRPD services [28].

Security Gateway (SeGW): The SeGW provides secure access for the femtocell access point to access services within the network [28].

Femtocell Gateway (FGW): The FGW provides aggregation, proxy, and signal routing functions for the FAP to access cdma2000 1x packet data and HRPD services within the cellular network [29].

Femtocell Authentication, Authorization, and Accounting Server (F-AAA): The Femtocell AAA is a server that sends authorization policy information to the SeGW in order to provide the FAP authentication and authorization functions within the cellular network [29].

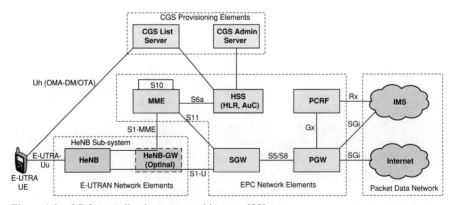

Figure 1.8   3G femtocell sub-system architecture [28].

Femtocell Management System (FMS): The FMS facilitates the auto-configuration of the FAP within the network [29].

Packet Data Serving Node (PDSN): The PDSN is a network entity that routes the HRPD AT based packet data traffic through the Internet.

Home Agent/Local Mobility Anchor (HA/LMA): This network entity acts as a mobility anchor within the network to provide the mobility management based on the location information of the MS and HRPD ATs and supports Proxy Mobile IP (PMIP) Version 6 (IPv6) [29].

Access Network Authentication, Authorization, and Accounting (AN-AAA) Server: The AN-AAA is a network entity that performs access authentication and authorization functions for the FAP. It also provides authorization function for local IP access to allow an MS to access either local IP networks or the Internet through the local interface [29].

1.7.2 Air Interface

The interface that carries user traffic between the FAP and the FGW or the PDSN is called A10 interface. On the other hand, the A11 interface is used for carrying signaling information between the FAP and the FGW or the PDSN. The A12 interface carries signaling information for access control and authentication between the FAP and the AN-AAA [29]. The A13 interface is used to provide HRPD idle session handoff between the FAP and the HRPD AN [28]. The A16 interface carries signaling information between the FAP and the HRPD AN for HRPD active session handoff (known as hard handoff) [28]. The A24 interface carries buffered data traffic for the user during idle session handoff between the FAP and the HRPD AN. The Fm interface is used for the auto-configuration of the FAP by femto user equipment (FUE). Fx4(AAA) interface is used to enable the authorization procedure for the FAP with in the network. The Fx5(AAA) interface is used for remote IP access authentication for FAP.

Note that the air-interface performance of 3G femtocells primarily depends on the power control method. However, in a large-scale deployment scenario, centralized power control would be infeasible for these femtocells.

1.8 CHANNEL MODELS FOR SMALL CELL NETWORKS

The propagation channel model can be represented by a combination of path-loss, log-normal shadowing, and multipath fading. For example, for an outdoor link (e.g., link between a macro base station and a macro user) or an indoor link (e.g., link between a FAP and a femto user), the channel gain g_{ji} of user j to BS i can be modeled as

$$g_{ji} = K_{ji} d_{ji}^{-\alpha_j} \chi_{ji} f_{ji} \qquad (1.1)$$

where K_{ji} is a constant factor (depends on carrier frequency, antenna gains, reference distance), d_{ji} is the distance from user j to BS i, α_j is the path-loss

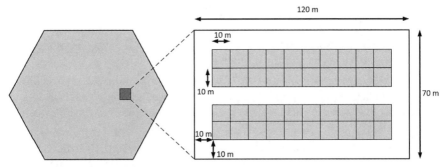

Figure 1.9   3GPP dual-stripe model.

exponent corresponding to the propagation environment (e.g., cellular, indoor, and indoor–outdoor), χ_{ji} is the log-normal shadowing between the BS and the user (i.e., $10 \log \chi \sim N(0, \sigma^2)$), and f_{ji} is the channel power gain due to multipath fading.

For an outdoor-to-indoor link (e.g., link between a macro base station and a femto user), the penetration loss (L_w) needs to be taken into account, in which case the link gain can be given by

$$g_{ji} = K_{ji} d_{ji}^{-\alpha_j} \chi_{ji} L_w^{-1} f_{ji}. \tag{1.2}$$

For an indoor-to-indoor link (e.g., link between an FAP and a femto user in another femtocell), the link gain can be given by

$$g_{ji} = K_{ji} d_{ji}^{-\alpha_j} \chi_{ji} L_w^{-2} f_{ji}. \tag{1.3}$$

In the following sections, we briefly outline the channel models defined by 3GPP for hierarchical networks and by ITU (International Telecommunication Union) for indoor propagation channels.

1.8.1   3GPP Model

For system simulations, 3GPP has defined the different physical-layer aspects and channel parameters for different deployment scenarios (e.g., homogeneous and heterogeneous deployment[3] scenarios) [30]. The path-loss between a small cell base station (SBS) and users takes into account-free space loss, indoor path-loss, indoor wall penetration loss, floor penetration loss, and outdoor wall penetration loss.

The dense urban deployment model for HeNBs, which is useful to model environments with many femtocells, is referred to as the dual-stripe model (Figure 1.9), where each block of deployment represents two stripes of apartments. Each stripe has $2 \times N$ apartments ($N = 10$ in Figure 1.9) and each apartment is of size $10\,\text{m} \times 10\,\text{m}$, and there is a street of width $10\,\text{m}$ between the two stripes of apartments.

---

[3]In a heterogeneous deployment scenario, low power nodes such as femtocells, pico eNB, relays, and remote radio heads (RRHs) are placed throughout a macrocell.

For the dual-stripe model, the path-loss from a femto base station (FBS) to an FUE) is calculated as follows:

$$PL(dB) = 38.46 + 20\log_{10} D + 0.7d_{2D,indoor}$$
$$+ q \times L_{iw} + 18.3n^{\frac{n+2}{n+1}-0.46} \tag{1.4}$$

where D (in m) is the distance between femtocell and user, $d_{2D,indoor}$ (in m) is the indoor distance, L_{iw} is the penetration loss of the indoor walls (5 dB), q is the number of indoor walls, and n is the number of penetrated floors.

The path-loss from FBS to FUE in another femtocell is calculated as

$$PL(dB) = \max(15.3 + 37.6\log_{10} D, 38.46 + 20\log_{10} D) + 0.7d_{2D,indoor}$$
$$+ q \times L_{iw} + 18.3n^{\frac{n+2}{n+1}-0.46} + L_{ow,1} + L_{ow,2} \tag{1.5}$$

where $L_{ow,1}$ and $L_{ow,2}$ represent the penetration loss of outdoor walls (e.g., 20 dB).

The path-loss from FBS to MUE is calculated as

$$PL(dB) = \max(15.3 + 37.6\log_{10} D, 38.46 + 20\log_{10} D) + 0.7d_{2D,indoor}$$
$$+ q \times L_{iw} + 18.3n^{\frac{n+2}{n+1}-0.46} + L_{ow,1}. \tag{1.6}$$

For macrocells, the path-loss from MBS to MUE is calculated as $15.3 + 37.6\log_{10}(D)$ [dB], where D (in m) is the distance. The path-loss from MBS to FUE/FBS can be calculated as $15.3 + 37.6\log_{10}(D)$ + Penetration loss (in dB).

For picocells, the distance-dependent path-loss from BS to UE is calculated as

$$PL_{LOS} = 103.8 + 20.9\log_{10} D$$
$$PL_{NLOS} = 145.4 + 37.5\log_{10} D \tag{1.7}$$

where D is in km and the carrier frequency is 2 GHz.

1.8.2 ITU Model

For indoor environments, the ITU indoor propagation model estimates the path-loss of radio propagation which is applicable for carrier frequency ranging from 900 MHz to 5.2 GHz and for buildings with one to three floors [31]. According to this model, the indoor propagation path-loss is given by

$$PL_{indoor}(dB) = 20\log f + N\log D + L_{floor}(n) - 28 \tag{1.8}$$

where N is the distance power loss coefficient (which denotes the loss of signal power with distance), D is the distance (in m), L_{floor} is the floor penetration loss factor (an empirical constant which depends on the number of floors the radio waves need to penetrate), n is the number of floors between the transmitter and the receiver, and f is the transmission frequency (in MHz). When installed outdoor, the path-loss can be calculated as

$$PL = PL_{indoor} + L_{wall} \tag{1.9}$$

where L_{wall} is the exterior wall loss.

1.9 MULTI-TIER CELLULAR WIRELESS NETWORK MODELING AND ABSTRACTION

In a wireless network, the coverage and interference statistics highly depend on the network geometry. That is, the locations of the network entities and the users with respect to each other highly impact their performances[4]. Therefore, to model and design multi-tier cellular wireless networks, the network topology should be abstracted to a general baseline model that is able to capture the interactions among the coexisting network entities. For instance, the hexagonal grid-based model has been used traditionally for modeling, analysis and design of cellular networks.

In the grid-based model, it is assumed that the locations of the MBSs are known, follow a deterministic hexagonal grid, and all MBSs have the same coverage area. For macrocells, the hexagonal grid model is easy to simulate along with outdoor channel model considering path-loss, shadowing, and fading. However, due to the variation of the capacity demand across the service area (i.e., downtowns, residential areas, parks, sub-urban areas, etc.), and the infeasibility of deployment of MBSs in some locations (i.e., rivers, hills, rails, buildings, etc.), there are spatial randomness in the locations of MBSs and all MBSs do not have the same coverage area [32]. Hence, the assumption of hexagonal grids is violated, and therefore, is considered to be very idealized. Moreover, the grid-based model does not provide tractable results for ICI (i.e., interference from different BSs) for the simple one-tier cellular networks and the performance metrics of interest are usually obtained via Monte Carlo simulations [33].

Recently, a new network model based on stochastic geometry has been used to model homogeneous cellular networks as well as multi-tier cellular networks [32, 34]. In that model, it is assumed that the locations of the network entities are drawn from some realizations of a stochastic point process in the $\mathbb{R}^2$ plane. Although the locations of the MBSs are planned through a sophisticated network planning procedure and hence their locations are known, the locations of the MBSs with respect to each other highly vary from one location to another. The stochastic geometric network models not only explicitly account for the random locations of the network entities, but also provide tractable yet accurate results for the performance metrics of interest [32, 34]. Moreover, the results from the stochastic geometric network models are general and topology independent. The simplest, the most tractable and well-understood stochastic point process in the literature is the Poisson point process (PPP). A point process in $\mathbb{R}^d$ is a PPP if and only if the number of points within any bounded region has a Poisson distribution with a mean directly proportional to the d-dimensional volume of that region, and the numbers of points within disjoint regions are independent [35]. That is, the PPP assumes that the positions of the points are uncorrelated. Although the assumption on uncorrelated MBS locations is not realistic, it was shown in [32] that the PPP assumption provides a lower bound on coverage probability (i.e., the complement of the outage probability) and average achievable rate that is as much tight as the upper bound provided by the idealized

---

[4]Network entity is a generic term that will be used to denote both the MBSs or the SBSs.

grid-based model. In [34], it was shown that the PPP assumption is accurate to within 1–2 dB of the performance of an actual LTE network overlaid by heterogeneous tiers modeled as PPP.

A two-tier network is shown in Figure 1.10. Figure 1.10(a) shows a coexistence scenario for 5 MBSs and 14 SBSs.[5] As shown in the figure, the coverage of each network entity highly depends on its type (i.e., an MBS or an SBS) and the network geometry (i.e., its location with respect to other network entities). That is, assuming that each user will associate with (i.e., covered by) the network entity that provides the highest signal power, the coverage of each network entity will depend on its transmission power as well as the relative positions of the neighboring network entities and their transmission powers. For instance, if two MBSs have the same transmission power, a line bisecting the distance between them will separate their coverage areas. However, for an MBS with 50 times higher transmission power than an SBS, a line dividing the distance between them with a ratio of 50:1 will separate their coverage areas, and so on. Hence, it can be seen how the coverage of the network entities highly depends on the network geometry. In the same manner, if we have a test receiver at a generic location, it can also be shown that the interference statistics at that receiver also depends on the network geometry. Stochastic geometry can mathematically model these location-dependent phenomena and provide tractable results for the network performance metrics. For instance, assuming that both MBSs and SBSs follow independent PPPs, the network can be modeled via a weighted Voronoi tessellation as shown in Figure 1.10(b), and performance metrics such as outage probability and minimum achievable rate can be analyzed [34].

1.10 TECHNICAL CHALLENGES IN SMALL CELL DEPLOYMENT

The mass deployment of small cells gives rise to several technical challenges which are outlined below.

- *Resource allocation and interference management*: One of the major challenges is interference management between neighboring small cells and between small cells and macrocell. In general, two types of interferences that occur in a two-tier network architecture (i.e., a central macrocell is underlaid/overlaid with 3G/OFDMA femtocells, respectively) are as follows.

  Co-tier interference: This type of interference occurs among network elements that belong to same tier in the network. In case of a two-tier macrocell–femtocell network, co-tier interference occurs between neighboring femtocells. For example, a femtocell UE (aggressor) causes UL co-tier interference to the neighboring femtocell base stations (victims). On the other hand, a femtocell base station acts as a source of DL co-tier interference to the neighboring

---

[5]This figure is drawn for deterministic channel gains. For random channel fading, the same concept applies but the boundaries will be randomly curved lines and the coverage area of each network entity may be disjoint.

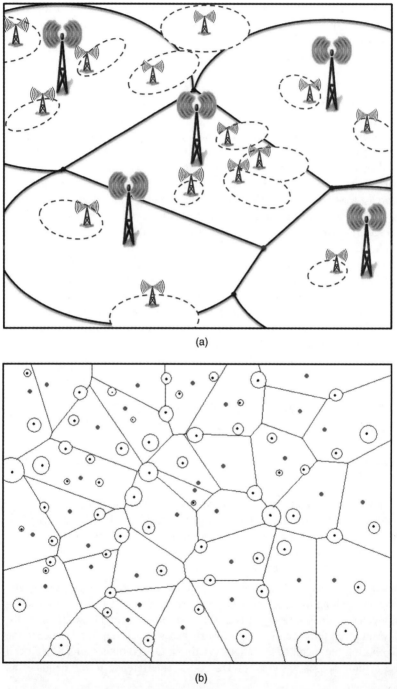

(a)

(b)

Figure 1.10   (a) A two-tier network model and (b) the network modeled as a weighted Voronoi tessellation (the larger dots represent the MBSs).

femtocell UEs. However, in OFDMA systems, the co-tier UL or DL interference occurs only when the aggressor (or the source of interference) and the victim use the same sub-channels. Therefore, efficient allocation of sub-channels is required in OFDMA-based femtocell networks to mitigate the co-tier interference.

Cross-tier interference (CTI): This type of interference occurs among network elements that belong to the different tiers of the network, that is, interference between femtocells and macrocells. For example, femtocell UEs and macrocell UEs (also referred to as MUEs) act as a source of UL CTI to the serving macrocell base station and the nearby femtocells, respectively. On the other hand, the serving macrocell base station and femtocells cause DL CTI to the femtocell UEs and nearby macrocell UEs, respectively. Again, in OFDMA-based femtocell networks, cross-tier UL or DL interference occurs only when the same sub-channels are used by the aggressor and the victim.

Distributed interference management schemes are required which satisfy the QoS requirements of the macrocell and small cell (e.g., femtocell) users and at the same time enhance the capacity and coverage of the network. In a CDMA femtocell network, interference caused to the macro users need to be controlled such that the QoS requirements of the macro users are satisfied while provisioning soft QoS for femtocell users. Also, the trade-off between the throughput and the power can be optimized for femtocell users. ICIC for femtocell networks is a major issue in 3GPP LTE-Advanced standardization. In LTE femtocells, backhaul-based coordination, dynamic orthogonalization, sub-band scheduling, and adaptive fractional frequency reuse (FFR) can be used for interference coordination. Advanced techniques such as interference cancellation and cooperative communication among BSs can be also used for ICI mitigation. In both open- and closed-access modes, signaling for coordinating CTI may be difficult. Since femtocells are not typically connected directly to the operator's core network, increased delay may occur in the backhaul signaling.

- *Cell association and admission control*: A simple approach for cell association is to assign each user to the "strongest" BS. With open-access and strongest cell selection, heterogeneous and multi-tier deployments may not worsen the overall interference conditions or even change the SINR statistics [5]. However, a better approach is *biasing* in which users are pushed into small cells. In OFDMA femtocells, biased users can be assigned orthogonal resources to the macrocell. The problems of optimal cell association, load balancing, optimal biasing, and admission control are open research issue. The optimal admission control decision should depend on factors such as distribution of users and locations of BSs, traffic patterns, information available to the mobiles and femtocell access points, etc.

- *Backward compatibility*: It may be infeasible to extensively modify the existing wireless infrastructure to accommodate the newly deployed SBSs. Hence, majority of the coexistence issues should be burdened to the SBSs.

Therefore, the protocol stacks at the SBSs need to take care of the convergence and backward compatibility issues.

- *Backhaul for small cells*: Designing a high capacity backhaul network (in particular, a wireless backhaul) for small cells is a big challenge. Lower cost backhaul solutions will be required for cost-effective deployment of small cells. The physical attributes of the backhaul solutions should be such that they are space-optimized. Also, due to the non-line-of-sight (NLOS) propagation environment, traditional point-to-point wireless backhaul solutions may not be suitable. In the backhaul network, the interference among the cell sites needs to be managed.

- *Limited cross-tier signaling and lack of completely centralized control*: Due to the large number of deployed SBSs, centralized control and optimization are infeasible. Moreover, cross-tier signaling via the IP network may be infeasible due to the large delay. Therefore, the resource allocation and interference management methods developed for multi-tier cellular networks with small cells should be developed such that they only involve limited signaling and take into consideration the limited capacity of the backhaul network.

- *Network performance analysis*: Precise characterization of interference in heterogeneous and multi-tier networks is important for the performance analysis and network dimensioning. By modeling the aggregate interference in a multi-tier set up under general fading scenarios, the coverage, outage, and capacity in the network can be analyzed. In this context, the analysis of UL SINR by modeling user location and BS location simultaneously, is a significant problem. Also, modeling and analysis of the effects of traffic burstiness (which affects interference) and packet-level performance analysis are crucial for design and optimization of multi-tier networks.

- *Handoff and mobility management*: An effective and efficient mobility management and handover scheme (macrocell-to-small cell, small cell-to-macrocell, and small cell-to-small cell) is necessary for mass deployment of small cells in UMTS and LTE networks [36]. The scheme should be able to reduce the network complexity and signaling cost, deal with different access modes and exhibit proper resource management beforehand to perform efficient handover. Also, vertical handoffs between licensed small cells and non-cellular access technologies such as Wi-Fi need to be considered. Lack of a low delay connection to the core network may result in significant handoff signaling delays. CDMA femtocells are typically unable to share a radio network controller (RNC) with a macrocell or other small cells for coordinating soft-handoffs. Therefore, architectural changes are required in the core network and small cell gateway functions.

- *Auto-configuration/self-organization, self-optimization, self-healing*: Since small cells are randomly deployed by the subscribers and are not location constrained, they must support a "plug-and-play" operation with automatic configuration and adaptation for their scalable deployment and maintenance

[37]. Therefore, efficient methods are required for automatic channel selection, power adjustment, and frequency assignment for autonomous interference coordination and coverage optimization. Also, the procedures for automatic registration and authentication, neighbor discovery, cell ID selection will be required for small cells. Self-organizing small cells will reduce the operational expenditure (OPEX). In this context, the *cognitive radio* concepts will be particularly useful to design these methods. The self-organizing and self-optimizing small cells will then be referred to as *cognitive small cells*. Cognitive small cells should be able to dynamically sense spectrum usage by the macrocell and adapt their transmissions accordingly. Cognitive small cells should be able to optimize the network parameters for transmission power, physical resources, access modes, admission control, handoff control, etc. For better convergence of the cognitive methods, semi-distributed schemes may be better. The small cells may also have the capability of shutting down and waking up autonomously to improve the energy efficiency (EE) performance of the FBSs. These type of small cells can be referred to as *green small cells*.

To deploy cognitive SBSs, there should be little or no modification required either in the existing MBSs or the mobile terminals so that backward compatibility is maintained. All the frequency bands are accessible to both the network tiers. Since the cognitive SBSs avoid interference from major interference sources by opportunistic access to orthogonal channels, topology-aware adaptive sub-channel allocation is achieved, and the spatial frequency reuse can be maximized via optimizing the spectrum sensing threshold for the cognitive SBSs. However, for the cognitive SBSs to be robust and adaptive to topological changes, the design parameters should be independent from the topology and account for the topological randomness. For instance, the spectrum sensing threshold is a very critical design parameter that should be tuned carefully while considering the topological randomness to achieve the required trade-off between the spatial frequency reuse and the experienced interference (which translates to outage probability).

- *Security*: The small cells (e.g., femtocells) need to be secured to prevent unwanted users from accessing them [38]. For example, femtocells are usually connected to the core network via DSL or broadband connection or wireless links (e.g., WiMAX) and vulnerable to a variant of malicious attacks such as masquerading, eavesdropping, and man-in-the-middle attack. Enhanced authentication and key agreement mechanisms are required to secure the femtocell networks [39].

- *Timing and synchronization*: SBSs such as the femto access points (FAPs) need to connect to the clock of the core network and use that clock for synchronization with the rest of the network. Also the packets need to be synchronized for transmission. Timing and synchronization is one of the major challenges for femtocells since synchronization over IP backhaul is difficult and inconsistent delays may occur due to varying traffic congestion [40].

There are a number of options available for synchronization which include the following: timing over packet, global positioning system (GPS), and network listening over the air. Each of the options has its merits and demerits. For example, the GPS signals and the BS signals may not penetrate indoors. Therefore, a combination of methods may be required.

REFERENCES

1. ITU Telecommunications Indicators Update-2009. Available at: http://www.itu.int/ITU-D/ict/statistics/
2. Cisco, "Cisco Visual Networking Index: Global Mobile Data Traffic Forecast Update, 2011–2016," Technical Report, Cisco systems Inc., San Jose, CA, White Paper, February 2012.
3. L. Lindbom, R. Love, S. Krishnamurthy, C. Yao, N. Miki, and V. Chandrasekhar, "Enhanced inter-cell interference coordination for heterogeneous networks in LTE-Advanced: A survey". Available at: http://arxiv.org/ftp/arxiv/papers/1112/1112.1344.pdf
4. "More Than 50 Billion Connected Devices", Ericsson White Paper, February 2011.
5. J. Andrews, H. Claussen, M. Dohler, S. Rangan, and M. Reed , "Femtocells: Past, present, and future," vol 30, no. 3, pp. 497–508, April 2012.
6. Qualcomm Document Cente,"LTE Advanced: Heterogeneous Networks," Jan 27, 2011. Available at: http://www.qualcomm.com/documents/lte-advanced-heterogeneous-networks-0.
7. R. Bendlin, V. chandrasekhar, R. Chen, A. Ekpenyong, and E. Onggosanusi, "From homogeneous to heterogeneous networks: A 3GPP Long Term Evolution Rel. 8/9 Case Study," in *Proceedings of Conference on Information Sciences and Systems*, March 2011.
8. V. Chandrasekhar and J. G. Andrews, "Femtocell networks: A survey," vol. 46, no. 9, pp. 59–67, September 2008.
9. Femto Forum, available at: http://www.femtoforum.org/femto. (http://www.smallcellforum.org)
10. A. Golaup, M. Mustapha, and L. B. Patanapongipibul, "Femtocell access control strategy in UMTS and LTE," vol. 47, no. 9, pp. 117–123, September 2009.
11. Presentations by ABI Research, Picochip, Airvana, IP access, Gartner, Telefonica Espana, *2nd International Conference on Home Access Points and Femtocells*. Available at: http://www.avrenevents.com/dallasfemto2007/purchase_presentations.htm.
12. J. Cullen, "Radioframe presentation," in *Femtocell Europe*, London, UK, June 2008.
13. D. Chambers, "Femtocell History". Available at: http://www.thinkfemtocell.com/FAQs/femtocell-history.html.
14. LTE, available at: http://www.3gpp.org/LTE.
15. E. Yaacoub and Z. Dawy, "A survey on uplink resource allocation in OFDMA wireless networks," *IEEE Communications Surveys and Tutorials*, vol. 14, no. 2, pp. 322–337, Second Quarter 2012.
16. M. Rumney, "Introducing LTE Advanced," Agilent Technologies, May 22, 2011. Available at: http://www.eetimes.com/design/microwave-rf-design/4212869/Introducing-LTE-Advanced
17. 3GPP Technical Specification 25.913: "Requirements for Evolved UTRA (E-UTRA) and Evolved UTRAN (E-UTRAN)," v8.0.0, December 2008. Available at: http://www.3gpp.org
18. A. Ghosh, J. Zhang, J. G. Andrews, and R. Muhamed, *Fundamentals of LTE*. 1st edition, Communications Engineering and Emerging Technologies Series, Prentice Hall, 2010.

19. N. A. Ali, A. M. Taha, and H. S. Hassanein, *LTE, LTE-Advanced and WiMAX: Towards IMT-Advanced Networks*. 1st edition, John Wiley & Sons, Ltd., 2012.

20. S. Sesia, I. Toufik, and M. Baker, *LTE - The UMTS Long Term Evolution: From Theory to Practice*, 2nd edition, John Wiley & Sons, Ltd., 2011.

21. 3GPP Technical Specification 36.300: "Evolved Universal Terrestrial Radio Access (E-UTRA) and Evolved Universal Terrestrial Radio Access Network (E-UTRAN); Overall description; Stage 2," v10.2.0, January 2011. Available at: http://www.3gpp.org

22. 3GPP Technical Specification 24.301: "Non-Access-Stratum (NAS) Protocol for Evolved Packet System (EPS); State 3," v9.5.0, December 2010. Available at: http://www.3gpp.org

23. 3GPP Technical Specification 23.401: "General Packet Radio Service (GPRS) enhancements for Evolved Universal Terrestrial Radio Access Network (E-UTRAN) access," v10.7.0, March 2012. Available at: http://www.3gpp.org

24. 3GPP Technical Specification 36.913: "Requirements for Further Advancement for E-UTRA," v8.0.1, March 2009. Available at: http://www.3gpp.org

25. A. Damnjanovic, J. Montojo, Y. Wei, T. Ji, T. Luo, M. Vajapeyam, T. Yoo, Q. Song, and D. Malladi, "A survey on 3GPP heterogeneous networks," *IEEE Wireless Comm.*, June 2011.

26. J. G. Andrews, "Interference cancellation for cellular systems: A contemporary overview," *IEEE Wireless Communications*, vol. 2, no. 3, pp. 19–29, April 2005.

27. 3GPP Technical Report 23.830, "Architecture Aspects of Home NodeB and Home eNodeB," v9.0.0, September 2009. Available at: http://www.3gpp.org

28. 3GPP2 X.S0059-000-A: "cdma2000 Femtocell Network: Overview," ver. 1.0, December 2011. Available at: http://www.3gpp2.org

29. 3GPP2 S.R0135-0, "Network Architecture Model for cdma2000 Femtocell Enabled Systems," ver. 1.0, April 2010. Available at: http://www.3gpp2.org

30. 3GPP TR 36.814 V9.0.0 (2010-03), 3rd Generation Partnership Project; Technical Specification Group Radio Access Network; Evolved Universal Terrestrial Radio Access (E-UTRA); Further advancements for E-UTRA physical layer aspects (Release 9), Technical Report.

31. Wikipedia, "ITU Model for Indoor Attenuation," http://en.wikipedia.org/wiki/ITU_Model_for_Indoor_Attenuation. Retrieved 2013–01–14.

32. J. Andrews, F. Baccelli, and R. Ganti, "A tractable approach to coverage and rate in cellular networks," *IEEE Transactions on Communications*, vol. 59, no. 11, pp. 3122–3134, November 2011.

33. J. Xu, J. Zhang, and J. Andrews, "On the accuracy of the Wyner model in cellular networks," *IEEE Transactions on Wireless Communications*, vol. 10, no. 9, September 2011.

34. H. Dhillon, R. Ganti, F. Baccelli, and J. Andrews, "Modeling and analysis of K-tier downlink heterogeneous cellular networks," *IEEE Journal on Selected Areas in Communications*, vol. 30, no. 3, pp. 550–560, April 2012.

35. D. Stoyan, W. Kendall, and J. Mecke, *Stochastic Geometry and Its Applications*. John Wiley & Sons, 1995.

36. L. Wang, Y. Zhang, and Z. Wei, "Mobility management schemes at radio network layer for LTE femtocells," in *Proceedings of IEEE 69th Vehicular Technology Conference*, pp. 1–5, 26-29 April 2009.

37. Y. J. Sang, H. G. Hwang, and K. S. Kim, "A self-organized femtocell for IEEE 802.16e system," in *Proceedings of IEEE Global Telecommunications Conference (GLOBECOM)*, pp. 1–5, November 30–December 4, 2009.

38. 3GPP Technical Specification 33.320: "Security of Home Node B (HNB) / Home evolved Node B (HeNB)," v9.2.1, June 2010. Available at: http://www.3gpp.org

39. C. K. Han, H. K. Choi, and I. H. Kim, "Building femtocell more secure with improved proxy signature," in *Proceedings of IEEE Global Telecommunications Conference (GLOBECOM)*, pp. 1–6, November 30–December 4, 2009.
40. S. Lee, "An enhanced IEEE 1588 time synchronization algorithm for asymmetric communication link using block burst transmission," *IEEE Communications Letters*, vol. 12, no. 9, pp. 687–689, September 2008.

CHAPTER 2

RESOURCE ALLOCATION APPROACHES IN MULTI-TIER NETWORKS

2.1 INTRODUCTION

In a two-tier macrocell–femtocell network, femtocells are deployed over the existing macrocell network and share the same frequency spectrum with macrocells. Due to spectral scarcity, the femtocells and macrocells have to reuse the total allocated frequency band partially or totally which leads to cross-tier or co-channel interference. At the same time, in order to guarantee the required QoS to the macro users, femtocells should occupy as little bandwidth as possible that leads to co-tier interference. As a result, the throughput of the network will decrease substantially due to such co-tier interference and CTI. In addition, severe interference may lead to "deadzones," that is, areas where the QoS degrades significantly. The deadzones are created due to asymmetric levels of transmission power within the network and the distance between macrocell UE and macrocell base station. For example, a macrocell UE located at a cell edge and transmitting at a high power will create a deadzone to the nearby femtocell UL transmission due to co-channel interference. On the other hand, in the DL transmission, due to high path-loss and shadowing, a cell-edge macrocell UE may experience severe co-channel interference from the nearby femtocells. Thus, it is essential to adopt an effective and robust interference management scheme that would mitigate the co-tier interference and reduce the CTI substantially in order to enhance the throughput of the overall network. In this chapter, we will review the existing literature on resource allocation and interference management in two-tier macrocell–femtocell networks.

In OFDMA-based small cell networks, due to the flexibility in spectrum allocation, orthogonal sub-carriers can be assigned to small cells and macrocells. This gives OFDMA-based small cells an edge over CDMA-based small cells in terms of utilizing the spectrum resources efficiently. In OFDMA-based wireless networks, radio resource allocation (e.g., sub-channel and power allocation), CAC, and power control algorithms are crucial to optimize the resource utilization while providing support to a wide range of multimedia broadband services with heterogeneous QoS requirements. In general, the radio resource management (RRM) strategies for OFDMA-based networks can be grouped into three categories [1]. The first category includes frequency/time resource allocation schemes such as channel allocation, scheduling, transmission rate control, and bandwidth reservation schemes. The second category consists of power allocation and control schemes, which control the transmission power of the UEs and the BSs. Note that the channel and power allocation schemes (i.e., the resource allocation schemes) can be referred to as interference management schemes. The third category comprises of CAC, BS assignment, and handoff algorithms, which control the accessibility of the UEs to the radio network. The CAC is responsible for admission or rejection of incoming request to the network based on predefined criterion while taking the network load conditions and QoS requirements of both incoming and ongoing users into account. Then, scheduling techniques assign available resources to the admitted users. CAC takes into account the amount of resource available in the cell, QoS requirements of the new users as well as their priority levels, and the currently provided QoS to the active sessions in the cell. A new request is only granted resources if it is estimated that the QoS requirement for the new request can be fulfilled, while still being able to provide acceptable QoS to the existing in-progress sessions in the cell. While we focus on the channel allocation and interference management issues in this chapter, we will deal with the CAC issue in multi-tier cellular networks in Chapter 6.

2.2 DESIGN ISSUES FOR RESOURCE ALLOCATION IN MULTI-TIER NETWORKS

In this section, we highlight the major performance measures and issues which are considered when designing radio resource allocation in traditional OFDMA-based networks in general. The same issues arise in the context of multi-tier OFDMA-based networks.

Minimum rate outage: When the transmission rate of a selected user is not satisfied, a rate failure (or rate outage) occurs. The minimum rate required for each user has to be achieved with a predefined rate outage.

Channel utilization: Channel utilization, a term related to the channel usage, includes not only the data bits but also the overhead that makes use of the channel. A resource allocation method should minimize the overhead to maximize the channel utilization.

Bit rate-related parameters: The maximum sustained and the minimum reserved traffic rates (measured in bits per second) are bit rate related QoS parameters. The former is the peak information rate of the service flow whereas the latter is the minimum information rate promised to the service flow.

Fairness: It is one of the challenging problems for RRM. In general, there is trade-off between channel utilization and fairness. Increasing the channel utilization, for example, by serving users with good channel conditions, results in unfairness.

Delay-related parameters: These include maximum latency and tolerated jitter. The maximum latency measures the maximum time elapsed or delay between when a sending data unit enters the medium access control (MAC) layer and when it enters the air interface. On other hand, the tolerated jitter determines the maximum variation in delay.

Call-level and packet-level QoS parameters: Both call-level and packet-level QoS parameters need to be considered for resource allocation. Call dropping probability and handoff call dropping probability are considered as call-level QoS metrics while packet throughput, packet delay, packet dropping rate, and delay jitter are considered as packet-level QoS metrics.

Policy-related parameters: Both traffic priority and request/transmission policy relate to policy-based parameters. Traffic priority is the priority given to a service flow, for instance, if two services have identical QoS parameters, preference is given to the service flow that has higher priority. The request/transmission policy parameters specify the service flow attributes.

Pre-emption strategy: This strategy is used to pre-empt the assigned bandwidth to admitted users with low priority services for new high priority service users. Cut-off and service degradation are two basic schemes used for pre-emption. In the former case, when pre-emption is executed, the total bandwidth assigned to low priority traffic is abruptly pre-empted and as a result its QoS can no longer be sustained. In the latter scheme, to pre-empt the bandwidth that has been assigned to admitted low priority service users, bandwidth degradation is performed with some predefined degradation level.

Differentiated services: One of the most challenging issues for wireless networks is to support multiple classes of services with different QoS requirements. Efficient resource management techniques are the keys to support multimedia applications with QoS provisioning.

Adaptive bandwidth/power allocation: Adaptive bandwidth/power allocation is essential due to time-varying nature of wireless channel, diversity of applications and QoS requirements to improve utilization of wireless network resources.

Energy efficiency (EE): EE is the overall number of bits transmitted per unit energy and is equivalent to the sum capacity of users per unit power. To calculate EE, in addition to the energy consumption for transmission and reception, the circuit energy consumption should also be taken into account. It depends on network architectures, transmission schemes, and resource allocation strategies.

2.3   INTERFERENCE MANAGEMENT APPROACHES

Figure 2.1 illustrates all possible interference scenarios in an OFDMA-based two-tier macrocell–femtocell network. In this network, each UE communicates on a specific sub-channel with the BS from which it receives the strongest signal strength, while

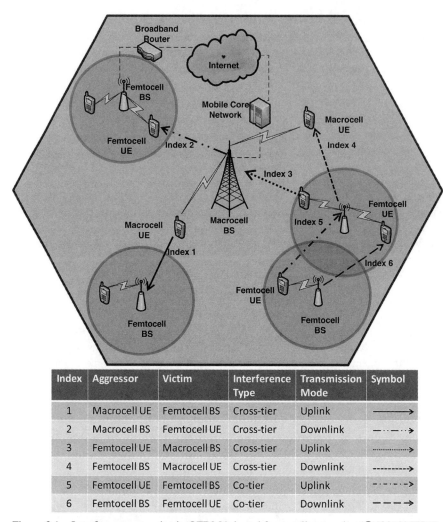

Index	Aggressor	Victim	Interference Type	Transmission Mode	Symbol
1	Macrocell UE	Femtocell BS	Cross-tier	Uplink	⟶
2	Macrocell BS	Femtocell UE	Cross-tier	Downlink	—··—·⇢
3	Femtocell UE	Macrocell BS	Cross-tier	Uplink	··········⇢
4	Femtocell BS	Macrocell UE	Cross-tier	Downlink	---------⇢
5	Femtocell UE	Femtocell BS	Co-tier	Uplink	—·—·—·⇢
6	Femtocell BS	Femtocell UE	Co-tier	Downlink	———⟶

Figure 2.1   Interference scenarios in OFDMA-based femtocell networks. (© [2012] IEEE)

the signals received from other BSs on the same sub-channel are considered as interference. Two types of interferences occur in such a multi-tier network.

Co-tier interference: This type of interference occurs between neighboring femtocells. For example, a femtocell UE (aggressor) causes UL co-tier interference to the neighboring femtocell base stations (victims) (e.g., index 5 in Figure 2.1). On the other hand, a femtocell base station acts as a source of DL co-tier interference to the neighboring femtocell UEs (e.g., index 6 in Figure 2.1).

Cross-tier interference (CTI): This type of interference occurs between femtocells and macrocells. For example, femtocell UEs (referred to as FUEs) and macrocell UEs (referred to as MUEs) act as sources of UL CTI to the serving MeNB (e.g., index 3 in Figure 2.1) and the nearby HeNBs (e.g., index 1 in Figure 2.1), respectively. On the other hand, the serving MeNB and HeNBs cause DL CTI to the FUEs (e.g., index 2 in Figure 2.1) and nearby MUEs[1] (e.g., index 4 in Figure 2.1), respectively.

If an effective interference management scheme can be adopted, then the co-tier interference can be mitigated and the CTI can be reduced which would enhance the throughput of the overall network.

Different techniques such as cooperation among macrocell BSs (i.e., MeNBs) and femtocell BSs (i.e., HeNBs), formation of groups of HeNBs and exchange of information (such as path-loss, geographical location, etc.) among neighboring HeNBs, accessing the spectrum intelligently, etc. can be considered to reduce co-tier interference and CTI. In the following, we provide an overview of the different approaches for interference mitigation in two-tier OFDMA networks. These approaches consider UL and/or DL transmissions as well as co-tier interference and/or CTI. In the following sections, we give an overview of the different interference management techniques for two-tier macrocell–femtocell networks presented in the recent literature. To this end, we will provide a qualitative comparison among these techniques based on some important criteria. We will conclude the chapter and outline some open challenges related to interference management in multi-tier networks.

2.3.1   Femto-Aware Spectrum Arrangement Scheme

In [2], Yi Wu *et al.* propose a femto-aware spectrum arrangement scheme to avoid *UL* CTI between a macrocell and femtocells. In this scheme, the allocated frequency spectrum for any macrocell coverage area is divided into two parts: the macrocell dedicated spectrum and macrocell–femtocell shared spectrum. It is assumed that shared spectrum allocated to femtocells (i.e., HeNBs) is configured by the mobile operator. Thus, the macrocell base station (i.e., MeNB) has adequate knowledge of the shared frequency spectrum. Based on this knowledge, the MeNB develops an interference pool which includes the macrocell UEs that pose a threat to the nearby HeNBs. These macrocell UEs are thus assigned a portion of the spectrum dedicated for macrocell usage which reduces/mitigates the UL CTI and solves the UL deadzone problem.

Figure 2.2 illustrates the femto-aware spectrum arrangement scheme, where macrocell UE4, macrocell UE5, and macrocell UE6 pose potential threat of CTI on their prospective nearby HeNBs. Therefore, these macrocell UEs are put into the femtocell-interference pool by the MeNB and are assigned a dedicated portion of the total frequency spectrum in order to mitigate co-channel interference. On the other hand, since other macrocell UEs (i.e., macrocell UE1, macrocell UE2, and macrocell UE3) are not close to any HeNB, they share the rest of the frequency spectrum along with the femtocell UEs (i.e., femtocell UE1, femtocell UE2, and femtocell UE3). However, this scheme does not consider inter-HeNB interference and may be inefficient if the number of macrocell UEs near the HeNB increases.

---

[1] The MUEs are considered as outdoor users and the FUEs are considered as indoor users.

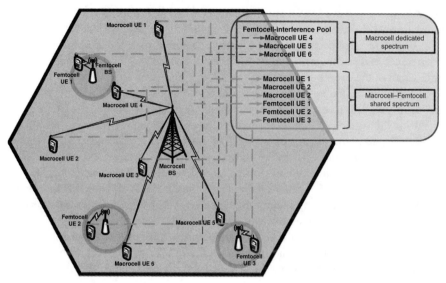

Figure 2.2   Femto-aware spectrum arrangement scheme. (© [2012] IEEE)

2.3.2   Clustering of Small Cells

In [3], a framework is presented to reduce *DL* interference (*both cross-tier and co-tier*) and enhance the spectral efficiency for an OFDMA-based closed access femtocell network. In this framework, a Femtocell System Controller (FSC) per macrocell obtains all the necessary knowledge of HeNB system configuration (i.e., position information of HeNBs and macrocell UEs) and performs the necessary computations. To mitigate interference, the scheme encompasses a combination of dynamic frequency band allocation among HeNBs and MeNB, and clustering of HeNBs based on their geographical locations. In this scheme, a portion of the entire frequency band is dedicated to the MeNB users and the rest is reused by the MeNB and HeNBs. The advantage of allocating a portion of the frequency band strictly for MeNB users is that it can solve the MeNB UE DL deadzone problem and guarantee users' QoS requirement. However, the portion of the frequency band to be shared, is determined by the total number of HeNB clusters obtained through a clustering algorithm.

The clustering algorithm allocates HeNBs into different frequency reuse clusters and UEs of different HeNBs in the same cluster use the same sub-channels allocated from the shared frequency band. Based on the geographical locations of the HeNBs, the threshold distance for clustering is calculated. If the Euclidean distance between any two HeNBs is less than the threshold distance, then they are assigned to different clusters to avoid co-tier interference and CTI. Simulation results show that high spectrum efficiency is achieved as the probability of cross-tier spectrum reuse becomes high. This signifies that the problem regarding a macrocell UE DL deadzone around HeNBs is effectively solved. For the proposed scheme, simulation results also show a significant improvement of the femtocell user capacity.

In [4], an energy-efficient interference mitigation scheme is presented for closed access HeNBs grouped in a neighborhood area based on their geographical locations. In this scheme, inter-femtocell or *co-tier* interference among neighboring HeNBs is minimized by reducing the unnecessary Available Intervals (AI) in Low Duty Operation (LDO) mode for HeNBs, which is determined based on the pattern of Low Duty Cycle (LDC). According to the IEEE 802.16m standard, an HeNB in the operation state may enter the LDO mode if no UE exists in its coverage zone, or if all the UEs in the coverage are in sleep/idle mode. In the LDO mode, an HeNB switches alternately between available interval (AI) and unavailable interval (UAI) modes. During UAI, an HeNB becomes inactive on the air interface. During AI, the HeNB may become active on the air interface by transmitting preambles to the new incoming UE for synchronization purposes. However, the HeNB in the LDO mode still has AIs even though there is no UE that will access the HeNB in near future. These unnecessary AIs cause co-tier interference for CGS HeNBs. In the proposed scheme, the unnecessary AIs are reduced which results in reduction of co-tier interference among neighboring HeNBs. The main idea behind reducing such interference is to cluster/group the neighboring HeNBs based on their geographical locations. In each cluster, one HeNB is designated as the leader and the adjacent HeNBs are referred to as members.

According to the IEEE 802.16m standard, a newly installed HeNB scans the surrounding area to search for neighboring HeNBs in its initialization state. Since it is assumed that the global knowledge about the topology of the network is available, the scanning report may include the group configuration in the network (i.e., the leader and the members of the group based on the HeNB ID). If a newly installed HeNB receives the preamble signal from the leader above a defined threshold then it becomes a member of the group. Otherwise, the newly installed HeNB will form a new group and assign itself as the leader of the group. The leader requires AIs in its LDC pattern so that the arrival of a UE at the group can discover the existence of the group by detecting the leader, even though the members in the group stay in UAI. As soon as the leader senses the arrival of the UE, it sends a message to the target HeNB to activate its AI in the LDC pattern so that the UE can detect the target HeNB and connect to it. In this pattern, unnecessary AI in the LDO mode of HeNB is reduced resulting in power conservation of HeNB and at the same time the co-tier interference is minimized. Through analysis and simulation, it is shown that for the proposed scheme, the gain in terms of co-tier interference reduction time and energy saving is up to 90% in comparison with conventional LDO scheme in the IEEE 802.16m standard.

In [5], another clustering-based scheme is proposed to mitigate the intra-tier interference among CA-based femtocells. Note that in the LTE-Advanced system, two or more CCs belonging to contiguous or non-contiguous frequency bands can be aggregated to support a wider transmission bandwidth. Femtocells that are likely to have an interfering relationship are grouped into the same cluster and femtocells in the same cluster are allowed to exchange local information with each other. In the proposed scheme, for selection of CC, the femtocells consider interference experienced in that CC, and they cooperatively adjust their transmission power in order to optimize the total throughput on each CC.

2.3.3 Beam Subset Selection Strategy

The authors in [6] propose an orthogonal random beamforming-based CTI reduction scheme in closed access two-tier femtocell networks. The macrocell beam subset selection strategy is based on the number of macrocell UEs and the intensity of HeNBs in the network. The MeNB selects the beam subset and the users for each channel based on the signal-to-interference-plus-noise (SINR) information for all the channels which is fed back by the macrocell UEs. The main objective is to enhance network throughput by optimizing the trade-off between multiplexing gain and multiuser interference (cross-tier) based on adaptive selection of optimal number of beams using max-throughput scheduler at the MeNB. The adaptive selection of the number of beams decreases CTI, and provides spatial opportunity to HeNBs to access the spectrum in an opportunistic manner. In addition, distributed power control mechanism for HeNBs integrated with the proposed scheme reduces CTI significantly.

2.3.4 Collaborative Frequency Scheduling

Co-channel *UL* and *DL* CTI can be mitigated if an HeNB can avoid using the macrocell RBs that belong to its nearby macrocell UEs through efficient spectrum sensing. However, the spectrum sensing results for HeNB may be impaired due to misdetection, false alarm, and improper timing synchronization. To deal with this problem, a framework for OFDMA-based HeNBs is provided in [7] where the scheduling information for macrocell UEs' (both UL and DL) is obtained from the MeNB through a backhaul or air interface. This information is used to improve the spectrum sensing results for HeNB and to utilize the RBs associated with a far-away macrocell UE in the UL and DL transmissions. The key features of the proposed framework are as follows.

- HeNB receives the macrocell UEs scheduling information for UL and DL transmissions from the MeNB.
- HeNB performs spectrum sensing for finding the occupied parts of the spectrum. The occupied parts of the UL spectrum can be determined through energy detection.
- HeNB compares the spectrum sensing results with the obtained scheduling information to decide about the spectrum opportunities.

Since the HeNB accesses the spectrum in an opportunistic manner, the authors analyze the impact of Inter-carrier Interference (ICI) from macrocell UEs to femtocells which is severe in the UL transmission. The ICI is basically due to the asynchronous arrival of macrocell UE signals at the femtocell. Through simulations (using Okumura–Hata model of radio propagation), it is shown that the variation of ICI power depends on center frequency, height of the femtocell, and length of the Cyclic Prefix (CP). A lower center frequency and a higher femto BS height increase the received ICI power at the HeNB. In addition, if the macrocell UEs' signal arrival time at HeNB exceeds the CP duration then the orthogonality between the sub-carriers is disrupted leading to ICI. Also, different sub-carrier assignment schemes result in different ICI.

2.3.5   Power Control Approach

Power control methods for CTI mitigation generally focus on reducing transmission power of HeNBs. These methods are advantageous in that the MeNB and HeNBs can use the entire bandwidth with interference coordination. Dynamic or adjustable power setting, which is preferred over fixed HeNB power setting, can be performed either in proactive or reactive manner and each of which again can be executed either in open-loop power setting (OLPS) or closed-loop power setting (CLPS) mode. In the OLPS mode, an HeNB adjusts its transmission power based on its measurement results or predetermined system parameters (i.e., in a proactive manner). In the CLPS mode, the HeNB adjusts its transmission power based on the coordination with MeNB (i.e., in a reactive manner). Also, a hybrid mode can be used where the HeNB switches between the two modes according to the operation scenarios [9].

Another related concept is power control for HeNBs on a cluster basis in which the initial power setting for the HeNBs is done opportunistically based on the number of active femtocells in a cluster (Figure 2.3) [8]. For this, centralized sensing can be used by which an MeNB can estimate the number of active femtocells per cluster and broadcast the interference allowance information to femtocells for their initial power setting. Alternatively, distributed sensing can be used where each cell senses if the others are active in the same cluster and adjusts its initial power setting accordingly.

In [10], a joint frequency bandwidth dynamic division, clustering and power control algorithm (JFCPA) for DL transmission is proposed to control both intra-tier and inter-tier interferences and to improve the system efficiency. The overall system bandwidth is divided into three bands. The macro-cellular coverage is divided into two areas taking the intensity of the interference from the macro base station to the femtocells into account. Both of these are dynamically computed using the JFCPA. A cluster is defined as the unit for frequency reuse among femtocells. The problem of clustering is mapped to the MAX k-CUT problem to eliminate the inter-femtocell collision interference and a graph-based heuristic algorithm is developed to solve it.

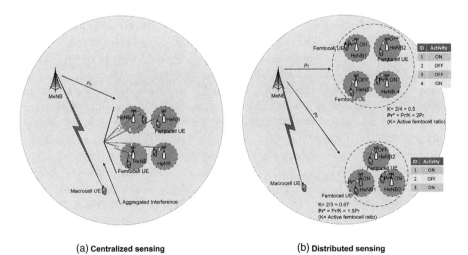

(a) **Centralized sensing**   (b) **Distributed sensing**

Figure 2.3   Sensing-based opportunistic power control (from [8]).

Frequency bandwidth sharing or splitting between the femtocell tier and the macrocell tier is determined by a step-migration-algorithm (SMA)-based power control method which allocates the minimal transmission power to every FAP such that the FUEs' SINR requirement is satisfied in any position of the femtocell.

For a single-carrier two-tier cellular system, in [11], the authors presented an estimation-based distributed UL power allocation approach for the FUEs to maximize their capacity in a fair manner. The UL power allocation problem for the FUEs in a macrocell–femtocell network can be modeled as a non-linear program (NLP) with an objective to maximizing the weighted sum-capacity of FUEs in different femtocells subject to the individual maximum power constraint at each FUE. However, this NLP can only be solved in a centralized manner. Therefore, the authors propose a distributed method in which, in each time slot, in each femtocell, the active FUE selects its power based on interference prediction and the SINR evolution. There is no exchange of any information among the FAPs, and among FAPs and macro base station. At every time slot, each FAP reports to its active FUE some interference related measurements to help the FUE select the transmission power.

Game theoretic models can be used to design and analyze distributed power control methods in a heterogeneous cellular wireless network with macrocells and femtocells. Two broad categories of game theoretic models are noncooperative and cooperative game models. In [12], a distributed power control problem is formulated for DL transmission in OFDMA-based femtocells overlaid upon a macrocell network. The problem is modeled as a noncooperative game, namely, a Stackelberg game, where the throughput of each station in the network is maximized under power constraints. In this game, the macrocell UEs are referred as the leaders and the femtocell UEs are considered to be the followers. The game is divided into two sub-games: a sub-game comprised of the set of leaders, referred to as the upper sub-game, and another sub-game comprised of the set of followers, referred to as the lower sub-game. The players in each sub-game compete with each other in a noncooperative manner to reach a sub-game Nash equilibrium, which is the solution of the power control game.

In [13], the EE problem for green communications is solved by using a power allocation approach to maximize BS EE when macrocells and femtocells share the same frequency spectrum. The telecom energy efficiency ratio (TEER) is defined as the ratio between effective system throughput and energy consumption of both macro BS and FAPs. An optimization problem is formulated to obtain power allocation for DL transmission in order to maximize local TEER while considering the reference FUE and the closely located interfered MUE (i.e., the CTI). Due to non-convex nature of the problem, a local optimum value is obtained by using the interior-point method for nonconvex NLPs.

2.3.6 Cognitive Radio Approach

Cognitive radio approach based on distributed spectrum sensing can be used for interference mitigation in femtocell networks. In [14], an efficient DL co-tier interference management scheme for an OFDMA-based LTE system is proposed where the path-loss information is shared among HeNB neighbors. In addition, adjacent HeNBs share the information related to the usage of LTE CCs, which are formed based on CA, in a distributed manner. The exchange of information between HeNBs may be done via femtocell gateway (HeNB GW) or over-the-air (OTA). The HeNB GW is an intermediate node between HeNBs and mobile core network that manages the inter-HeNB coordination messages via the S1 connection. On the other hand, the OTA method includes a direct link between HeNB and MeNB.

In the proposed scheme, when an HeNB is turned on, it identifies the adjacent neighbors and obtains the knowledge of the CCs used by the neighbors. The main idea of the scheme is that, each HeNB estimates the co-tier interference based on the path-loss information, capitalizes the knowledge of the usage of CCs by the neighbors, and accesses the spectrum intelligently to minimize interference. The selection of CC is done in such a way that, each HeNB selects the CC which is not used by the neighbor, or the CC that is occupied by the furthest neighbor, or the CC that is occupied by the least number of neighbors (in a chronological order as mentioned). Simulation results show a significant reduction in co-tier interference and signaling overhead within the network when compared with another cognitive based HeNB co-tier interference management technique.

Figure 2.4 illustrates a scenario of co-tier interference management (DL) of HeNBs through cognitive approach. In this scenario, the available CCs for HeNBs are CC1, CC2, CC3, and CC4. In Figure 2.4(a), since HeNB1 and HeNB3 are adjacent to each other, they select different CCs. On the other hand, since HeNB2 is a neighbor of neither HeNB1 nor HeNB3, it selects any one pair of available CCs (e.g., CC1 and CC2). Now, under such femtocell deployment, when HeNB4 is turned on, it discovers its adjacent neighbors, that is, HeNB1 and HeNB2. Through inter-HeNBs coordination mechanism, HeNB4 obtains the information related to the usage of CCs

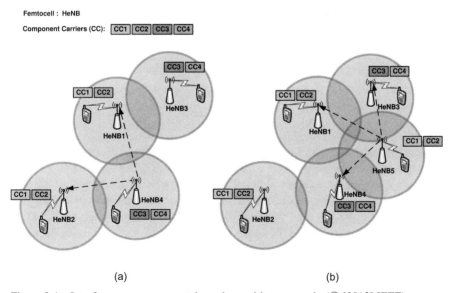

(a)   (b)

Figure 2.4   Interference management through cognitive approach. (© [2012] IEEE)

of its adjacent neighbors. Thus, in order to avoid co-tier interference, when HeNB4 selects CCs for the DL transmission, it selects the CCs (i.e., CC3 and CC4) which are different from those used by HeNB1 and HeNB2. Furthermore, in Figure 2.4(b), when HeNB5 is turned on, it identifies the adjacent neighbors (i.e., HeNB1, HeNB3, and HeNB4) and obtains the knowledge of the CCs used by the neighbors. Under these circumstances, HeNB5 selects the CCs that are occupied by the furthest neighbor, that is, HeNB1. To this end, HeNB5 selects CC1 and CC2 for DL transmission to reduce co-tier interference.

2.3.7 Fractional Frequency Reuse (FFR) and Resource Partitioning

The basic mechanism of this method divides the entire frequency spectrum into several sub-bands. Afterward, each sub-band is differently assigned to each macrocell or sub-area of the macrocell. Since the resources for the MeNB and HeNBs are not overlapped, interference between MeNB and HeNB can be mitigated.

In [15], the authors propose a frequency sharing mechanism that uses frequency reuse coupled with pilot sensing to reduce *cross-tier*/co-channel interference between macrocell and femtocells. In this scheme, FFR of three or above is applied to the macrocell. When an HeNB is turned on, it senses the pilot signals from the MeNB and discards the sub-band with the largest received signal power, and thus uses the rest of the frequency sub-bands resulting in an increased SINR for macrocell UEs. The overall network throughput is enhanced by adopting high order modulation schemes.

In [16], another interference management scheme for LTE femtocells is presented based on FFR. The scheme avoids *DL CTI* by assigning sub-bands from the entire allocated frequency band to the HeNBs that are not being used in the macrocell sub-area. In the proposed scheme, the macrocell is divided into a center zone (corresponding to 63% of the total macrocell coverage area) and an edge region including three sectors per each region. The reuse factor of one is applied in the center zone, while the edge region adopts the reuse factor of three. The entire frequency band is divided into two parts and one of them is assigned to the center zone. The rest of the band is equally divided into parts and assigned to the three edge regions.

Figure 2.5 illustrates the allocation of frequency sub-bands within the macrocell sub-areas. The sub-band A is used in the center zone ($C1$, $C2$, and $C3$), and sub-bands B, C, and D are used in regions $X1$, $X2$, and $X3$, respectively. Now, when an HeNB is turned on, it senses the neighboring MeNB signals, compares the Received Signal Strength Indication (RSSI) values for the sub-bands, and chooses the sub-bands which are not used in the macrocell sub-area. In addition, if the HeNB is located in the center zone, then it excludes the sub-band that is used in the center zone as well as the one that is used by the macrocell in the edge region of the current sector. For example, if an HeNB is located in edge region $X1$, then it would exclude sub-band B which is used by macrocell UEs, and select sub-band A, C, or D. However, if an HeNB is located in center zone $C1$, then it avoids sub-band A and at the same time sub-band B since the received signal strength indicator (RSSI) for this sub-band is comparatively higher for that HeNB. In this way, this scheme mitigates co-tier interference and CTI. Simulation results show that, the scheme offers throughput gains of 27% and 47% on

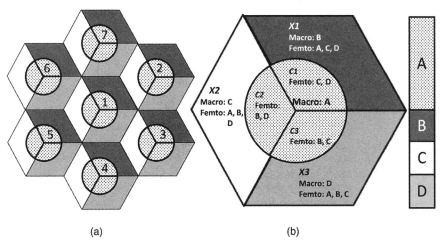

Figure 2.5   Interference management scheme using FFR. (© [2012] IEEE)

average, when compared with the FFR-3 scheme (with no center zone) and a scheme with no FFR, respectively.

To improve spectrum utilization when compared to that for orthogonal spectrum assignment for macrocell and femtocells, the authors in [17] use partially shared spectrum which is a solution between orthogonal and shared assignments. In this scheme, the macrocell and femtocell tiers access their own spectrum bands which overlap partially with each other.

The schemes described above use a fixed partitioning, which would cause a loss in throughput performance due to inefficient use of the bandwidth resources. A dynamic partitioning scheme (in both time and frequency domain) can be used for bandwidth sharing which minimizes CTI. In [18], an adaptive FFR scheme is presented to minimize DL interference caused by the HeNBs in the vicinities of a macrocell. The proposed scheme adopts FFR radio resource hopping or orthogonal FFR radio resource allocation based on the density (e.g., high or low) and location information (e.g., inner region or outer region) of the HeNBs. The location information of the HeNBs may be obtained and maintained within the network through using registered physical address associated with the broadband IP address that an HeNB uses. The proposed scheme only deals with the CTI posed by the HeNBs located (inner region) near the MeNB. If the HeNB is situated in a highly dense inner region, then orthogonal sub-channels are adopted by the HeNBs. Otherwise, the HeNB selects a sub-channel arbitrarily, utilizes it for a certain period of time, and then hops to other sub-channels. The proposed scheme reduces DL CTI.

The effect of spectrum allocation in two-tier networks is investigated in [19], where the macrocells employ closed access policy and the femtocells can operate in either open or closed access mode. By introducing a tractable model, the success probability for each tier under different spectrum allocation and femtocell access policies is derived. Both joint sub-channel allocation, in which the entire spectrum is shared by both tiers, as well as disjoint sub-channel allocation, whereby disjoint sets of sub-channels are assigned to both tiers, are considered. When femtocells are configured as closed access, each FAP is only accessible by its femtocell users, but when the femtocells are configured as open access, a FAP is accessed by both its users and all cochannel macro users. The throughput maximization problem is formulated subject to QoS constraints in terms of success probabilities and per-tier minimum rates.

Note that resource partitioning method can be used along with power control (thus resulting in a hybrid approach) to achieve *co-tier interference* and *CTI*. In [20], a joint sub-band, rate and power allocation scheme is proposed according to the interference statistics. Each femtocell determines its best sub-band, rate and power in a distributed manner in order to reach the maximum throughput on average.

2.3.8   Resource Scheduling Strategies

According to the different QoS requirements of the users, the resource scheduling schemes in OFDMA systems are generally classified into the following two categories: rate adaptive (RA) schemes and margin adaptive (MA) schemes [21, 22]. In an RA scheme, it is assumed that the overall transmission power is fixed and the objective is to allocate resources (sub-carrier and power) so that the total power does not exceed the overall transmission power limit. The fairness of resource allocation is achieved by using different techniques such as maximizing the minimum capacity among all users, using a proportional fairness constraint or using an objective function called utility function. In an MA scheme, each user has a fixed QoS demand (e.g., data rate and bit error rate [BER]) and the objective is to minimize the total transmission power while satisfying the QoS requirements of these users. MA-based schemes consider networks where all users have QoS demands.

Rate-adaptive scheduling schemes: In [23], a genetic algorithm is used to solve the resource allocation problem for macrocell–femtocell networks. The optimization of resource allocation is equivalent to the maximization of the total achievable throughput in the system, which is calculated using the Shannon's formula subject to the constraints on total available bandwidth, power, and capacity in macrocell and femtocells. The algorithm uses two stages: BS selection and bandwidth and power allocation. Some users are allowed to be connected to femtocells together with the femtocell subscribers. A decision on BS selection is made for each user to determine whether the user gets access to the core network via a macrocell or a nearby femtocell. A subscriber is given priority to connect to its own femtocell while a non-subscribing user can join it if the user is located inside the femtocell's coverage as long as the connection of the user does not affect the transmissions of the femtocell subscribers. After BS selection, a dual bandwidth and power assignment is implemented to distribute the available resources among users to maximize the overall system throughput. A portion of the available resource is allocated to each user depending on the demand and location of the user. The bandwidth assigned to each user in macro tier is less than or equal to demand but lower than the maximum allowed capacity in the zone the user is located at. In femto tier, bandwidth allocated to each user is kept less than or equal to the demand. The fitness function employed in the genetic algorithm is the objective function of the optimization problem which is performed after the random generation of the first population.

In [24], the resource allocation problem is formulated as a linear program (LP) which aims to maximize the system sum-rate under the proportional user rate constraint. The proportional rate constraint for users is regarded as a fairness criterion to balance the trade-off between the system sum-rate and user fairness.

A graph-based channel allocation scheme is proposed in [25] to maximize the system throughput while ensuring proportional fairness in rate among femtocells. The received signal-to-interference-plus-noise-ratio (SINR) is explicitly used by this scheme to generate the interference graph for the OFDMA femtocell network so as to guarantee the acceptable ICI for all the links. First, all the femtocells are partitioned into different groups by applying a greedy graph coloring algorithm to maximize the sum throughput of each group. The femtocells in the same group are given permission to share the assigned sub-channels while those in different groups are assigned orthogonal sub-channels. Then, an optimization problem is formulated to estimate the number of sub-channels assigned to each group. An approximation method is suggested to solve the optimization problem.

Margin-adaptive scheduling schemes: A distributed cell selection and resource allocation (CS–RA) mechanism is suggested in [26], where the CS–RA processes are performed by the UEs independently. The CS–RA problem is formulated as a two-tier game, namely, an inter-cell game and an intra-cell game. In the first tier, each UE selects the best cell according to its cell selection strategy which is stochastic, that is, there is a set of probabilities each representing the probability of selecting a particular cell. From the expected payoff, the optimal strategies of the UEs are derived, which depend not only on the channel qualities, but also on the load distribution of all cells and the strategies of other MSs. In the second tier, that is, in the intra-cell game, the UEs within the same cell choose the proper radio resource, typically sub-channels and power, to achieve their maximum payoff. Distributed algorithms, namely, the CS and RA algorithms are proposed to enable the independent UEs to converge to the Nash equilibria.

In [27], the available spectrum resource is dynamically used by each femtocell according to the macrocell resource usage. The proposed scheme gives priority to macro users and the resource allocation to the macro users is performed based on traditional optimization formulations. The concept of neighboring area around a femtocell is used which the macro users are vulnerable to CTI due to transmissions from the femto access points. A femtocell only uses the sub-carriers which are not used by the macro users inside the neighboring area. The resource allocation among the femtocells is modeled as a noncooperative game and a utility function is adopted that considers the benefits from sub-carrier and power allocation to the femto users.

In [28], a channel allocation scheme for two-tier OFDMA femtocell networks is recommended, where the soft frequency reuse (SFR) strategy is adopted by macro BSs. That is, a macrocell is divided into two regions: inner region and outer (or edge) region. For the MUEs located in the outer regions in macrocells, the frequency spectrum is equally partitioned across cells based on a reuse factor N. The MUEs in the inner regions of the macrocells use the sub-bands allocated to the outer regions of the neighboring cells. Considering that the macro user equipments (MUEs) have priority over the femtocell users (FUEs), the sub-bands are first assigned to MUEs utilizing SFR such that the ICI is mitigated. Subsequently, sub-bands are allocated to the femtocells such that CTI is mitigated. The scheme is optimized in terms of EE of the entire network (i.e., throughput per unit energy consumption), while guaranteeing that both the MUEs and FUEs attain at least a given minimum data rate.

A more detailed exposition to the resource scheduling problem in OFDMA networks will be provided in Chapter 3.

2.4 QUALITATIVE COMPARISON AMONG INTERFERENCE MANAGEMENT APPROACHES [30]

The "efficiency" of a scheme depends on whether it (i) mitigates/significantly reduces both co-tier interference and CTI; (ii) is applicable for both UL and DL transmissions; (iii) considers coordination among HeNBs and MeNB, or capitalizes on minimal amount of information, that is, path-loss, geographical location, or usage of the spectrum or sub-band among nearby HeNBs and/or among HeNBs and MeNB; (iv) handles ICI (e.g., by using frequency scheduling or any other method); (v) adopts an adaptive power control mechanism; (vi) corresponds to opportunistic access of the spectrum by the HeNBs based on RSSI value from MeNB signals; (vii) reduces the unnecessary AIs of LDO mode for HeNBs; (viii) is scalable and robust, that is, implementable for mass deployment of HeNBs; and (ix) is applicable for all three types of access modes (i.e., closed, open, and hybrid).

For example, the efficiency of *cognitive approach* can be considered to be moderate since it is capable of handling both CTI and co-tier interference with minimal amount of information (i.e., information about usage of sub-bands) exchange among neighboring HeNBs, applicable for all types of access modes of HeNBs, and more importantly, it accesses the spectrum in an opportunistic manner causing minimal harm to the nearby macrocell UEs. The *collaborative frequency scheduling* scheme can be considered to be highly efficient since it significantly reduces CTI and co-tier interference for mass deployment of HeNBs in both UL and DL transmission, handles ICI problem, and allows the HeNBs to opportunistically access the spectrum based only on the scheduling information of macrocell UEs that is exchanged between HeNBs and MeNB.

The "complexity" of each scheme increases with (i) the amount of information exchanged among neighboring HeNBs, (ii) the amount of information exchanged between HeNBs and MeNB, (iii) formation of clusters among HeNBs, (iv) algorithm executed in the HeNBs and/or in the MeNB to allow the HeNBs to access the spectrum opportunistically, etc. The more information exchanged among HeNBs or between HeNBs and MeNB, the more signaling overhead is introduced, and more processing is done in both HeNBs and MeNB, increasing the complexity of the scheme. For example, the complexity of the *beam subset selection strategy* scheme can be considered to be *high* since it requires the channel state information from all macrocell UEs to determine the optimal number of beams every time along with extensive coordination between HeNBs and MeNB regarding the spectrum access (thus increasing the signaling overhead). Also, the HeNBs have to run iterative power control algorithm to minimize interference.

Selection of an interference management scheme depends on the desired trade-off between complexity and efficiency. FFR can be considered to be a viable interference management scheme for two-tier femtocell networks due to the following reasons. First, the FFR requires minimal/no coordination among HeNBs and MeNB (and hence reduces the signaling overhead, and thus the complexity of the system), opportunistically accesses the spectrum based on only RSSI value from MeNB signals. Second, the FFR effectively solves the problem of CTI and co-tier interference in UL and DL transmissions for different access modes of HeNBs. Consequently, the FFR can increase the throughput of the network by a large margin, and can be used when the average number of HeNBs per macrocell is very high (about 180–200) while maintaining the QoS requirements of macrocell UEs. The FFR has been recommended as an effective interference management scheme for OFDMA-based two-tier networks [29].

2.5 FUTURE RESEARCH DIRECTIONS [30]

To enable mass deployment of femtocells, it is essential to develop distributed interference management schemes which primarily satisfy the QoS requirements of macrocell and femtocell UEs and at the same time enhance the capacity and coverage of the network. Such schemes should incur low overhead for coordination among macrocell BSs (i.e., MeNBs), and should also be able to integrate mobility management with different access modes and to address synchronization issues while keeping the complexity as minimal as possible. The interference management solution would strongly depend on the employed radio access technology (e.g., CDMA or OFDMA) and access mode (i.e., closed, open, or hybrid). In particular, adaptive admission control, power control, and advanced communication strategies such as interference cancellation and beamforming for multiple-antenna transceivers are important techniques to mitigate co-tier interference and CTI. For example, by using beamforming techniques femtocells can form antenna beams toward their UEs while nulling interference caused to macrocell UEs. In addition, macrocells would have higher priority in accessing the spectrum; therefore, suitable admission control mechanisms should be activated when femtocells create intolerable interference for macrocell UEs.

For OFDMA-based femtocell networks, if different sets of sub-channels are assigned to macrocells and femtocells, the CTI can be completely eliminated. However, to improve the spectrum utilization, a more efficient spectrum assignment method can be adopted.

Also, hybrid interference management schemes which combine power control with resource partitioning are promising. Power control schemes are advantageous in that MeNB and HeNB can use the entire bandwidth with interference coordination for both control and data channels. However, for this, the HeNB measurement scheme for power setting would need to be standardized. Also, such a scheme may not be fully effective when a macro UE is located very close to an HeNB. With resource partitioning schemes, interference between MeNB and HeNB can be eliminated.

However, multiple frequency bands are required. The merits of both the approaches can be exploited in a hybrid scheme, the design of which is not trivial.

An important consideration in the design of resource management algorithms for multi-tier networks is minimizing the coordination among the small cells and that between the small cells and macrocells. Also, the limited capacity of the backhaul network needs to be considered.

Energy consumption in the cellular wireless networks has become a critical issue and EE has become a very important system design parameter. With the deployment of a large number of small cells, this EE issue will become even more significant. Therefore, the resource allocation algorithms for multi-tier networks should be designed to improve the EE of the overall network. For example, the resource allocation schemes can be designed for small cells such that the overall power consumption is minimized while satisfying the QoS requirements and maintaining fairness among users.

The provisioning of CA in the evolving LTE-Advanced systems adds another dimension to the resource allocation problem. With CA, two or more CCs can be aggregated to support transmission bandwidths up to 100 MHz. In a multi-tier network, the RBs corresponding to the different CCs can be shared among the macro and the small cell users considering the transmission characteristics of the different CCs as well as the CA capabilities of the users (e.g., due to power limitations at the UEs). Effective resource allocation schemes need to be designed for these scenarios.

REFERENCES

1. M. Salem, A. Adinoyi, H. Yanikomeroglu, and D. Falconer, "Opportunities and challenges in OFDMA-based cellular relay networks: A radio resource management perspective," *IEEE Transactions on Vehicular Technology*, vol. 59, no. 5, pp. 2496–2510, 2010.

2. W. Yi, Z. Dongmei, J. Hai, and W. Ye, "A novel spectrum arrangement scheme for femtocell deployment in LTE macrocells," in *Proceedings of IEEE 20th Symposium on Personal, Indoor and Mobile Radio Communications*, pp. 6–11, 13–16 September 2009.

3. H. Li, X. Xu, D. Hu, X. Qu, X. Tao, and P. Zhang, "Graph method based clustering strategy for femtocell interference management and spectrum efficiency improvement," in *Proceedings of IEEE 6th International Conference on Wireless Communications Networking and Mobile Computing (WiCOM)*, pp. 1–5, 23–25 September 2010.

4. H. Widiarti, S. Pyun, and D. Cho, "Interference mitigation based on femtocells grouping in low duty operation," in *Proceedings of IEEE 72nd Vehicular Technology Conference Fall (VTC'10-Fall)*, pp. 1–5, 6–9 September 2010.

5. B. Wang, Y. Zhang, W. Wang, M. Lei, and L. Jiang, "A cooperative downlink power setting scheme for CA-based femtocells," in *Proceedings of 2012 IEEE 75th Vehicular Technology Conference (VTC-Spring)*, 2012.

6. S. Park, W. Seo, Y. Kim, S. Lim, and D. Hong, "Beam subset selection strategy for interference reduction in two-tier femtocell networks," *IEEE Transactions on Wireless Communications*, vol. 9, no. 11, pp. 3440–3449, November 2010.

7. M. E. Sahin, I. Guvenc, Moo-Ryong Jeong, H. Arslan, "Handling CCI and ICI in OFDMA femtocell networks through frequency scheduling," *IEEE Transactions on Consumer Electronics*, vol. 55, no. 4, pp. 1936–1944, November 2009.

8. M. S. Jin, S. Chae, and D. I. Kim, "Per cluster based opportunistic power control for heterogeneous networks," in *Proceedings of IEEE VTC'11-Spring*, Budapest, Hungary, May 2011.

9. 3GPP R1-105238, "Further Discussion on HeNB Downlink Power Setting in HetNet," 3GPP RAN1 Meeting, Xian, China, October 2010.

10. H. Li, X. Xu, D. Hu, X. Tao, P. Zhang, S. Ci, and H. Tang, "Clustering strategy based on graph method and power control for frequency resource management in femtocell and macrocell overlaid system," *Journal of Communications and Networks*, vol. 13, no. 6, December 2011, pp. 664–677.

11. N. Chakchouk and B. Hamdaoui, "Estimation-based non-cooperative power allocation in two-tier femtocell networks," in *Proceedings of 2011 IEEE Global Telecommunications Conference (GLOBECOM 2011)*, 5–9 December, 2011, pp. 1–6.

12. S. Guruacharya, D. Niyato, D. I. Kim, and E. Hossain, "Hierarchical competition for downlink power allocation in OFDMA femtocell networks," *IEEE Transactions on Wireless Communications*, vol. 12, no. 4, pp. 1543–1553, April 2013.

13. B. Han, W. Wang, and M. Peng, "A power allocation scheme for achieving high energy efficiency in two-tier femtocell networks," in *Proceedings of 2011 IEEE 13th International Conference on Communication Technology (ICCT)*, 2011.

14. L. Zhang, L. Yang, and T. Yang, "Cognitive interference management for LTE-A femtocells with distributed carrier selection," in *Proceedings of IEEE 72nd Vehicular Technology Conference Fall (VTC 2010-Fall)*, pp. 1–5, 6–9 September 2010.

15. T. Kim and T. Lee, "Throughput enhancement of macro and femto networks by frequency reuse and pilot sensing," in *Proceedings of IEEE International Performance, Computing and Communications Conference (IPCCC)*, pp. 390–394, December 2008.

16. L. Poongup, L. Taeyoung, J. Jangkeun, and S. Jitae, "Interference management in LTE femtocell systems using fractional frequency reuse," in *Proceedings of 12th International Conference on Advanced Communication Technology*, vol. 2, pp. 1047–1051, 7–10 February 2010.

17. L. Yang, L. Zhang, T. Yang, and W. Fang, "Location-based hybrid spectrum allocation and reuse for tiered LTE-A networks," in *Proceedings of IEEE 73rd Vehicular Technology Conference (VTC Spring)*, 2011.

18. R. Juang, P. Ting, H. Lin, and D. Lin, "Interference management of femtocell in macrocellular networks," in *Proceedings of Wireless Telecommunications Symposium (WTS'10)*, pp. 1–4, 21–23 April 2010.

19. W. C. Cheung, T. Q. S. Quek, and M. Kountouris, "Throughput optimization, spectrum allocation, and access control in two-tier femtocell networks," *IEEE Journal on Selected Areas in Communications*, vol. 30, pp. 561–574, 2012.

20. Z. Zhang, K. Wu, and A. Huang, "Optimal distributed subchannel, rate and power allocation algorithm in OFDM-based two-tier femtocell networks," in *Proceedings of IEEE 71st Vehicular Technology Conference (VTC 2010-Spring)*, 16–19 May 2010, pp. 1–5.

21. S. Bashar and Z. Ding, "Efficient algorithms for resource allocation in heterogeneous OFDMA networks," in *Proceedings of IEEE Global Telecommunications Conference (Globecom)*, November 30–December 4, 2008.

22. C. Liu, A. Schmeink, and R. Mathar, "Dual optimal resource allocation for heterogeneous transmission in OFDMA systems," in *Proceedings of IEEE Global Telecommunications Conference (Globecom)*, November 30–December 4, 2009.

23. H. Otrok, H. Barada, R. Estrada, A. Jarray, and Z. Dziong, "Resource allocation in macrocell-femtocell network using genetic algorithm," in *Proceedings of 2012 IEEE 8th International Conference on Wireless and Mobile Computing, Networking and Communications (WiMob)*, 2012.

24. P. Xue, P. Gong, J. H. Park, D. Park, and D. K. Kim, "Radio resource management with proportional rate constraint in the heterogeneous networks," *IEEE Transactions on Wireless Communications*, vol. 11, pp. 1066–1075, 2012.

25. K. Zheng, Y. Yang, C. Lin, X. Shen, and J. Wang, "Graph-based interference coordination scheme in orthogonal frequency-division multiplexing access femtocell networks," *IET Communications*, vol. 5, pp. 2533–2541, 2011.

26. L. Gao, X. Wang, G. Sun, and Y. Xu, "A game approach for cell selection and resource allocation in heterogeneous wireless networks," in *Proceedings of 2011 8th Annual IEEE Communications Society Conference on Sensor, Mesh and Ad Hoc Communications and Networks (SECON)*, 2011, pp. 530–538.

27. S. Han, B.-H. Soong, and Q. D. La, "Interference mitigation in resource allocation for OFDMA-based macro/femtocell two-tier wireless networks," in *Proceedings of 2012 IEEE International Symposium on Broadband Multimedia Systems and Broadcasting (BMSB)*, 2012, pp. 1–6.

28. W. Li, W. Zheng, H. Zhang, T. Su, and X. Wen, "Energy-efficient resource allocation with interference mitigation for two-tier OFDMA femtocell networks," in *Proceedings of 2012 IEEE 23rd International Symposium on Personal Indoor and Mobile Radio Communications (PIMRC)*, 2012.

29. Femto Forum, available at: http://www.femtoforum.org/femto. (http://www.smallcellforum.org)

30. 3GPP R1-106052, "Per Cluster Based Opportunistic Power Control," 3GPP RAN1 Meeting, Jacksonville, USA, November 2010.

CHAPTER *3*

RESOURCE ALLOCATION IN OFDMA-BASED MULTI-TIER CELLULAR NETWORKS

With the adoption of OFDMA as the radio access technology in the current and next-generation wireless networks such as the WiMAX and LTE/LTE-Advanced networks, radio resource allocation for OFDMA-based cellular networks has become a very important research topic. Early works in this area have focused on the single-cell scenario where resource allocations under different fairness criteria such as max–min and proportional fairness were considered for DL [1–8], and UL [9, 10]. Since these resource allocation problems are mixed integer programs, various low-complexity algorithms have been developed most of which are based on decomposed sub-channel and power allocation solution. More recently, there have been some works that proposed resource allocation algorithms considering multi-cell coordination and interference management where sub-channels are reused over the cells [11–16]. In general, resource and interference management problems for OFDMA-based networks are NP-hard under most fairness criteria [17]. Therefore, finding optimal solutions for these resource allocation problems typically involves exponential computational complexity. As a result, most low complexity resource allocation algorithms for OFDMA-based wireless networks seek to achieve sub-optimal but efficient solutions considering various QoS and fairness constraints.

In the multi-tier context, simple spectrum allocation methods [18–20] have been proposed. Again, RRM in the multi-tier heterogeneous cellular network (HetNet) must address the differentiated spectrum access priorities, more diverse QoS requirements of different tiers, and also the tier association problem.

In this chapter, we first review some existing works on resource allocation for OFDMA-based homogeneous cellular wireless networks. Then, we present an exemplary resource allocation design for OFDMA-based HetNets. Specifically, a max–min fair resource allocation framework with macrocell protection is described for OFDMA-based two-tier HetNets. We then reveal the optimal structure of the problem and present an optimal algorithm to resolve it. A distributed and low complexity

Radio Resource Management in Multi-Tier Cellular Wireless Networks, First Edition.
Ekram Hossain, Long Bao Le, and Dusit Niyato
© 2014 John Wiley & Sons, Inc. Published 2014 by John Wiley & Sons, Inc.

algorithm using an adaptive weight-based sub-channel allocation is also described. Then, we summarize the chapter and discuss some potential research directions.

3.1   RESOURCE ALLOCATION FOR OFDMA-BASED HOMOGENEOUS NETWORKS

We present some important resource allocation techniques and existing algorithms for single-cell OFDMA networks. In such networks, the key difference between the DL and UL resource allocation lies in the power constraints where there is a single- power constraint at the BS for the DL while each individual user imposes one corresponding power constraint in the UL. Moreover, the employment of an appropriate solution approach to solve the underlying problem usually depends on the fairness constraint imposed in the resource allocation problem. Solving a resource allocation problem requires to determine how to allocate sub-channels and the corresponding power levels to users considering the fairness and power constraints. To perform such resource allocation, either continuous or discrete rate models are assumed, which are described in the following.

3.1.1   Physical-Layer Model

The rate model describes how the achievable rate on a particular sub-channel varies with the allocated power or the corresponding SINR. One common rate model widely used in the resource allocation literature specifies the relationship between rate and SINR as follows:

$$r = \log_2 (1 + \beta \gamma) \tag{3.1}$$

where r is the normalized rate in b/s/Hz (i.e., the spectral efficiency) or the number of "bits" per OFDM symbol for discrete rates, γ denotes the SINR (or SNR if there is no interference), and β denotes the gap to capacity. In particular, the parameter β can be approximated when the adaptive modulation using MQAM is employed as $\beta = -1.5/\ln(5\overline{P}_e)$ where $\overline{P}_e$ denotes the target BER value [21]. Using (3.1) and this relationship for β, we can express the required SINR for a particular normalized rate r as follows:

$$\overline{\gamma}(r) = -\frac{\ln(5\overline{P}_e)}{1.5} \left(2^r - 1\right). \tag{3.2}$$

Although this formula can be used for both continuous and discrete rate, more accurate calculation can be performed if the explicit BER expression for the discrete rate adaptation is given. For example, suppose that MQAM modulation is adopted for communication on each sub-channel where the constellation size s for s-QAM is chosen from a predetermined set $\mathfrak{M}$. Let $\overline{\gamma}(s)$ be the required SINR value to guarantee acceptable performance in terms of BER when a constellation size s is adopted on a particular sub-channel. To determine $\overline{\gamma}(s)$, let $f_s(\gamma(s))$ be the expression of BER with respect to SINR $\gamma(s)$ when the constellation size s is adopted. Then, the SINR must satisfy the following requirement when the constellation size s is employed

$$\gamma(s) \geq \bar{\gamma}(s) = f_s^{-1}\left(\overline{P}_e\right) \tag{3.3}$$

where $f_s^{-1}(\cdot)$ denotes the inverse function of the BER function $f_s(\gamma(s))$, and $\overline{P}_e$ is the target BER value.

3.1.2 Downlink Single-Cell Resource Allocation

DL resource allocation for single-cell OFDMA systems with user rate requirements to minimize the total transmission power is one of earliest problems considered in this context [1]. In fact, the DL resource allocation for single-cell OFDMA under max–min fairness criterion is closely related to this problem. These two problems are referred to as margin adaptation (MA) and rate adaptation (RA) problems, respectively [5]. To describe formulations for these problems, suppose there are N users and K sub-channels in the considered cell. Let us denote the channel gain from the BS to user i on sub-channel n as h_i^n and the noise power at user i on each sub-channel n as σ_i^n. Then, the SNR achieved by user i on sub-channel n can be expressed as follows:

$$\gamma_i^n = \frac{h_i^n p_i^n}{\sigma_i^n} \tag{3.4}$$

where p_i^n is the transmission power at the BS to user i on sub-channel n. Using this relationship, we can find the required transmission power to achieve a particular rate r_i^n as follows:

$$p_i^n\left(r_i^n\right) = \overline{\gamma}\left(r_i^n\right) \frac{\sigma_i^n}{h_i^n} \tag{3.5}$$

where $\overline{\gamma}(r_i^n)$ is the required SNR to support rate r_i^n given in Section 3.1.1. Let a_i^n represent the sub-channel allocation variables where $a_i^n = 1$ if sub-channel n is allocated for user i and $a_i^n = 0$, otherwise. For brevity, we use a vector **a** to represent all sub-channel allocation variables for all users over all sub-channels.

Margin adaptive optimization: In the MA optimization problem, each user i requires to achieve a particular rate R_i. The MA optimization problem can be formulated as follows:

$$\min \sum_{i=1}^{N} \sum_{n=1}^{K} a_i^n \overline{\gamma}\left(r_i^n\right) \frac{\sigma_i^n}{h_i^n} \tag{3.6}$$

$$\text{subject to} \sum_{n=1}^{K} a_i^n r_i^n = R_i, \quad \forall i, \tag{3.7}$$

$$\sum_{i=1}^{N} a_i^n = 1, \quad \forall n. \tag{3.8}$$

In this formulation, we aim to minimize the total transmission power in (3.6) subject to the rate constraints in (3.7). In addition, it is required that each sub-channel can only be allocated to exactly one user (3.8).

Rate adaptive optimization: The RA optimization can be formulated as follows:

$$\max_{\mathbf{a}} \min_{i} \sum_{n=1}^{K} a_i^n r_i^n \tag{3.9}$$

$$\text{subject to} \sum_{i=1}^{N} \sum_{n=1}^{K} a_i^n \overline{\gamma}(r_i^n) \frac{\sigma_i^n}{h_i^n} \leq P_{\text{BS}} \tag{3.10}$$

$$\sum_{i=1}^{N} a_i^n = 1, \quad \forall n \tag{3.11}$$

where we aim to maximize the minimum rate achieved by all users in (3.9) (i.e., max–min fairness) and P_{BS} represents the power budget at the BS. Here, $\sum_{n=1}^{K} a_i^n r_i^n$ represents the rate achieved by user i. Moreover, it is required that the total transmission power at the BS is smaller or equal to its power budget in (3.10). Finally, we also impose the sub-channel allocation constraints in (3.11) as in the MA optimization problem.

Relationship Between MA and RA optimization problems: The optimal solution of the RA problem can be obtained by using the optimal solution of the MA problem. Specifically, suppose we know how to find the optimal solution of the MA problem for any set of required rates of all users. Now, suppose all users require to achieve the same rate R. Then, we can find the optimal solution of the RA problem by finding the maximum common rate R that can be achieved with the maximum power budget P_{BS} of the BS. This can be obtained by iteratively applying the solution of the MA problem. We describe an iterative algorithm for the discrete and integer rate adaptation in the following algorithm [5]. We assume that users can achieve only integer normalized rate on each allocated sub-channel, which can be realized by adaptive modulation in practice.

For the continuous rate, a similar algorithm can be developed by slowly increasing the common rate R in step 4 of the above algorithm as long as the power constraint

---

Algorithm 3.1 RA OPTIMIZATION VIA RECURSIVELY SOLVING MA PROBLEM

1: Initialization: Set common rate $R = 1$.
2: Solve the MA optimization problem with current common rate requirements $R_i = \sum_{n=1}^{K} a_i^n r_i^n = R$ and find minimum required power $P_{\text{tot}}(R) = \sum_{i=1}^{N} \sum_{n=1}^{K} a_i^{*n} p_i^n(r_i^{*n})$ where a_i^{*n} and r_i^{*n} denote the optimal sub-channel and rate allocations, respectively.
3: **while** $P_{\text{tot}}(R) < P_{\text{BS}}$ **do**
4:    Increase the common rate $R = R + 1$.
5:    Solve the MA optimization problem with current common rate requirements $R_i = \sum_{n=1}^{K} a_i^n r_i^n = R$ and find minimum required power $P_{\text{tot}}(R)$ accordingly.
6: **end while**
7: Optimal solution of the RA problem is the optimal solution of the MA problem with common rate $R - 1$.

---

at the BS is still satisfied. Specifically, we update the common rate in each step as $R = R + \epsilon$ where ϵ is a predetermined small number. Due to the above relationship between the MA and RA optimization problems, it is sufficient to consider only the MA optimization problem.

Efficient and low complexity algorithm for MA problem: The MA problem can be transformed into an integer linear program (ILP) by combing the sub-channel and rate allocation variables a_i^n and r_i^n into one single variable [5]. Therefore, we can obtain the optimal solution of the MA problem by using any available ILP solver such as CPLEX. However, computation time needed to solve such ILP can be very large, which may prevent it from being useful in practice. Therefore, a low complexity algorithm that can find an efficient solution for the MA problem is more desirable. There are several such algorithms that have been proposed in the literature [1,3–5]. In fact, a low complexity algorithm developed by relaxing integer sub-channel allocation variables to real variables and solving the Karush–Kuhn–Tucker (KKT) conditions of the corresponding relaxed problem was the first proposed in the literature [1]. Other greedy algorithms with comparable performance but lower complexity were later proposed in [3,4].

Low complexity algorithms typically aim to determine an efficient sub-channel allocation solution in the first step and power allocation is performed in the second step for the given sub-channel allocation. In the following, we describe the sub-channel allocation technique proposed in [5]. This solution approach is developed based on the observation that the number of bits (i.e., normalized discrete rate) loaded on different sub-channels for any particular user is roughly the same in optimality. This observation was indeed verified through numerical studies in [5]. Now, suppose that user i is allocated m_i sub-channels and the rate achieved on each allocated sub-channel is equal to r_i. Then, the MA optimization problem (3.6–3.8) can be rewritten as follows:

$$\min \sum_{i=1}^{N} \sum_{n=1}^{K} p_i^n(r_i) a_i^n \tag{3.12}$$

$$\text{subject to} \sum_{n=1}^{K} a_i^n = \frac{R_i}{r_i}, \quad \forall i \tag{3.13}$$

$$\sum_{i=1}^{N} a_i^n = 1, \quad \forall n. \tag{3.14}$$

The constraints (3.13) come from the constraints (3.7) where we have used the following relationship $R_i = m_i r_i$. For given channel gains, the quantities $p_i^n(r_i)$ in the objective function (3.12) can be considered as known sub-channel assignment weights. Hence, the optimization problem (3.12–3.14) is a transportation problem whose optimal solution can be obtained by solving the corresponding relaxed linear program (LP) [22,23]. Therefore, we can obtain the sub-channel allocation solution for all users if we can determine the value r_i of each user i. To fulfil this task, we estimate the average required transmission power to support a particular rate r_i for user i as follows:

$$\overline{p}_i(r_i) = \overline{\gamma}(r_i) \frac{\overline{\sigma}_i}{\overline{h}_i} \tag{3.15}$$

where $\overline{h}_i = 1/K \sum_{n=1}^{K} h_i^n$ and $\overline{\sigma}_i = 1/K \sum_{n=1}^{K} \sigma_i^n$ are the average channel gain and noise power of each user i over all sub-channels, respectively. Then, the MA optimization problem (3.6–3.8) can be simplified to

$$\min \sum_{i=1}^{N} \overline{p}_i(r_i) \frac{R_i}{r_i} \qquad (3.16)$$

$$\text{subject to} \quad \sum_{i=1}^{N} \frac{R_i}{r_i} = K. \qquad (3.17)$$

Note that the rate on every assigned sub-channel for user i is assumed to be equal to r_i. Therefore, R_i/r_i gives the number of sub-channels required to support rate R_i for user i. The constraint (3.17) holds since there are K sub-channels to be shared among all users. To solve the problem (3.16–3.17), we define $x_i = 1/r_i$ and obtain the following equivalent optimization problem

$$\min \sum_{i=1}^{N} \overline{p}_i\left(\frac{1}{x_i}\right) R_i x_i \qquad (3.18)$$

$$\text{subject to} \quad \sum_{i=1}^{N} R_i x_i = K. \qquad (3.19)$$

For practical systems, the power-rate function $p_i(\cdot)$ is convex. Therefore, the optimization problem (3.18–3.19) is convex, whose optimal solution can be found from the KKT conditions. Specifically, we can write the Lagrangian for this problem as follows:

$$L(x_i, \lambda) = \sum_{i=1}^{N} \overline{p}_i\left(\frac{1}{x_i}\right) R_i x_i - \lambda \left(\sum_{i=1}^{N} R_i x_i - K \right). \qquad (3.20)$$

where λ represents the Lagrange multiplier. Differentiate $L(x_i, \lambda)$ with respect to x_i and set the result to zero we obtain

$$\overline{p}_i\left(\frac{1}{x_i}\right) - \overline{p}_i'\left(\frac{1}{x_i}\right)\frac{1}{x_i} - \lambda = 0, \quad \forall i. \qquad (3.21)$$

Using this set of N equations together with the constraint on the number of sub-channels, that is,

$$\sum_{i=1}^{N} R_i x_i = K \qquad (3.22)$$

we can find the optimal values of x_i^* and λ^*. It is difficult to find closed-form expressions for these optimal variables since these equations involve non-linear functions. However, we can employ the vector-based Newton's method to find its solution [24]. Once we have determined x_i^*, we are able to calculate $r_i^* = 1/x_i^*$. Then, we can find the sub-channel allocation solution by solving the transportation problem (3.12–3.14). Note that the right-hand side of the constraints (3.13) must be integer numbers; therefore, we must round the quantities R_i/r_i^* to the corresponding integer number.

The procedure presented above allows us to determine a feasible sub-channel allocation for all users. To complete the resource allocation, we have to perform power allocation for all sub-channels based on the obtained sub-channel allocation solution. In fact, optimal power allocation can be performed by using the well-known bit-loading algorithm if the rate is discrete [1,5]. Recall that R_i denotes the required rate of user i. Also, let S_i be the set of sub-channels allocated for user i. The bit-loading algorithm is presented in the following. For the continuous rate, the power allocation can be performed by a water-filling type algorithm. Examples of such power allocation algorithms can be found in [6,8].

Algorithm 3.2   BIT LOADING ALGORITHM

1: Initialization: For all users i, set $r_i^n = 0$ and $\Delta P_i^n(r_i^n) = p_i^n(r_i^n + 1) - p_i^n(r_i^n)$, $\forall n \in S_i$

2: For each user i, repeat the following operation R_i times

  - Find sub-channel: $\widehat{n} = \mathrm{argmin}_{n \in S_i} \Delta P_i^n$
  - Update rate: $r_i^{\widehat{n}} = r_i^{\widehat{n}} + 1$
  - Evaluate new values of $\Delta P_i^{\widehat{n}}(r_i^{\widehat{n}})$

DL Resource Allocation Under Other Fairness Criteria: While max–min fairness provides essentially the same throughput performance for all users, it may harm the total network throughput if there are some users in very poor channel conditions. There are some existing works that attempt to perform DL resource allocation under other fairness constraints. In [6], the total throughput maximization problem is considered subject to the constraints that achievable rates of all users satisfy predetermined ratios as well as the BS power constraint. A greedy iterative sub-channel allocation algorithm is developed under uniform power allocation over sub-channels where the user with minimum ratio between its current rate and required rate is allocated the best available sub-channel in each iteration. Then, the transmission power is allocated for the obtained sub-channel allocation solution considering both power and rate constraints by solving the KKT conditions of the corresponding optimization problem. Again, a numerical method based on iterative Newton's algorithm is employed to solve the set of non-linear equations obtained from the KKT conditions.

In [8], a general DL resource allocation problem for utility maximization is considered. In the following, we describe a low complexity solution approach proposed in this work. Let $U_i(R_i)$ be the utility achieved by user i for a data rate R_i. Then, the utility maximization problem can be written as follows:

$$\max \sum_{i=1}^{N} U_i(R_i) \tag{3.23}$$

$$\text{subject to} \sum_{i=1}^{N} \sum_{n=1}^{K} a_i^n p_i^n(r_i^n) \leq P_{\mathrm{BS}} \tag{3.24}$$

$$\sum_{i=1}^{N} a_i^n = 1, \quad \forall n \tag{3.25}$$

where $p_i^n(r_i^n)$ is the required power to support rate r_i^n by user i on sub-channel n, which is given in (3.15), and $R_i = \sum_{n=1}^{K} a_i^n r_i^n$ is the achieved rate of user i. In general, the utility functions $U_i(\cdot)$ can be chosen to be a concave and increasing function that enables fair spectrum sharing among users under certain criteria such as α-fairness [35]. Define the total utility as $U(\mathbf{a}) = \sum_{i=1}^{N} U_i(R_i)$ where $\mathbf{a}$ denotes the sub-channel allocation vector whose elements are the sub-channel allocation variables a_i^n. For a certain power allocation solution such as uniform power allocation over sub-channels, the following must hold for concave utility functions

$$U(\mathbf{a}) - U(\mathbf{b}) \geq \nabla_{\mathbf{a}}^T (\mathbf{a} - \mathbf{b}) \tag{3.26}$$

where $\nabla_{\mathbf{a}}$ is the gradient of $U(\mathbf{a})$ with respect to $\mathbf{a}$, which can be expressed as follows:

$$\nabla_{\mathbf{a}} = \begin{bmatrix} U_1'(R_1)r_1^1 \\ \vdots \\ U_1'(R_1)r_1^K \\ U_N'(R_N)r_N^1 \\ \vdots \\ U_N'(R_N)r_N^K \end{bmatrix} \tag{3.27}$$

where

$$U_i'(R_i) = \frac{dU_i(R_i)}{dR_i}. \tag{3.28}$$

In [8], the sub-channel allocation is developed so that $\mathbf{a}^*$ satisfies the following condition

$$\nabla_{\mathbf{a}^*}^T (\mathbf{a}^* - \mathbf{a}) \geq 0, \text{ for all feasible solution } \mathbf{a}. \tag{3.29}$$

Then, the result in (3.26) implies that this sub-channel allocation solution $\mathbf{a}^*$ satisfies

$$U(\mathbf{a}^*) - U(\mathbf{a}) \geq 0 \tag{3.30}$$

which means that $\mathbf{a}^*$ is an optimal solution for the considered resource allocation problem. Let S_i be the set of sub-channels allocated for user i. Then, the optimality condition (3.26) implies the following

$$U_i'(R_i^*)r_i^n \geq U_j'(R_j^*)r_j^n, \ \forall n \in S_i, \ \forall i, j \tag{3.31}$$

where $R_i^* = \sum_{n \in S_i} r_i^n$. For the system with two users, an optimal sub-channel allocation based on this optimality condition can be obtained as follows. We first sort the sub-channel n in the increasing values of r_2^n/r_1^n to form a set of sub-channels $S = \{1, 2, \ldots, K\}$. Then, there exists an index $1 \leq m \leq K + 1$ so that the optimal sub-channel allocation sets for the two users are $S_1 = \{1, 2, \ldots, m - 1\}$ and $S_2 = \{m, \ldots, K\}$. By comparing the total utility achieved by all possible $K + 1$ allocation solutions, the optimal one can be obtained. Based on the optimal solution for two users, an efficient sub-channel allocation for $N > 2$ users can be developed by iteratively updating the sub-channel allocations for any two users.

Given a sub-channel allocation solution, we can perform power allocation again to enhance the system performance. For discrete-rate systems, a greedy bit-loading algorithm, which is similar to that used in the MA problem, can be developed as follows. We iteratively increase the current rate on one "best" sub-channel to the next rate (i.e., loading one more bit) in each iteration. Specifically, the best sub-channel n is the one that achieves maximum $\Delta U_n / \Delta P_n$ where ΔU_n and ΔP_n represent the increase in utility and transmission power due the bit-loading operation. In addition, only the rate updates that can maintain the power constraint at the BS are considered in each iteration. For concave utility functions, this greedy power allocation can be shown to be optimal. For continuous-rate systems, the power allocation must solve the following optimization problem

$$\max \sum_{i=1}^{N} U_i(R_i) \tag{3.32}$$

$$\text{subject to} \sum_{i=1}^{N} \sum_{n=1}^{K} a_i^n p_i^n(r_i^n) \leq P_{\text{BS}} \tag{3.33}$$

where no sub-channel allocation constraint is needed since the sub-channel allocation solution is assumed to be given. The KKT conditions of this non-linear and convex problem give the optimal power allocation for the users

$$p_i^{n*} = \left[\frac{U_i'(R_i^*)}{\lambda} - \frac{\sigma_i^n}{\beta h_i^n} \right]^+, \ n \in S_i \tag{3.34}$$

where $[x]^+ = \max \{0, x\}$; λ and R_i^* must satisfy the following

$$\sum_{n=1}^{K} p_{i(n)}^{n*} = P_{\text{BS}} \ R_i^* = \sum_{n \in S_i} \log_2 \left(1 + \beta \frac{h_i^n p_i^{n*}}{\sigma_i^n} \right). \tag{3.35}$$

This is a utility-based water-filling to obtain the optimal power solution [8], whose detailed implementation is described in the algorithm presented in the next page.

3.1.3   Uplink Single-Cell Resource Allocation

UL resource allocation problems in OFDMA-based systems is more challenging than the DL counterparts in general. This is because multiple power constraints for individual users must be imposed for the UL while there is a single BS power constraint for the DL. There are some existing works on the UL resource allocation for throughput and utility maximization, which are in [9, 10], respectively. Given that the throughput maximization problem is a special case of the utility maximization problem, in the following, we describe the low complexity algorithm proposed in [10].

Algorithm 3.3 ITERATIVE WATER-FILLING BASED POWER ALLOCATION

1: Initialization: Choose initial values for λ and allocated powers (e.g., uniform power allocation).

2: **while** $\sum_{i=1}^{N} U_i'(R_i^{(t)}) \left(R_i^{(t+1)} - R_i^{(t)} \right) > \epsilon$ do **do**

3:      Obtain new power allocation solution from the following constraints

$$p_i^n = \left[\frac{\omega_{i,n}^{(t)}}{\lambda} - \frac{\sigma_i^n}{\beta h_i^n} \right]^+, \quad n \in S_i \tag{3.36}$$

$$R_i^{(t+1)} = \sum_{n \in S_i} \log_2 \left(1 + \beta \frac{h_i^n p_i^{n*}}{\sigma_i^n} \right), \quad \forall i. \tag{3.37}$$

4:      Update $\omega_{i,n}^{(t)}$ using a step size μ as follows:

$$\omega_{i,n}^{(t+1)} = (1 - \mu) \omega_{i,n}^{(t)} + \mu U_i'(R_i^{(t+1)}), \quad \forall i. \tag{3.38}$$

5: **end while**

Denote by $U_i(R_i)$ the utility achieved by user i and P_i the maximum power budget of user i for UL transmission. Then, the UL resource allocation problem for utility maximization can be formulated as follows:

$$\max \sum_{i=1}^{N} U_i(R_i) \tag{3.39}$$

$$\text{subject to} \quad \sum_{n=1}^{K} a_i^n p_i^n \left(r_i^n \right) \leq P_i, \quad \forall i \tag{3.40}$$

$$\sum_{i=1}^{N} a_i^n = 1, \quad \forall n \tag{3.41}$$

where (3.40) describes the power constraint for each user and (3.41) captures the sub-channel allocation constraints. Here, $a_i^n \in \{0, 1\}$ and $p_i^n \geq 0$ denote the sub-channel and power allocation variables, respectively. Also, $R_i = \sum_{n=1}^{K} a_i^n r_i^n$ denotes the total achievable rate of user i over all allocated sub-channels. Since this is a mixed integer and non-linear optimization problem, finding its global optimal solutions would require exponential computation efforts. Therefore, the KKT necessary conditions are used to develop an efficient but sub-optimal sub-channel allocation solution [10]. Specifically, given a power allocation solution, it is necessary that a sub-channel n must be allocated to user $\widehat{i}$ as follows:

$$\widehat{i} = \underset{1 \leq i \leq N}{\operatorname{argmax}} \left\{ U_i'(R_i) \log_2 \left(1 + \beta \frac{h_i^n p_i^n}{\sigma_i^n} \right) \right\}. \tag{3.42}$$

In addition, if a sub-channel allocation solution for all users is given, then the optimal power allocation for continuous-rate systems is the standard water-filling algorithm, that is,

$$p_i^n = a_i^n \left[\lambda_i - \frac{\sigma_i^n}{\beta h_i^n} \right]^+ \tag{3.43}$$

where λ_i is determined so that

$$\sum_{n=1}^{K} p_i^n = P_i. \tag{3.44}$$

In fact, λ_i is called the water level of user i. By employing these sub-channel allocation criteria and the optimal water-filling power allocation, we can develop a low complexity resource allocation for UL utility maximization. This algorithm can be modified slightly for discrete-rate systems by using a standard bit-loading algorithm for power allocation in the last step.

Algorithm 3.4   UPLINK RESOURCE ALLOCATION FOR UTILITY MAXIMIZATION

1: Initialization: The set of available sub-channels $\mathcal{A}$ includes all sub-channels; the set of sub-channels allocated for each user i is set $S_i = \emptyset$
2: **while** $\mathcal{A} \neq \emptyset$ do **do**
3:    Each user i performs optimal power allocation over the set of sub-channels $\mathcal{A} \cup S_i$ using the above water-filling solution.
4:    Compute rate R_i for each user i using the above power allocation solution due to total rate in the set S_i.
5:    Among all unallocated sub-channels, find the user-sub-channel pair (i, n) that achieves the largest $U_i'(R_i) \log_2 \left(1 + \beta \frac{h_i^n p_i^n}{\sigma_i^n} \right)$. Then, allocate sub-channel n for user i.
6: **end while**
7: Perform optimal power allocation for the obtained sub-channel allocation solution.

3.1.4   Resource Allocation for Homogeneous Multi-Cell OFDMA Networks

Resource allocation problems for OFDMA-based cellular networks in the multi-cell environment are very difficult to solve even in a centralized setting. In fact, solving such problems requires to determine the joint sub-carrier (or sub-channel) and power allocation for all users in different cells to optimize a predetermined objective function subject to power and other constraints. There are closely related but simpler spectrum management problems in the context of DSL where each subscriber line can utilize all sub-carriers and power allocation for the sub-carriers must be determined to optimize the chosen objective function (e.g., weighted sum-rate maximization). These DSL spectrum management problems are challenging since the rate region is not convex, which implies that the underlying optimization problems are not convex optimization problems. However, it was discovered in [25, 26] that solving the corresponding dual problem can help obtain the optimal solution if the number of sub-carriers is sufficiently large. This result was proved rigorously later in [17]. Other efficient algorithm based on successive convex approximation has been recently proposed in [27] to tackle the non-convexity of the spectrum management problem.

The resource allocation for OFDMA-based multi-cell wireless networks involves optimization over integer sub-channel allocation and continuous power allocation variables. There are several low complexity algorithms that are proposed to solve the underlying mixed integer non-linear resource allocation problems in the literature [11–13]. In this section, we describe a dual-based low complexity algorithm proposed in [12] which is proved to converge and achieve local optimality.

We proceed by presenting the system model and problem formulation. We consider an OFDMA-based cellular network where users in different cells share K sub-channels using full frequency reuse. Suppose there are M cells where the set of users in cell m is denoted as $\mathcal{N}_m$ whose cardinality is N_m. Also, associations between users and their BSs are assumed to be fixed during the resource allocation. Denote the channel gain from BS m to user i and the noise power at user i over sub-channel n as $h_{m,i}^n$ and σ_i^n, respectively. Also, let p_m^n be the transmission power by BS m on sub-channel n. Then, we can express the SINR achieved by user i associated with BS m over sub-channel n as follows:

$$\gamma_{m,i}^n = \frac{h_{m,i}^n p_m^n}{\sum_{k \neq m} h_{k,i}^n p_k^n + \sigma_i^n}. \tag{3.45}$$

The corresponding achieved rate of user i in cell m on sub-channel n is

$$r_{m,i}^n = \log_2\left(1 + \gamma_{m,i}^n\right). \tag{3.46}$$

Then, the DL multi-cell resource allocation problem for weighted sum rate maximization can be formulated as follows:

$$\max \ f(\mathbf{p}, \mathbf{i}) = \sum_{m=1}^{M} \sum_{n=1}^{K} w_{i(m,n)} r_{m,i(m,n)}^n \tag{3.47}$$

$$\text{subject to} \sum_{n=1}^{K} p_m^n \leq P_{\text{BS}}^m, \quad \forall m \tag{3.48}$$

where $i(m, n)$ denotes the user in cell m that is allocated sub-channel n, $w_{i(m,n)}$ is the weight of user $i(m, n)$, and P_{BS}^m denotes the maximum power budget of BS m. For brevity, the optimization variables are expressed in the compact form $\mathbf{p}$ and $\mathbf{i}$, respectively.

Low complexity dual-based algorithm: We first obtain the necessary optimality condition of this optimization problem. First, optimal sub-channel allocation must satisfy the following for a given feasible power allocation vector $\mathbf{p}$:

$$\widehat{i}(m, n) = \underset{i \in \mathcal{N}_m}{\operatorname{argmax}} \ w_{i(m,n)} r_{m,i}^n, \quad \forall n, m. \tag{3.49}$$

To obtain additional optimality condition, we construct the Lagrangian as follows:

$$L\left(\mathbf{p}, \mathbf{i}, \lambda\right) = f(\mathbf{p}, \mathbf{i}) + \sum_{m=1}^{M} \lambda_m \left(P_{BS}^m - \sum_{n=1}^{K} p_m^n \right) \tag{3.50}$$

where λ_m are the Lagrange multipliers and λ is the corresponding vector capturing all individual λ_m. Then, the KKT conditions can be expressed as follows:

$$p_m^n + \frac{1 + \sum_{k \neq m} h_{k,i(m,n)}^n p_k^n}{h_{m,i(n,n)}^n} = \frac{w_{i(m,n)}}{\lambda_m \ln 2 + t_m^n} \tag{3.51}$$

$$\lambda_m \left(P_{BS}^m - \sum_{n=1}^{K} p_m^n \right) = 0 \tag{3.52}$$

$$\sum_{n=1}^{K} p_m^n \leq P_{BS}^m \tag{3.53}$$

$$\mathbf{p} \succeq 0; \lambda \succeq 0 \tag{3.54}$$

where

$$t_m^n = \sum_{k=1, k \neq m}^{M} \frac{w_{i(k,n)} h_{m,i(k,n)}^n \gamma_{k,i(k,n)}^n}{1 + \sum_{l=1}^{M} p_l^n h_{l,i(k,n)}^n}. \tag{3.55}$$

The conditions in (3.51) is obtained by setting the derivative of $L(\mathbf{p}, \mathbf{i}, \lambda)$ with respect to p_m^n to zero. To develop dual-based algorithm, we need to define the dual function as follows:

$$g\left(\lambda\right) = \max_{\mathbf{p}, \mathbf{i}} L\left(\mathbf{p}, \mathbf{i}, \lambda\right). \tag{3.56}$$

Then, the dual problem can be defined as follows:

$$\lambda^* = \operatorname*{argmax}_{\lambda \succeq 0} g\left(\lambda\right). \tag{3.57}$$

In general, the optimal dual objective function $g(\lambda^*)$ gives an upper bound of the objective function $f(\mathbf{p}, \mathbf{i})$ in (3.47). However, they are approximately the same if the number of sub-channels is sufficiently large [26]. Therefore, solving the dual problem (3.57) can provide an efficient solution for the considered resource allocation problem. The low complexity algorithm can be developed by iteratively and sequentially solving the primal problem (3.56) for fixed λ and the dual problem (3.57) for fixed $\mathbf{p}$ and $\mathbf{i}$. We describe how to solve these problems in the following. The dual function can be rewritten as follows:

$$g\left(\lambda\right) = \max_{\mathbf{p}, \mathbf{i}} \sum_{n=1}^{K} \left(\sum_{m=1}^{M} w_{i(m,n)} r_{m,i(m,n)}^n - \sum_{m=1}^{M} \lambda_m p_m^n \right) + \sum_{m=1}^{M} \lambda_m P_{BS}^m. \tag{3.58}$$

Therefore, the optimal solution of this primal problem can be obtained by solving K per-sub-channel problems as follows:

$$\max_{\mathbf{p}^n, \mathbf{i}^n} \left(\sum_{m=1}^{M} w_{i(m,n)} r^n_{m,i(m,n)} - \sum_{m=1}^{M} \lambda_m p^n_m \right) \tag{3.59}$$

where $\mathbf{p}^n$ and $\mathbf{i}^n$ capture power and sub-channel allocation decisions on sub-channel n. This per-sub-channel problem is still very complex to solve since it is a mixed integer non-linear optimization problem. To overcome this challenge, we optimize power and sub-channel allocation in separate steps. In addition, power allocation is performed by using a low complexity iterative coordinate ascent search, which is also employed in the DSL context [26]. Let $\widehat{\mathbf{p}}^n$ and $\widehat{\mathbf{i}}^n$ be the power and sub-channel allocation in a previous iteration, respectively. Then, the coordinate ascent search algorithm works as follows. We first keep $\widehat{p}^n_2, \widehat{p}^n_3, \ldots, \widehat{p}^n_M$ and $\widehat{\mathbf{i}}^n$ unchanged but optimize $\widehat{p}^n_1$. Then, we use the previous values of $\widehat{p}^n_1, \widehat{p}^n_3, \ldots, \widehat{p}^n_M$ and $\widehat{\mathbf{i}}^n$ to optimize $\widehat{p}^n_2$ and so on. For a particular cell m, this one-dimensional optimization problem can be written as follows:

$$\max_{p^n_m} \left(\sum_{m=1}^{M} w_{i(m,n)} r^n_{m,i(m,n)} - \sum_{m=1}^{M} \lambda_m p^n_m \right) \tag{3.60}$$

$$\text{subject to } p^n_j = \widehat{p}^n_j, \quad \forall j \neq m. \tag{3.61}$$

After the power optimization is performed based on the above coordinate ascent search, the sub-channel allocation solution is determined according to (3.49).

We now describe how to solve the dual problem in (3.57) by using the ellipsoid method. The idea is to form an ellipsoid which contains at least one optimal solution λ. Then, by updating the sub-gradient of the dual function iteratively based on which we can determine the new ellipsoids whose size reduces over iterations. First, it can be verified that the sub-gradient $\mathbf{d}$ of the dual function $g(\lambda)$ has its components that can be expressed as follows:

$$d_m = P^m_{\text{BS}} - \sum_{n=1}^{K} p^n_m, \quad m = 1, 2, \ldots, M. \tag{3.62}$$

An ellipsoid with center λ_0 and with the shape defined by a positive semidefinite matrix $\mathbf{A}_0$ can be defined as follows:

$$\text{Ellipsoid } (\mathbf{A}_0, \lambda_0) = \left\{ \mathbf{y} : (\mathbf{y} - \lambda_0)^T \mathbf{A}_0 (\mathbf{y} - \lambda_0) \leq 1 \right\}. \tag{3.63}$$

It is shown in [12] that the optimal λ must lie within an ellipsoid formed with the following parameters

$$\lambda_0 = \left[\frac{\lambda^{\text{single}}_1}{2} \max_{i \in \mathcal{N}_1} w_i, \ldots, \frac{\lambda^{\text{single}}_M}{2} \max_{i \in \mathcal{N}_M} w_i \right] \tag{3.64}$$

where $\lambda_m^{\text{single}}$ is the dual variable that solves (3.57) when only BS m is active and $w_i = 1$, $\forall i \in \mathcal{N}_m$. The ellipsoid algorithm then updates λ in iteration t as follows:

$$\widetilde{\mathbf{d}}_t = \frac{\mathbf{d}_t}{\sqrt{\mathbf{d}_t^T \mathbf{A}_t \mathbf{d}_t}} \tag{3.65}$$

$$\lambda_{t+1} = \left(\lambda_t - \frac{1}{M+1} \mathbf{A}_t^{-1} \widetilde{\mathbf{d}}_t \right)^+ \tag{3.66}$$

$$\mathbf{A}_{t+1}^{-1} = \frac{M^2}{M^2-1} \left(\mathbf{A}_t^{-1} - \frac{2}{M+1} \mathbf{A}_t^{-1} \widetilde{\mathbf{d}}_t \widetilde{\mathbf{d}}_t^T \mathbf{A}_t^{-1} \right) \tag{3.67}$$

where the sub-gradient vector $\mathbf{d}_t$ is calculated as in (3.62) at $\lambda = \lambda_t$. This low complexity resource allocation algorithm is summarized in the following algorithm.

Algorithm 3.5 DUAL-BASED MULTI-CELL RESOURCE ALLOCATION FOR WEIGHTED SUM-RATE MAXIMIZATION

1: Initialization: initial power and sub-channel allocations are set to equal $\mathbf{p}_{\text{start}}$ and $\mathbf{i}_{\text{start}}$, respectively; choose initial λ according to (3.64).
2: **repeat**
3:   **for** $n = 1 \rightarrow N$ **do**
4:     **repeat**
5:       **for** $m = 1 \rightarrow M$ **do**
6:         Update $\widehat{p}_m^n$ according to (3.60).
7:       **end for**
8:       Update sub-channel allocation $\widehat{i}^n$ according to (3.49).
9:     **until** $\widehat{\mathbf{p}}^n$ and $\widehat{\mathbf{i}}^n$ converge
10:   **end for**
11:   Update λ according to (3.65)–(3.67) and set $t = t + 1$
12: **until** convergence or $t = T_{\max}$

Algorithm initialization and signaling reduction: Since the underlying resource allocation problem is a mixed integer non-convex optimization problem, the proposed low complexity can only achieve a local optimal solution. Therefore, good initialization helps improve the performance achieved at convergence. In [12], an initial solution is chosen based on binary power control [28]. Specifically, under this binary power control scheme, BS m transmits with power $q_m^n = P_{\text{BS}}^m / K$ if it is active on sub-channel n or with power $q_m^n = 0$, otherwise. Then, the initial power and sub-channel allocations on sub-channel n are chosen as follows:

$$\left\{ \mathbf{p}_{\text{start}}^n, \mathbf{i}_{\text{start}}^n \right\} = \underset{\mathbf{q}^n, \mathbf{i}^n}{\operatorname{argmax}} \sum_{m=1}^{M} w_{i(m,n)} r_{m,i}^n \left(\mathbf{q}^n \right) . \tag{3.68}$$

In general, the resource allocation algorithm presented above requires centralized implementation since the channel gains between all BSs and users must be known by a central controller to run the algorithm. However, the required feedback and signaling can be reduced by employing simpler sub-channel allocation as follows [12]. Each user i in cell m reports to BS m the following SINR information

$$\overline{\gamma}_{m,i}^n = \frac{h_{m,i}^n P_{BS}^m}{\sum_{k \neq m} h_{k,i}^n P_{BS}^k + \sigma_i^n}. \tag{3.69}$$

Then, the sub-channel allocation is determined based on these estimated SINR as follows:

$$\overline{i}(m, n) = \underset{i \in \mathcal{N}_m}{\operatorname{argmax}} \; w_{i(m,n)} \log_2 \left(1 + \overline{\gamma}_{m,i}^n\right), \quad \forall n, m. \tag{3.70}$$

This sub-channel allocation is fixed and the proposed power control algorithm can be employed to determine transmission powers on all sub-channels. Under this reduced signaling scheme, sub-channel allocation is determined locally by each BS and only scheduled users on each sub-channel need to send their corresponding channel gains to the controller to perform power control. Therefore, the signaling overhead can be reduced significantly. It is shown in [12] that this simpler algorithm only results in slightly lower performance compared to the algorithm presented before.

An UL resource allocation problem for OFDMA cellular networks in the multi-cell context is more complicated to solve than the DL counterpart. This is because power constraints for individual users in any cell must be satisfied besides other fairness and sub-channel allocation constraints. There are quite few existing low complexity algorithms developed for UL resource allocation that can guarantee to achieve even local optimality performance. An example of such existing works is the one given in [16] where several auction based resource allocation algorithms are proposed.

3.2 FAIR RESOURCE ALLOCATION FOR TWO-TIER OFDMA NETWORKS

In general, resource allocation for OFDMA-based HetNets requires to determine joint sub-channel and power allocation as well as to protect macro users from the transmissions of femto users. Moreover, it is desirable to achieve fair resource sharing among femto users by exploiting the wireless capacity beyond what is needed to support QoS requirements of macro users. In this section, we will illustrate a max–min radio resource allocation framework for two-tier OFDMA HetNets with robust macro users' protection [29].

We consider an OFDMA-based two-tier macrocell–femtocell network where users of both tiers share the spectrum comprising K sub-channels. We assume that there are N_f femto users served by $(M - 1)$ FBSs, which are underlaid by one macrocell serving N_m macro users. Let $\mathcal{N}_k$ be the set of users in the kth cell, that is, they are served by BS k of the corresponding tier. For convenience, let $\mathcal{N}_m = \mathcal{N}_1$ represent the set of macro users and $\mathcal{N}_f = \cup_{k=2}^M \mathcal{N}_k = \{N_m + 1, \ldots, N_m + N_f\}$ denote the set of all femto users. In addition, let $\mathcal{N}$ and $\mathcal{M}$ be the sets of all users and BSs, respectively. Then, we have $\mathcal{N} = \mathcal{N}_m \cup \mathcal{N}_f = \{1, 2, \ldots, N\}$ and $\mathcal{M} = \{1, 2, \ldots, M\}$ and $\mathcal{M}_f = \{2, \ldots, M\}$ where BS 1 is assumed to be the MBS.

We consider fixed BS association for all users of both tiers in the network (i.e., each user is served by a fixed BS in the corresponding tier). Let $b_i \in \mathcal{M}$ denote the BS serving user i and let $\mathcal{K} = \{1, 2, \ldots, K\}$ be the set of all orthogonal sub-channels. We assume that there is no interference among transmissions on different sub-channels. We consider a system with full-frequency reuse where all K sub-channels are allocated for users in all cells of either tier. To describe the sub-channel assignments, let $\mathbf{A} \in \mathfrak{R}^{N \times K}$ be the sub-channel assignment matrix for all N users over K sub-channels where

$$\mathbf{A}(i, n) = a_i^n = \begin{cases} 1, & \text{if sub-channel } n \text{ is assigned for user } i \\ 0, & \text{otherwise.} \end{cases} \tag{3.71}$$

We assume that a sub-channel can be allocated to at most one user in any cell. Then, we have

$$\sum_{i \in \mathcal{M}_k} a_i^n \leq 1, \qquad \forall k \in \mathcal{M} \text{ and } \forall n \in \mathcal{K}. \tag{3.72}$$

These constraints will be applied to the UL resource allocation problem studied in the following.

To model the rate and SINR relationship, suppose that MQAM modulation with square signal constellations (i.e., s-QAM where $s \in \mathfrak{M} \triangleq \{4, 16, 64, 256, \ldots, s_{\max}\} = \{2^{2\upsilon} | \upsilon \text{ is a positive integer and } \upsilon \leq 1/2 \log_2 s_{\max}\}$) is employed. According to [30], the BER of the s-QAM scheme with Gray encoding can be approximated as follows:

$$f_s(\gamma(s)) = x_s \mathbb{Q}(\sqrt{y_s \gamma(s)}), \qquad s \in \mathfrak{M} \tag{3.73}$$

where $\mathbb{Q}(\cdot)$ stands for the Q-function, $x_s = \frac{2(1 - 1/\sqrt{s})}{\log_2 s}$, $y_s = \frac{3}{2(s-1)}$ and $\gamma(s)$ is the SINR. Using the result in (3.3), the target SINR for the s-QAM modulation scheme can be calculated as follows:

$$\bar{\gamma}(s) = \frac{\left[\mathbb{Q}^{-1}(\overline{P}_e / x_s)\right]^2}{y_s}, \qquad s \in \mathfrak{M} \tag{3.74}$$

where recall that $\overline{P}_e$ is the target BER value. Suppose ideal Nyquist data pulses are used where the symbol rate on each sub-channel is equal to B where B is the bandwidth of one sub-channel. Then, if s-QAM modulation scheme is employed, the spectral efficiency per one Hz of the system bandwidth is $\frac{\log_2 s}{K}$ (b/s/Hz).

3.2.1 Uplink Resource Allocation Problem

Let p_i^n represent the transmission power of user i over sub-channel n in the UL where $p_i^n \geq 0$. We impose the following constraints on the total transmission power of each individual user

$$\sum_{n=1}^{K} p_i^n \leq P_i, \qquad i \in \mathcal{N} \tag{3.75}$$

where P_i is the maximum transmission power of user i. Similar to the sub-channel assignments, we define $\mathbf{P}$ as an $N \times K$ power allocation matrix where $\mathbf{P}(i, n) = p_i^n$. For convenience, we also define partitions of sub-channel assignment and power allocation matrices $\mathbf{A}$ and $\mathbf{P}$ as follows. In particular, let $\mathbf{A}_k, \mathbf{P}_k \in \mathfrak{R}^{|\mathcal{N}_k| \times K}$ represent the sub-channel assignment and power assignment matrices for users in cell k over K sub-channels, respectively.

Let $h_{i,j}^n$ and σ_i^n be the channel gain from user j to BS i and the noise power at BS i over sub-channel n, respectively. Then, for a given sub-channel assignment and power allocation solution, that is, given $\mathbf{A}$ and $\mathbf{P}$, the SINR achieved at BS b_i due to the transmission of user i over sub-channel n can be written as follows:

$$\gamma_i^n(\mathbf{A}, \mathbf{P}) = \frac{a_i^n h_{b_i,i}^n p_i^n}{\sum_{j \notin \mathcal{N}_{b_i}} a_j^n h_{b_i,j}^n p_j^n + \sigma_{b_i}^n} = \frac{a_i^n p_i^n}{R_i^n(\mathbf{A}, \mathbf{P})} \tag{3.76}$$

where $R_i^n(\mathbf{A}, \mathbf{P})$ is the effective interference corresponding to user i on sub-channel n, which is defined as follows:

$$R_i^n(\mathbf{A}, \mathbf{P}) = \frac{\sum_{j \notin \mathcal{N}_{b_i}} a_j^n h_{b_i,j}^n p_j^n + \sigma_{b_i}^n}{h_{b_i,i}^n}. \tag{3.77}$$

We assume that sub-channel assignments for macro users have been determined by a certain mechanism and fixed while power allocation over the corresponding sub-channels for macro users are updated to cope with the CTI due to transmissions of femto users. That means $\mathbf{A}_1$ is fixed while we need to determine $\mathbf{A}_k$, $2 \leq k \leq M$ and the corresponding power allocation solution.

To protect the QoS of the licensed macro users, we wish to maintain a predetermined target SINR $\bar{\gamma}_i^n$ for each of its assigned sub-channel n. Specifically, we have the following constraints for the macro users

$$\gamma_i^n(\mathbf{A}, \mathbf{P}) \geq \bar{\gamma}_i^n, \quad \text{if } a_i^n = 1, \ \forall i \in \mathcal{N}_m \tag{3.78}$$

where $\bar{\gamma}_i^n = \bar{\gamma}(s^m)$, which is calculated as in (3.74).

For femto users, we assume that they all employ s^f-QAM for a predetermined s^f. Then, the spectral efficiency (bits/s/Hz) achieved by femto user i on one sub-channel is

$$r_i^n(\mathbf{A}, \mathbf{P}) = \begin{cases} 0, & \text{if } \gamma_i^n(\mathbf{A}, \mathbf{P}) < \bar{\gamma}_i^n, \\ r_f, & \text{if } \gamma_i^n(\mathbf{A}, \mathbf{P}) \geq \bar{\gamma}_i^n \end{cases} \tag{3.79}$$

where $r_f = \frac{\log_2 s^f}{K}$ and $\bar{\gamma}_i^n = \bar{\gamma}(s^f)$, which is calculated as in (3.74). Note that the femto users in all femtocells are assumed to employ the same constellation size s^f although this assumption can be relaxed to the adaptive rate scenario. In practice, we would choose $s^f \geq s^m$ since femto users can achieve higher transmission rates on each sub-channel than that of macro users thanks to their shorter distance to the associated FBS. This is indeed equivalent to $\bar{\gamma}(s^f) \geq \bar{\gamma}(s^m)$. In the analysis, we use the same notation $\bar{\gamma}_i^n$ to refer to the required target SINR of femto users or macro users where $\bar{\gamma}_i^n = \bar{\gamma}(s^f)$ for $i \in \mathcal{N}_f$ and $\bar{\gamma}_i^n = \bar{\gamma}(s^m)$ for $i \in \mathcal{N}_m$. Now, we can express the total spectral efficiency achieved by user i for given sub-channel assignment and power allocation matrices $\mathbf{A}$ and $\mathbf{P}$ as follows:

$$r_i^{\text{tot}}(\mathbf{A}, \mathbf{P}) = \sum_{n=1}^{K} r_i^n(\mathbf{A}, \mathbf{P}). \tag{3.80}$$

To impose the max–min fairness for femto users associated with the same FBS, we define the following minimum spectral efficiency for femtocell k

$$\mathbf{r}^{(k)}(\mathbf{A}, \mathbf{P}) = \min_{i \in \mathcal{N}_k} r_i^{\text{tot}}(\mathbf{A}, \mathbf{P}). \tag{3.81}$$

The UL resource allocation problem for femto users can be formulated as follows:

$$\max_{(\mathbf{A}, \mathbf{P})} \sum_{2 \le k \le M} \mathbf{r}^{(k)}(\mathbf{A}, \mathbf{P}) \tag{3.82}$$

$$\text{subject to} \quad (3.72), (3.75), (3.78) \tag{3.83}$$

where the minimum spectral efficiency at femtocell k, that is, $\mathbf{r}^{(k)}(\mathbf{A}, \mathbf{P})$, is given in (3.81). Therefore, this UL resource allocation problem aims to maximize the total min spectral efficiency of all femtocells subject to sub-channel assignment constraints, maximum power constraints, and protection constraints for macro users. Note that the SINR constraints for femto users are embedded in the calculation of femto spectral efficiency in (3.79). The resource allocation problem (3.82)–(3.83) can be transformed into a standard mixed integer program. Therefore, this resource allocation problem is NP-hard. In the following, we study the feasibility of a particular sub-channel assignment solution, which reveals the optimal structure of the optimization problems (3.82) and (3.83) based on which we develop optimal and suboptimal low complexity algorithms.

3.2.2 Feasibility of a Sub-Channel Assignment Solution

The resource allocation problem (3.82)–(3.83) involves finding a joint sub-channel assignment and power allocation solution. For a certain sub-channel assignment solution represented by matrix $\mathbf{A}$ satisfying the sub-channel assignment constraints (3.72), we can indeed find its "best" power allocation and verify its feasibility with respect to the constraints in (3.83). In particular, we wish to satisfy the SINR constraints for femto users and macro users in (3.79) and (3.78), respectively. Specifically, for each sub-channel n we need to satisfy the SINR constraints $\gamma_i^n(\mathbf{A}, \mathbf{P}) \ge \bar{\gamma}_i^n$ for both macro users and femto users who are allocated this sub-channel.

Now, let $S^n = \{i \in \mathcal{N} | a_i^n \ne 0\} = \{n_1, \dots, n_{c^n}\}$ denote the set of users of both tiers who are assigned sub-channel n and $c^n = |S^n|$ be the number of elements in set S^n. Then, the SINR constraints for users in set S^n, that is, $\gamma_i^n(\mathbf{A}, \mathbf{P}) \ge \bar{\gamma}_i^n$, $i \in S^n$, can be rewritten in a matrix form by using the SINR expression in (3.76) as follows:

$$(\mathbf{I}^n - \mathbf{G}^n \mathbf{H}^n) \mathbf{p}^n \ge \mathbf{g}^n \tag{3.84}$$

where $\mathbf{g}^n = \left[g_{n_1}^n, \ldots, g_{n_{c^n}}^n \right]^T$ with $g_i^n = \frac{\sigma_{b_i}^n \bar{\gamma}_i^n}{h_{b_i,i}^n}$, $\mathbf{I}^n$ is $c^n \times c^n$ identity matrix, $\mathbf{G}^n = \mathrm{diag}\{\bar{\gamma}_{n_1}^n, \ldots, \bar{\gamma}_{n_{c^n}}^n\}$, $\mathbf{p}^n = [p_{n_1}^n, \ldots, p_{n_{c^n}}^n]^T$ and $\mathbf{H}^n$ is a $c^n \times c^n$ matrix defined as follows:

$$\left[\mathbf{H}_{i,j}^n \right] = \begin{cases} 0, & \text{if } j = i, \\ \dfrac{h_{b_{n_i},n_j}^n}{h_{b_{n_i},n_i}^n}, & \text{if } j \neq i. \end{cases} \tag{3.85}$$

According to the Perron–Frobenius theorem, there exists a non-negative power allocation solution for users that are allocated sub-channel n if and only if the maximum eigenvalue of $\mathbf{G}^n \mathbf{H}^n$, that is, spectrum radius $\rho(\mathbf{G}^n \mathbf{H}^n)$, is less than 1 [30]. In addition, the Pareto-optimal power allocation for users in S^n (i.e., minimum power vector in the element-wise sense) can be expressed as [30]

$$\mathbf{p}^n = \begin{cases} (\mathbf{I}^n - \mathbf{G}^n \mathbf{H}^n)^{-1} \mathbf{g}^n, & |\rho(\mathbf{G}^n \mathbf{H}^n)| < 1 \\ +\infty, & \text{otherwise.} \end{cases} \tag{3.86}$$

In addition, this Pareto-optimal power allocation can be achieved at equilibrium by employing the following iterative Foschini–Miljanic [30] power updates

$$p_i^n(t+1) = p_i^n(t) \frac{\bar{\gamma}_i^n}{\gamma_i^n(t)} = R_i^n(t) \bar{\gamma}_i^n \tag{3.87}$$

where $\gamma_i^n(t)$ and $R_i^n(t)$ are the SINR and effective interference achieved by user i in iteration t, respectively. We state one important result for a given sub-channel assignment solution $\mathbf{A}$ that satisfies the sub-channel assignment constraints (3.72) in the following lemma.

Lemma 3.1.   Suppose that we can find a finite Pareto-optimal power allocation solution $\mathbf{p}^n$ on each sub-channel n for a particular sub-channel assignment matrix $\mathbf{A}$, which is given in (3.86) (i.e., the spectrum radius $\rho(\mathbf{G}^n \mathbf{H}^n) < 1$, $\forall n$). Then, the underlying sub-channel assignment represented by $\mathbf{A}$ is feasible if and only if these Pareto-optimal power allocation vectors $\mathbf{p}^n$ satisfy the power constraints in (3.75).

Proof: The Pareto-optimal power allocation solution $\mathbf{p}^n$ on each sub-channel n requires the minimum powers in the element-wise sense for all users, who are assigned sub-channel n, to meet their SINR requirements. Therefore, the considered sub-channel assignment solution $\mathbf{A}$ is feasible if and only if the corresponding Pareto-optimal power allocation solutions $\mathbf{p}^n$ on all sub-channels n satisfy the power constraints in (3.75). Therefore, we have completed the proof of the lemma. ∎

Since the number of possible sub-channel assignments is finite, the result in this lemma paves the way for developing an optimal exhaustive search algorithm, which is presented in the following.

3.2.3 Optimal Algorithm and Its Complexity Analysis

Based on the results in the previous section, we can find the optimal solution for the resource allocation problem (3.82)–(3.83) by performing exhaustive search as follows. For a fixed and feasible $\mathbf{A}_1{}^1$, let $\Omega\{\mathbf{A}\}$ be the list of all potential sub-channel assignment solutions that satisfy the sub-channel assignment constraints (3.72) and the fairness condition: $\sum_{n\in\mathcal{N}} a_i^n = \sum_{n\in\mathcal{N}} a_j^n = \tau_k$ for all femto users $i, j \in \mathcal{N}_k$. We sort the list $\Omega\{\mathbf{A}\}$ in the decreasing order of $\sum_{k=2}^{M} \tau_k$ to obtain the sorted list $\Omega^*\{\mathbf{A}\}$. Then, the feasibility of each sub-channel assignment solution in the list $\Omega^*\{\mathbf{A}\}$ can be verified as presented in Section 3.2.2. Among all feasible sub-channel assignments, the feasible solution achieving the highest value of the objective function (3.82) and its corresponding power allocation solution given in (3.86) correspond to the optimal solution of the optimization problem (3.82)–(3.83).

    The complexity of the exhaustive search algorithm can be determined by finding the number of the elements in the list $\Omega^*\{\mathbf{A}\}$ and the complexity involved in the feasibility verification for each of them. The number of the elements in $\Omega^*\{\mathbf{A}\}$ (i.e., the number of potential sub-channel assignments) is the product of the number of potential sub-channel assignments for all femtocells that satisfy the "fairness condition." Therefore, the number of potential sub-channel assignments can be calculated as follows:

$$
\begin{aligned}
T &= \prod_{k=2}^{M} \sum_{\tau_k=0}^{\eta_k} \prod_{i=0}^{N_k-1} C_{K-i\tau_k}^{\tau_k} \\
&= \prod_{k=2}^{M} \sum_{\tau_k=0}^{\eta_k} \frac{K!}{(\tau_k!)^{N_k}(K-N_k\tau_k)!} \approx O\left((K!)^{(M-1)}\right)
\end{aligned}
\tag{3.88}
$$

where $\eta_k = \lfloor K/N_k \rfloor$ represents for the largest integer less than or equal to K/N_k and $C_m^n = \frac{m!}{n!(m-n)!}$ denotes the *"m-choose-n"* operation. According to Section 3.2.2, the complexity of the feasibility verification for each potential sub-channel assignment mainly depends on the eigenvalue calculation of the corresponding matrix and solving the linear system to find the power allocation solution for each sub-channel. It requires computation complexity of $O(M^3)$ to calculate the eigenvalues of $\mathbf{G}^n\mathbf{H}^n$ [31] and $O(M^3)$ to obtain $\mathbf{p}^n$ by solving a system of linear equations [32]. Therefore, the complexity of the optimal exhaustive search algorithm is $O\left(M^3 \times K \times (K!)^{(M-1)}\right)$, which is exponential in the numbers of sub-channels and FBSs.

3.2.4 Sub-Optimal and Distributed Algorithm

To resolve the exponential complexity of the centralized optimal algorithm presented in Section 3.2.3, we develop a low complexity resource allocation algorithm which can be implemented in a distributed manner. As implied by (3.79) and (3.80), the total spectral efficiency achieved by each femto user i is equal to the product of r_f and the number of assigned sub-channels given that the femto user can find

---

[1]The sub-channel assignment solution for macro users corresponding to $\mathbf{A}_1$ is feasible if there exists a power allocation solution that satisfies the constraints in (3.78) when there is no femto tier.

feasible power allocation solutions on all assigned sub-channels. Since we wish to maximize the total minimum spectral efficiency of all femtocells, an efficient resource allocation algorithm would attempt to assign the maximum and equal number of sub-channels to femto users in each femtocell and to perform Pareto-optimal power allocation for all users on each sub-channel so that they meet the SINR constraints in (3.79) and (3.78).

Note that if an efficient sub-channel assignment solution can be determined then one can employ the distributed Foschini–Miljanic power control updates (3.87) on all sub-channels to find the power allocation solution. However, there can be two scenarios where a particular sub-channel assignment is not feasible. The first scenario is for the case where there not exist any power allocation that can support the SINR constraints (3.79) and (3.78) (i.e., we have $|\rho(\mathbf{G}^n \mathbf{H}^n)| \geq 1$ for some "infeasible" sub-channel n). In this case the Foschini–Miljanic power control updates (3.87) will increase the transmission power on the "infeasible" sub-channel n to infinity [31, 32] (i.e., the power update diverges). The second scenario corresponds the case where Foschini–Miljanic power control updates (3.87) converge but the power constraints for some users (3.75) are violated.

Based on these observations, we develop an iterative resource allocation algorithm in which we update the sub-channel assignment for femto users over iterations based on estimated transmission powers on their sub-channels given by the Foschini–Miljanic power control updates (3.87). Specifically, we represent the quality of a sub-channel assignment for a pair of sub-channel and femto user by a weight, which is proportional to the estimated transmission power. Moreover, the sub-channel assignment for all femto users in each femtocell is performed so that the minimum total weight is achieved. This sub-channel assignment attempts to attain an efficient and low power sub-channel assignment solution. In addition, we reduce the number of sub-channels allocated for femto users in the femtocell in which the total minimum weight is too large since this implies that the SINR constraints (3.79) and (3.78) or power constraints (3.75) are likely to be violated. This aims to prevent the system from falling into the two infeasible sub-channel assignment scenarios described above.

The proposed resource allocation algorithm employs an iterative weight-based sub-channel assignment that is performed in parallel at all femtocells. The weight for each sub-channel and femto user pair is defined as the multiplication of the estimated transmission power and a scaling factor capturing the "quality" of the corresponding sub-channel assignment. In addition, the scaling factor is updated over iterations so that it becomes larger if the corresponding assignment results in violations of the SINR constraints for femto users and/or macro users.

Each user i in cell k estimates the transmission power on sub-channel n in each iteration t of the algorithm. User i first estimates the effective interference on sub-channel n at the beginning of iteration t, denoted as $R_i^n(t)$, which is given in (3.77). Then, it calculates the required transmission power by using the Foschini–Miljanic power update formula given in (3.87) as follows:

$$p_i^{n,\min} = R_i^n(t)\bar{\gamma}_i^n. \tag{3.89}$$

Then, femto user i in cell k updates the assignment weight as $w_i^n = \chi_i^n p_i^{n,\min}$ where the scaling factor χ_i^n is defined as follows:

$$\chi_i^n = \begin{cases} \alpha_i^n, & \text{if } p_i^{n,\min} \leq P_i/\tau_k \\ \alpha_i^n \theta_i^n, & \text{if } P_i/\tau_k < p_i^{n,\min} \leq P_i \\ \alpha_i^n \delta_i^n, & \text{if } P_i < p_i^{n,\min} \end{cases} \tag{3.90}$$

where τ_k denotes the number of sub-channels assigned for each femto user in femto-cell k, α_i^n is a factor, which helps maintain SINR constraints of macro users (i.e., it is increased if assigning sub-channel n to femto user i tends to result in violation of the SINR constraint of the corresponding macro user), θ_i^n and δ_i^n are another factors which are set to larger values if the assignment of sub-channel n for femto user i tends to require transmission power larger than the average power per sub-channel (i.e., P_i/τ_k) and the maximum power budget (i.e., P_i), respectively. We will set δ_i^n as $\delta_i^n = \mu_i \theta_i^n$ where $\mu_i = K$ in the proposed algorithm. Given the weights defined for each femto user i, femtocell k finds the sub-channel assignment for its femto users by solving the following problem

$$\begin{align} \min_{\mathbf{A}_k} \quad & \sum_{i \in \mathcal{N}_k} \sum_{n \in \mathcal{K}} a_i^n w_i^n \\ \text{subject to} \quad & \sum_{n \in \mathcal{K}} a_i^n = \tau_k, \quad \forall i \in \mathcal{N}_k. \end{align} \tag{3.91}$$

This problem aims to find a sub-channel assignment matrix $\mathbf{A}_k$ for which each femto user in femtocell k is assigned τ_k sub-channels with the minimum total weight. Therefore, by assigning a larger weight w_i^n for bad sub-channel assignments, the sub-channel assignment solution returned by (3.91) would be efficient.

The optimization problem in (3.91) can be transformed into the standard matching problem (say between "jobs" and "employees") as follows. Suppose we create τ_k virtual "employees" for each femto user in femtocell k, then we can consider the matching problem between $\tau_k N_k$ virtual "employees" (virtual femto users) and K "jobs" (sub-channels). In particular, femto user i is equivalent to τ_k virtual femto users $\{i_1, \ldots, i_{\tau_k}\}$. Let the edge $v_{i_u}^n$ ($u \in \{1, \ldots, \tau_k\}$) between sub-channel n and virtual femto user i_u represent the assignment of that sub-channel to the corresponding femto user. Then, the weight $w_{i_u}^n$ of the edge $v_{i_u}^n$ is equal to w_i^n. After performing this transformation, the sub-channel assignment solution of the problem in (3.91) can be found by using the standard Hungarian algorithm (i.e., Algorithm 14.2.3 given in [33]). After running the Hungarian algorithm, if there exists a virtual femto user i_u, $u \in \{1, \ldots, \tau_k\}$, being matched with sub-channel n, then we set $a_i^n = 1$; otherwise, we set $a_i^n = 0$. For further interpretation of the proposed algorithm, let $W_k(t)$ denote the total minimum weight due to the optimal solution of (3.91). Sub-channel assignment for femtocell k is illustrated in Figure 3.1, which is also described in Figure 3.2.

The main operations of the proposed algorithm in one iteration can be summarized as follows.

- Each macro user calculates its transmission power based on current effective interference levels by using the Foschini–Miljanic power update formula (3.89).

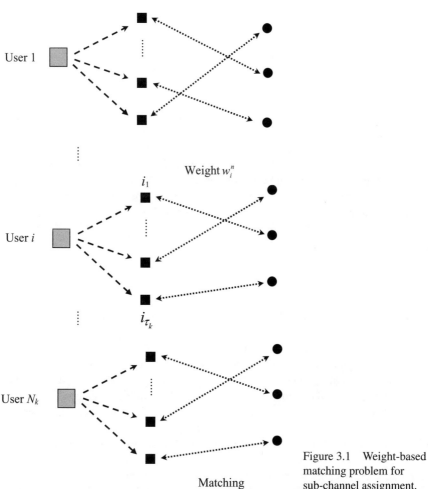

Figure 3.1 Weight-based matching problem for sub-channel assignment.

If its power constraint is still satisfied, the macro user updates its transmission powers on assigned sub-channels accordingly (steps 5–6). In contrast, if the power constraint of any macro user is violated, the MBS will inform all FBSs which find the femto user creating the largest interference for the victim macro user and increase the parameter $\alpha_{m_i^*}^{n_i^*}$ by a factor of 2 (steps 7–10).

- Each femtocell which has any femto power constraints being violated in the previous iteration solves the sub-channel assignment problem (3.91) with the current weight values. If the total minimum weight is larger than V times the total maximum powers of all femto users, the target number of sub-channels τ_k is decreased by one (steps 19–20).

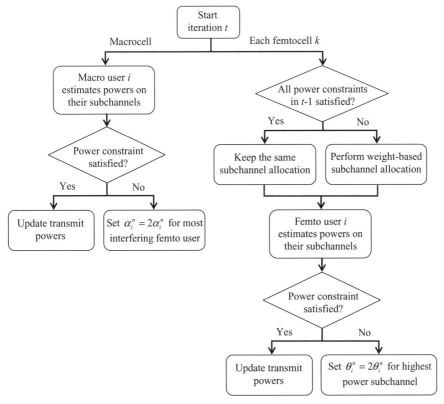

Figure 3.2   Operations in one iteration of the distributed algorithm.

- Each femto user that has its SINR constraint satisfied updates its transmission power on the assigned sub-channels (steps 24–25). In contrast, any femto user that has its SINR constraint violated increases its $\theta_i^{n_i^*}$ parameter by a factor of 2 for the assigned sub-channel requiring the largest transmission power (steps 26–29). In addition, we set $\varrho_{k,i} = 0$ for this femto user (step 30) so that the algorithm will update the sub-channel assignments in the next iteration.

The proposed algorithm can be implemented distributively. In fact, except for steps 7–10, each BS of either tier can collaborate with its associated users to conduct all required tasks in other steps. Moreover, the required tasks in steps 7–10 can be performed by allowing the MBS and FBSs coordinate with each other by using the wired backhaul link.

Convergence Analysis: The convergence of the proposed algorithm is stated in the following proposition.

Proposition 3.1.   The proposed algorithm converges to a feasible solution $(\mathbf{A}, \mathbf{P})$ of the optimization problem (3.82)–(3.83).

Algorithm 3.6 DISTRIBUTED UPLINK RESOURCE ALLOCATION

1: Initialization
   - Set $p_i(0) = 0$ for all user i, $i \in \mathcal{N}_f$, feasible $\mathbf{A}_1$.
   - Set $\tau_k = \lfloor \frac{N}{|N_k|} \rfloor$ and $\varrho_k = 0$ for all $k \in \mathcal{M}_f$.
   - Set $\alpha_i^n = \theta_i^n = 1$, $\mu_i = K$, $\forall i \in \mathcal{N}_f$, $n \in \mathcal{K}$.
2: **For the macrocell:**
3: Each macro user i estimates $R_i^n(t)$ and calculates $p_i^{n,\min}$ as in (3.89).
4: Let $\beta_i = \sum_{n \in \mathcal{K}} a_i^n(t) p_i^{n,\min} / P_i$.
5: **if** $\beta_i \leq 1$ **then**
6:    Set $p_i^n(t+1) = a_i^n(t) p_i^{n,\min}$, $\forall n \in \mathcal{K}$.
7: **else if** $\beta_i > 1$ **then**
8:    Set $p_i^n(t+1) = \frac{a_i^n(t) p_i^{n,\min}}{\beta_i}$, $\forall n \in \mathcal{K}$
9:    Find $n_i^* = \underset{n \in \mathcal{K}, c_n > 1}{\arg\max}\, a_i^n(t)\, p_i^{n,\min}$ and $m_i^* = \underset{m \in \mathcal{N}_f}{\arg\max}\, a_m^{n_i^*}(t) p_m^{n_i^*}(t) h_{1m}^*$
10:    Set $\alpha_{m_i^*}^{n_i^*} = 2\alpha_{m_i^*}^{n_i^*}$ and $\varrho_{b_{m_i^*}} = 0$.
11: **end if**
12: **For each femtocell $k \in \mathcal{M}_f$:**
13: Each femto user i estimates $R_i^n(t)$ and calculates $p_i^{n,\min}$ as in (3.89).
14: **if** $\varrho_k = 1$ **then**
15:    Keep $\mathbf{A}_k(t) = \mathbf{A}_k(t-1)$
16: **else if** $\varrho_k = 0$ **then**
17:    Define weight edges $w_{i_u}^n$ between $\mathcal{K}$ and $\cup_{i \in \mathcal{N}_k}\{i_1, \ldots, i_{\tau_k}\}$ as in (3.90)
18:    Run Hungarian algorithm with $\{w_{i_u}^n\}$ to obtain $W_k(t)$ and $\mathbf{A}_k(t)$.
19:    **if** $W_k(t) > V \sum_{i \in \mathcal{N}_k} P_i$ **then**
20:       Set $\tau_k := \tau_k - 1$.
21:    **end if**
22: **end if**
23: Let $\beta_i = \sum_{n \in \mathcal{K}} a_i^n(t) p_i^{n,\min} / P_i$.
24: **if** $\beta_i \leq 1$ **then**
25:    Set $p_i^n(t+1) = a_i^n(t) p_i^{n,\min}$, $\forall n \in \mathcal{K}$ and $\varrho_{k,i} = 1$.
26: **else if** $\beta_i > 1$ **then**
27:    Set $p_i^n(t+1) = \frac{a_i^n(t) p_i^{n,\min}}{\beta_i}$, $\forall n \in \mathcal{K}$
28:    Find $n_i^* = \underset{n \in \mathcal{K}}{\arg\max}\, a_i^n(t)\, p_i^{n,\min}$
29:    Set $\theta_i^{n_i^*} = 2\theta_i^{n_i^*}$.
30:    Set $\varrho_{k,i} = 0$.
31: **end if**
32: Set $\varrho_k = \prod_{i \in \mathcal{N}_k} \varrho_{k,i}$.
33: Let $t := t + 1$, return to step 2 until convergence.

Proof: To prove this proposition, we will consider two possible scenarios. For the first scenario, there exists an iteration after which the scaling factors χ_i^n for sub-channel assignment weights given in (3.90) do not change. Then it can be verified that after this iteration the sub-channel assignment solution will remain unchanged. In addition, the proposed algorithm simply updates transmission powers $p_i^{n,\min}$ for all users on their assigned sub-channels by using the Foschini–Miljanic power updates (steps 6 and 25). Therefore, the authors in [30] have shown that these power updates indeed converge, which implies the convergence of the proposed algorithm.

Now suppose that the scaling factors χ_i^n vary over iterations. We will prove that the system will ultimately evolve into the first scenario discussed above. First, it can be verified that the power $p_i^{n,\min}$ given in (3.89) is lower bounded by $\bar{\gamma}_i^n \sigma_{b_i}^n / h_{b_i,i}^n$. Therefore, if the scaling factors χ_i^n keep increasing over iterations, then the total weight $W_k(t)$ returned by the assignment problem in (3.91) will increase over iterations as well. Therefore, there exist some femtocells k that decrease their number of assigned sub-channels τ_k over iterations (steps 19–21). Since the initial values of all τ_k are finite, this process will terminate after a finite number of iterations. Then, the system will be in the first scenario discussed above; therefore, the proposed algorithm converges. Therefore, we have completed the proof of the proposition. ∎

Complexity Analysis

It is observed that the major complexity of the proposed algorithm is involved in solving the sub-channel assignment by using the Hungarian method in step 18. According to Theorem 14.2.4. of [33], the complexity of the Hungarian algorithm is $O(K^3)$. Therefore, the complexity of the proposed algorithm is $O(M \times K^3)$ for each iteration. However, the local sub-channel assignments can be performed in parallel at all $(M - 1)$ femtocells. Therefore, the run-time complexity of the algorithm is $O(K^3)$ multiplied by the number of required iterations, which is quite moderate according to the simulation results (i.e., tens of iterations).

The considered resource allocation problem and its resulting distributed algorithm can be extended for DL communication as follows. Specifically, the max–min fair resource allocation for the DL can be formulated similarly where the UL power constraints in (3.75) for individual users can be replaced by the corresponding power constraints at BSs of both network tiers. The optimal and suboptimal algorithms proposed in this section can be modified to account for these new power constraints. In particular, the sub-channel assignment weights defined in (3.90) used in the suboptimal algorithm may need to be modified to capture these new power constraints accordingly.

3.2.5 Numerical Examples

We present illustrative numerical results for the proposed algorithms using two different networks, which have a small and large number of femtocells and users, respectively. The network setting and user placement for the simulations are illustrated in Figure 3.3, where macro and femto users are randomly located inside circles of radii of $r_1 = 1000$ m and $r_2 = 30$ m, respectively. The channel gains

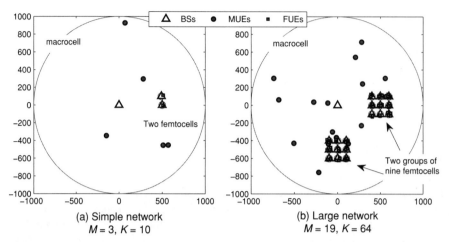

Figure 3.3 Macrocell–femtocell networks used in simulation (MUE, macro users; FUE, femto users; network dimension in horizontal and vertical axes is in meter (m), M denotes the number of BSs, K denotes the number of subchannels). © [2013] IEEE.

h_{ij}^n are generated by considering both Rayleigh fading, which is represented by an exponentially distributed random variable with the mean value of one, and the path-loss $L_{ij} = A_i \log_{10}(d_{ij}) + B_i + C\log_{10}(\frac{f_c}{5}) + W_l n_{ij}$, where d_{ij} is the distance from user j to BS i, (A_i, B_i) are set as $(36, 40)$ and $(35, 35)$ for MBS and FBSs, respectively, $C = 20$ and $f_c = 2.5$ GHz. W_l is the wall-loss parameter and n_{ij} is the number of walls between BS i and user j. The noise power is set as $\eta_i = 10^{-13}$ W, $\forall i \in \mathcal{K}$.

To obtain simulation results, we use two modulation schemes (4- and 16-QAM) for macro users and five modulation schemes (4-, 16-, 64-, 256-, and 1024-QAM) for femto users. Moreover, we choose the target BER $\overline{P}_e = 10^{-3}$ whose target SINRs corresponding to different modulation schemes, which can be calculated by using (3.74), are given in Table 3.1. Each simulation result is obtained by taking the average over 20 different runs where for each run, users are randomly located and a feasible sub-channel allocation matrix for macro users $\mathbf{A}_1$ is chosen so that each macro user is assigned $\frac{N}{M_1}$ sub-channels. The maximum powers of femto and macro users are denoted as $P_f^{\max}$ and $P_m^{\max}$ in the figures, respectively.

In Figure 3.4, we show the total minimum spectral efficiency of all femtocells (i.e., the optimal objective value of (3.82)) versus the constellation size of femto users (s^f) for the small network due to both optimal and sup-optimal algorithms. Clearly, for the low constellation sizes (i.e., low target SINRs), the low complexity algorithm can achieve almost the same spectral efficiency as the optimal one while for higher values of s^f, it results in just slightly lower spectral efficiency than that due to the

TABLE 3.1 Target SINRs for different constellation sizes with target BER $\overline{P}_e = 10^{-3}$

Constellation size	4	16	64	256	1024
bits/symbol	2	4	8	6	10
$\bar{\gamma}(s)$	9.55	45.11	179.85	694.17	2667.32

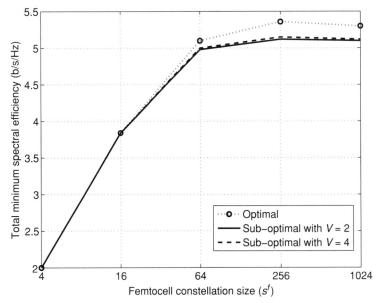

Figure 3.4   Total minimum spectral efficiency for small network under optimal and sub-optimal algorithms for $s^m = 4$, $W_l = 5$ dB, $P_m^{\max} = P_f^{\max} = 0.01$ W. © [2013] IEEE.

optimal one. Moreover, when we increase the value of parameter V, a slightly better performance can be achieved.

In Figures 3.5 and 3.6, we plot the total femtocell minimum spectral efficiency versus the maximum power of femto users ($P_f^{\max}$) and macro users ($P_m^{\max}$), respectively, for different modulation levels of macro users (the modulation scheme of femto users is 256-QAM). These figures show that the total minimum spectral efficiency increases with the increases in maximum power budgets, $P_f^{\max}$ or $P_m^{\max}$. However, this value is saturated as the maximum power budgets $P_f^{\max}$ or $P_m^{\max}$ become sufficiently large. In addition, as the number of macro users increases, the total femtocell minimum spectral efficiency increases thanks to the better diversity gain offered by the macro tier.

3.3   CHAPTER SUMMARY AND OPEN RESEARCH DIRECTIONS

In this chapter, we have discussed various OFDMA-based resource allocation problems and described some important design techniques that can be employed to devise efficient algorithms to solve the underlying problems. In particular, DL and UL resource allocations under different design objectives and fairness criteria have been described for both single-cell and multi-cell scenarios of single-tier wireless cellular networks. Then, we have presented an example of resource management design for a two-tier OFDMA-based macrocell–femtocell network. In general, adaptive radio

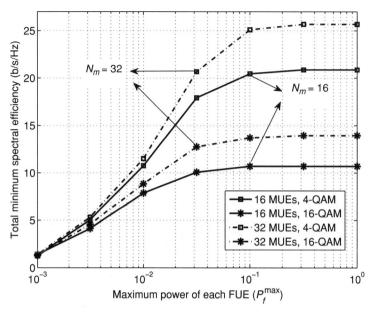

Figure 3.5   Total minimum spectral efficiency versus $P_f^{\max}$ for the large network with $s^f = 256$, $W_l = 5$ dB, $P_m^{\max} = 0.01$ W. © [2013] IEEE.

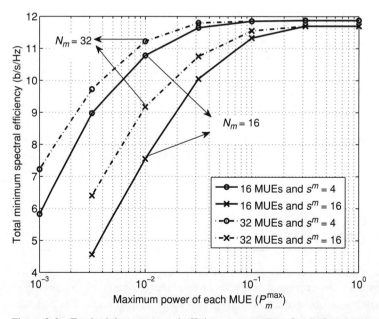

Figure 3.6   Total minimum spectral efficiency versus $P_m^{\max}$ for the large network with $s^f = 256$, $W_l = 5$ dB, $P_f^{\max} = 0.01$ W. © [2013] IEEE.

resource allocation for OFDMA HetNets is a fertile research area with many open research problems. Some of the potential research directions are presented in the following.

- The resource allocation formulation and algorithms presented for the two-tier OFDMA HetNets in the last section of this chapter restrict the same rate on all sub-channels in each femtocell. In general, better performance can be achieved if optimal rate adaptation is performed on each sub-channel where transmission rates on different sub-channels can be varied depending on their channel gains. Here, either continuous or discrete rate adaptation can be considered. Development of efficient and low complexity resource allocation algorithms for DL and UL transmissions considering various fairness and protection constraints for macro users is an important but challenging area, which requires further studies. Moreover, if the BSs and users of different tiers are equipped with multiple antennae, then joint beamforming and resource allocation can be performed to further enhance the network performance. There are some existing algorithms, which are proposed for multi-antenna OFDMA single-tier wireless cellular networks [33, 34]. For the HetNet setting, it would be desirable to have more flexible resource allocation algorithms that can protect macro users and enable controllable capacity sharing among different network tiers.

- In OFDMA-based two-tier HetNets, development of a flexible joint BS association and resource allocation framework is an important research problem. In particular, it is desirable that such a framework enables off-loading of any desirable amount of traffic from the macro-tier to the femto-tier for efficient load balancing and capacity sharing between the two network tiers. In practice, the femto tier may be able to provide much larger capacity than that of the macro-tier thanks to its small cell structure. In addition, user connections to the femto-tier may involve much cheaper or even free services. Therefore, there can be great benefits in off-loading more traffic to the femto-tier as long as minimum QoS requirements of femto users can be maintained. Also, distributed BS association and resource allocation algorithms, which are scalable and require only low signaling overhead, are strongly desired.

REFERENCES

1. C. Y. Wong, R. S. Cheng, K. B. Letaief, and R. D. Murch, "Multiuser OFDM with adaptive subcarrier, bit, and power allocation," *IEEE Journal on Selected Areas in Communications*, vol. 17, no. 10, pp. 1747–1758, October 1999.
2. J. Jang and K. B. Lee, "Transmit power adaptation for multiuser OFDM systems," *IEEE Journal on Selected Areas in Communications*, vol. 21, no. 2, pp. 171–178, February 2003.
3. D. Kivanc, G. Li, and H. Liu, "Computationally efficient bandwidth allocation and power control for OFDMA," *IEEE Transactions on Wireless Communications*, vol. 2, no. 6, pp. 1150–1158, November 2003.
4. M. Ergen, S. Coleri, and P. Varaiya, "QoS aware adaptive resource allocation techniques for fair scheduling in OFDMA based broadband wireless access systems," *IEEE Transactions on Broadcasting*, vol. 49, no. 4, pp. 362–370, December 2003.

5. I. Kim, I. S. Park, and Y. H. Lee, "Use of linear programming for dynamic subcarrier and bit allocation in multiuser OFDM," *IEEE Transactions on Vehicular Technology*, vol. 35, no. 4, pp. 1195–1207, July 2006.

6. Z. Shen, J. G. Andrews, and B. L. Evans, "Adaptive resource allocation in multiuser OFDM systems with proportional rate constraints," *IEEE Transactions on Wireless Communications*, vol. 4, no. 6, pp. 2726–2737, November 2005.

7. G. Song and Y. Li, "Cross-layer optimization for OFDM wireless networks – Part I: Theoretical framework," *IEEE Transactions on Wireless Communications*, vol. 4, no. 2, pp. 614–624, March 2005.

8. G. Song and Y. Li, "Cross-layer optimization for OFDM wireless networks – Part II: Algorithm development," *IEEE Transactions on Wireless Communications*, vol. 4, no. 2, pp. 625–634, March 2005.

9. K. Kim, Y. Han, and S.-L. Kim, "Joint subcarrier and power allocation in uplink OFDMA systems," *IEEE Communications Letters*, vol. 9, no. 6, pp. 526–528, June 2005.

10. C. Y. Ng and C. W. Sung, "Low complexity subcarrier and power allocation for utility maximization in uplink OFDMA systems," *IEEE Transactions on Wireless Communications*, vol. 7, no. 5, pp. 1667–1675, May 2008.

11. G. Li and H. Liu, "Downlink radio resource allocation for multi-cell OFDMA system," *IEEE Transactions on Wireless Communications*, vol. 5, no. 12, pp. 3451–3459, December 2006.

12. L. Venturino, N. Prasad, and X. Wang, "Coordinated scheduling and power allocation in downlink multicell OFDMA networks," *IEEE Transactions on Vehicular Technology*, vol. 58, no. 6, pp. 2835–2848, July 2009.

13. R. Y. Chang, Z. Tao, J. Zhang, and C.-C. J. Kuo, "Multicell OFDMA downlink resource allocation using a graphic framework," *IEEE Transactions on Vehicular Technology*, vol. 58, no. 7, pp. 3494–3507, September 2009.

14. Z. Han, Z. Ji, and K. J. R. Liu, "Non-cooperative resource competition game by virtual referee in multi-cell OFDMA networks," *IEEE Journal on Selected Areas in Communications*, vol. 25, no. 6, pp. 1079–1090, August 2007.

15. T. Wang and L. Vanderdorpe, "Iterative resource allocation for maximizing weighted sum min-rate in downlink cellular OFDMA systems," *IEEE Transactions on Signal Processing*, vol. 59, no. 1, pp. 223–234, January 2011.

16. K. Yang, N. Prasad, and X. Wang, "An auction approach to resource allocation in uplink OFDMA systems," *IEEE Transactions on Signal Processing*, vol. 57, no. 11, pp. 4482–4496, November 2009.

17. Z.-Q. Luo and S. Zhang, "Dynamic spectrum management: Complexity and duality," *IEEE Journal of Selected Topics in Signal Processing*, vol. 2, no. 1, pp. 57–73, February 2008.

18. Y. Sun, R. P. Jover, and X. Wang, "Uplink interference mitigation for OFDMA femtocell networks," *IEEE Transactions on Wireless Communications*, vol. 11, no. 2, pp. 614–625, February 2012.

19. Z. Lu, T. Bansal, and P. Sinha, "Achieving user-level fairness in open-access femtocell based architecture," *IEEE Transactions on Mobile Computing*, vol. 12, no. 10, pp. 1943–1954, October 2013.

20. Y.-S. Liang, W.-H. Chung, G.-K. Ni, I.-Y. Chen, H. Zhang, and S.-Y. Kuo, "Resource allocation with interference avoidance in OFDMA femtocell networks," *IEEE Transactions on Vehicular Technology*, vol. 61, no. 5, pp. 2243–2255, June 2012.

21. A. J. Goldsmith and S. G. Chua, "Variable-rate variable-power MQAM for fading channels," *IEEE Transactions on Communications*, vol. 45, no. 10, pp. 1218–1230, October 1997.

22. W. L. Winston, *Operations Research: Applications and Algorithms*. 4th edition. Belmont, CA: Duxbury, 2004.

23. L. A. Wolsey, *Integer Programming*. New York: Wiley, 1998.

24. K. E. Atkinson, *An Introduction to Numerical Analysis*. New York: Wiley, 1998

25. R. Cendrillon, W. Yu, M. Moonen, J. Verlinden, and T. Bostoen, "Optimal multiuser spectrum balancing for digital subscriber lines," *IEEE Transactions on Communications*, vol. 54, no. 5, pp. 922–933, May 2006.

26. W. Yu and R. Liu, "Dual methods for non-convex spectrum optimization of multicarrier systems," *IEEE Transactions on Communications*, vol. 54, no. 7, pp. 1310–1322, July 2006.

27. J. Papandriopoulos and J. S. Evans, "SCALE: A low-complexity distributed protocol for spectrum balancing in multiuser DSL networks," *IEEE Transactions on Information Theory*, vol. 55, no. 8, pp. 3711–3724, November 2009.

28. A. Gjendemsj, D. Gesbert, G. E. Oien, and S. G. Kiani, "Binary power control for sum rate maximization over multiple interfering links," *IEEE Transactions on Wireless Communications*, vol. 7, no. 8, pp. 3164–3173, August 2008.

29. V. N. Ha and L. B. Le, "Fair Resource Allocation for OFDMA Femtocell Networks with Macrocell Protection," *IEEE Transactions on Vehicular Technology*, to appear.

30. G. J. Foschini and Z. Miljanic, "A simple distributed autonomous power control algorithm and its convergence," *IEEE Transactions on Vehicular Technology*, vol. 42, no. 4, pp. 641–646, November 1993.

31. N. Bambos, S. C. Chen, and G. J. Pottie, "Channel access algorithms with active link protection for wireless communication networks with power control," *IEEE/ACM Transactions on Networking*, vol. 8, no. 5, pp. 583–597, October 2000.

32. M. Andersin, Z. Rosberg, and J. Zander, "Gradual removals in cellular PCS with constrained power control and noise," *Wireless Networks*, vol. 2, pp. 27–43, 1996.

33. G. Zheng, K. K. Wong, and T. S. Ng, "Throughput maximization in linear multiuser MIMO-OFDM downlink systems," *IEEE Transactions on Vehicular Technology*, vol. 57, no. 3, pp. 1993–1998, May 2008.

34. E. Bjornson, N. Jalden, M. Bengtsson, and B. Ottersten, "Optimality properties, distributed strategies, and measurement-based evaluation of coordinated multicell OFDMA transmission," *IEEE Transactions on Signal Processing*, vol. 59, no. 12, pp. 6086–6101, December 2011.

35. J. Mo and J.Walrand, "Fair end-to-end window-based congestion control," *IEEE/ACM Transactions on Networking*, vol. 8, no. 5, pp. 556–567, October 2000.

CHAPTER 4

RESOURCE ALLOCATION FOR CLUSTERED SMALL CELLS IN TWO-TIER OFDMA NETWORKS

4.1 INTRODUCTION

As we have discussed before, in an OFDMA-based two-tier cellular network with spectrum sharing among small cells and macrocells, co-tier interference and CTI significantly affect network performance [1]. Hence, interference mitigation techniques need to be developed to manage the radio resources of small cells in order to achieve the QoS requirements of all the users. In this chapter, we study the problem of sub-channel and power allocation in a dense deployment of femtocells in an OFDMA-based two-tier network with constraints on co-tier interference and CTI as well as minimum data rate requirements. We formulate the problem of joint sub-channel and power allocation for a clustered femtocell network with interference constraints on co-tier interference and CTI as well as data rate constraints as an optimization problem. The optimization problem aims at maximizing the sum-rate of the femtocells in a cluster of cooperating femtocells. This problem turns out to be a Mixed Integer Non-Linear Program (MINLP). Therefore, to solve this problem sub-optimally, different approaches are proposed. The first approach works by reformulating the MINLP problem as a convex one by using the time sharing idea, where sub-channels can be time shared among femtocells. Joint sub-channel and power allocation using the reformulated problem gives an upper bound to the original problem. In the other approaches, joint sub-channel and power allocation is done in two phases. In the first phase, given power allocation, sub-channels are allocated to femtocells. For sub-channel allocation, different schemes are proposed. The first scheme uses Linear Programming (LP) and the time sharing concept. This approach also gives an upper bound to the optimal solution. Other schemes that preserve the integrality constraint (i.e., a sub-channel is allocated to only one femtocell) are proposed. Those schemes are Branch and Bound (BnB), LP with rounding, Lagrangian Relaxation, and a heuristic scheme. Knowing the sub-channel allocation, in the second phase, we perform power allocation. The power allocation problem is

Radio Resource Management in Multi-Tier Cellular Wireless Networks, First Edition.
Ekram Hossain, Long Bao Le, and Dusit Niyato
© 2014 John Wiley & Sons, Inc. Published 2014 by John Wiley & Sons, Inc.

a non-linear convex problem that can be efficiently solved by interior point method. We compare the proposed schemes in terms of closeness to the optimal solution (which is obtained by an exhaustive search) and complexity. The obtained results show that the proposed schemes have performance that is close to that of the optimal scheme and have reduced complexity. In addition, we study the effect of clustering (i.e., placing femtocells into organized disjoint groups, where co-tier interference is eliminated) on the femtocell network performance. We compare different cluster configurations to a distributed scheme (i.e., without clustering) and a centralized scheme (i.e., with all femtocells in one cluster). Results show that in a dense deployment of femtocells with tight interference constraints, clustering is a very efficient technique.

In Section 4.2, a review of related work is presented. The system model and assumptions are discussed in Section 4.3. In Section 4.4, the main problem for joint sub-channel and power allocation is formulated. Section 4.5 reformulates joint sub-channel and power allocation as a convex problem. Section 4.6 then presents the different proposed schemes for sub-channel allocation. Section 4.7 discusses the power allocation step. Finally, Section 4.8 presents the numerical results before we conclude the chapter in Section 4.9.

4.2 RELATED WORK

4.2.1 Clustering and Coalition Formations of Femtocells

In [2–5], the authors propose a coalition formation games where, femtocells cooperate by forming coalitions. A coalition here is a group of femtocells organized by a coalition level scheduler such that interference within a coalition is eliminated by time-division duplex (TDD) mode of operation. Therefore, femtocells in a coalition can use the same sub-channels but on orthogonal time slots. However, there is inter coalition interference. Interference in this work is considered among femtocells only due to the assumption of split spectrum operation with macrocells. Each femtocell has four sub-channels (independent of the coalition size). In forming the coalitions, there is a cost in terms of transmission power that prevents forming coalitions with distant femtocells, and hence grand coalitions. Each femtocell evaluates the options of joining available coalitions in the form of iterations. From the results, it was shown that the number of iterations increases exponentially as the number of femtocells increases. Obtaining the optimal coalition structure is very challenging, and hence the authors proposed an algorithm that is based on recursive core to obtain the coalitions in a distributed manner.

In [6], the authors propose a scheme for macro and femtocell interference mitigation. In their work, femtocells exist in groups in the form of (5×5) grid. Each group of FAPs is managed by a femtocell controller; hence, we have a centralized operation. Interference exists between FAPs and macrocells. To reduce the amount of interference introduced to the MUEs, the FAPs rely on cognitive capabilities and sensing, where they employ both overlay and underlay modes of operation. Each of the FAPs and MUEs requires at most one sub-channel. No power control is considered for the macrocell but power control is considered for the FAPs. Femtocells are grouped into clusters, where a cluster is a group of co-channel FAPs that use the same sub-channel. Femtocells suffering from strong mutual interference are placed in different clusters. Formation of the clusters is done by using a graph approach relying on the max k-cut problem. A sub-optimal algorithm is proposed to place the femtocells in the clusters. Then each cluster is assigned a sub-channel using graph colouring. The only aspect considered by the authors here to maximize the throughput is by minimizing the co-tier interference. After that power control is performed for femtocells.

In [7], the authors formulate and solve the problem of femtocell deployment in a two-tier cellular network. Given a certain number of femtocells, the target is how to optimally deploy them to serve a set of house holds (HH) with the minimum data rate requirements and set their transmission powers. The number of femtocells is not enough to have a femtocell installed in every HH. Therefore, one femtocell will be installed in an HH that will act as a cluster head (CH) and will serve neighboring HHs acting as cluster members (CMs) in a TDMA fashion. The CH with its CMs form a cluster, where interference is eliminated through TDMA operation. The authors divide the joint optimization problem at hand into two sub-problems, namely, the cluster formation sub-problem and the power control subproblem. In their work, the number of clusters is known a priori and the maximum number of HHs that a FAP can serve is known as well. Therefore, starting with some initial coalition structure, the power control sub-problem is solved to obtain the social welfare. After that, using the simulated annealing algorithm, another coalition structure is proposed for which a new power control sub-problem is solved. The new social welfare is compared to that of the previous coalition structure. Iterations proceed up to the best coalition structure. A shared spectrum operation is assumed where macrocell and femtocells operate on the same channel. Hence, no channel allocation is considered and closed access FAPs are assumed.

4.2.2 Resource Allocation in Clustered Femtocells

In [8], the authors propose a joint power control and resource allocation algorithm in an OFDMA femtocell network, where femtocells are grouped into disjoint clusters. However, the authors only consider orthogonal channel assignment between femtocells and macrocells; hence, no CTI exists. In addition, optimal selection of the cluster size is not discussed. In [9], the authors propose a dynamic clustering based sub-band allocation scheme in a dense femtocell environment. However, the authors only consider frequency allocation. Also, they treat the interference from the macrocell network as additive white Gaussian noise (AWGN). The authors in [10] propose a coalition formation game approach to form cooperative groups among interfered femtocells by sharing the radio spectrum.

In [11], a DL power control method is proposed to mitigate interference in a macrocell–femtocell network. The QoS for both MUE and FUE is guaranteed by limiting interference from femtocells to nearby MUEs. A centralized and a distributed approach for power control are proposed and compared. However, no frequency allocation is considered.

In [12], a joint power control and resource allocation scheme for a co-channel femtocell–macrocell network is proposed with QoS guarantees for both MUE and FUE. Both centralized and distributed approaches are introduced.

In [13], a joint power and sub-channel allocation scheme is proposed to maximize the system capacity for indoor dense mobile communication systems. The authors assume that the femtocells are densely deployed and propose a centralized resource allocation framework. The authors prove that the optimal power allocation in such an environment has a special form known as binary power allocation, where for every sub-channel, a single femtocell only loads power. However, in [11–13], the effect of clustering is not studied for the femtocell resource allocation problem. In addition, [13] considers the interference from macrocell base stations (MBSs) as additive white Gaussian noise (AWGN).

A resource allocation and admission control algorithm, called QoS-based femtocell resource allocation (Q-FCRA), is proposed in [8], which is based on clustering and taking into account QoS constrained of high priority user and best effort users. The Q-FCRA algorithm is comprised of three main phases: (i) Cluster formation, (ii) Intra-cluster resource allocation, and (iii) Inter-cluster resource contention resolution. When powered on, a FAP listens to its surrounding transmissions and gathers information through measurements collected from users attached to it or via a receiver function within the FAP. Based on this information, the FAP computes the number of interfering femtocells and transmits it along with its Physical Cell Identity (PCI) to each of them. The FAP with the highest interference degree among its one-hop neighbors is selected as a CH. After partitioning the femtocell network into clusters, resources are jointly allocated to all FAPs within each cluster taking into account QoS requirements of attached users. The resource allocation problem is formulated as a multi-objective optimization problem. The objective is to find the optimal resource allocation of a set of tiles in each FAP to deliver user data, while minimizing the interference among the FAPs and at the same time providing QoS guarantees for high priority users as well as maximizing the throughput for best-effort users.

4.3   SYSTEM MODEL AND ASSUMPTIONS

We consider a multi-tier wireless network where femtocells are deployed in a dense manner to cover an indoor area and are overlaid by a single macrocell. In such an environment, FUEs will have very good channel conditions to their femtocells. Signals received from outdoor macrocells, however, are highly attenuated. We denote by k_f a user served by femtocell f. The MUEs exist indoor as well and are served by macrocell m, where we denote by k_m a user served by macrocell m. We denote by $\mathbf{F}$ the set of femtocells where, $F = |\mathbf{F}|$ and by $\mathbf{m}$ the macrocell. Let $\mathbf{N}$ be the set of indices of the sub-channels in the system and $N = |\mathbf{N}|$. We assume channel states of sub-carriers are the same within a sub-channel n of bandwidth Δf. Define $g_{i,j}^n$ as the channel gain between user i and BS j on sub-channel n. In our model, channel gain includes path-loss, log-normal shadowing, and Rayleigh fading.

The unit power SINR of an FUE k_f served by femtocell f on sub-channel n is as follows:

$$\gamma_{k_f,f}^n = \frac{g_{k_f,f}^n}{\sum_{j \neq f, j \in \mathbf{F}} p_{k_j,j}^n g_{k_f,j}^n + p_{k_m,m}^n g_{k_f,m}^n + N_o} \tag{4.1}$$

where $p_{i,j}^n$ is the power allocated to the link between user i and BS j on sub-channel n and N_o is the noise power.

Closed access femtocells are assumed where the access is restricted to registered UEs only. Femtocells and the macrocell operate in a shared spectrum environment. Femtocells have unique Cell-IDs. All the FUEs are capable of performing interference measurements and reporting them to the FAPs. Femtocells are equipped with GPS devices to obtain the accurate estimate of femtocell locations so that femtocell clusters can be formed. Communication among femtocells is possible via air or backhaul.

The femtocells are divided into disjoint clusters. The idea behind clustering is to divide the joint sub-channel and power allocation problem in the femtocell network into smaller sub-problems. We define $\mathbf{C}$ as the set of clusters of femtocells. A cluster $c_l \in \mathbf{C}$ is the lth set of femtocells such that $c_l \subseteq \mathbf{F}, \forall l \in \{1, 2, \dots, |\mathbf{C}|\}$, $\bigcup_l c_l = \mathbf{F}$, and $\bigcap_l c_l = \varnothing$. Within each cluster, one femtocell is elected as a **CH** and it takes the responsibility of performing sub-channel and power allocation within the cluster. Clustering is simply done by grouping femtocells closely located to each other. Therefore, it is basically done based on distance. Note that the entire set of sub-channels $\mathbf{N}$ is available to each cluster and within a cluster, no two femtocells transmit simultaneously on the same sub-channel. Therefore, no co-tier interference exists within a cluster.

Various configurations are available for cluster sizes. One extreme is to have a grand cluster, where all femtocells are in one cluster. One benefit of this configuration is that it eliminates co-tier interference among femtocells. However, in addition to the huge burden imposed on the **CH** performing resource allocation, the share of each femtocell in the available spectrum is small. We refer to this scheme as the *centralized scheme*. Another extreme is to have femtocells acting independently, that is, no clustering. In this case, each femtocell has the entire spectrum available and the complexity of resource allocation is the lowest. However, co-tier interference is severe in this case. We refer to this scheme as the *distributed scheme*.

4.4 JOINT SUB-CHANNEL AND POWER ALLOCATION IN FEMTOCELL CLUSTERS

For each femtocell cluster c_l, we require to maximize the sum-rate of all femtocells within the cluster, given that there are interference constraints for co-channel FUEs in neighboring clusters as well as co-channel MUEs. In addition, we have a data rate requirement for each femtocell. The **CH** within each cluster c_l solves the following optimization problem:

$$\max_{\Gamma_{k_f,f}^n, P_{k_f,f}^n} \sum_{f \in c_l} \sum_{n=1}^{N} \Gamma_{k_f,f}^n \Delta f \log_2 \left(1 + P_{k_f,f}^n \gamma_{k_f,f}^n\right) \tag{4.2}$$

subject to

$$\sum_{n=1}^{N} \Gamma_{k_f,f}^n \Delta f \log_2 \left(1 + P_{k_f,f}^n \gamma_{k_f,f}^n\right) \geq R_f, \quad \forall f \qquad (4.3)$$

$$\sum_{n=1}^{N} \Gamma_{k_f,f}^n P_{k_f,f}^n \leq P_{f\max}, \quad \forall f \qquad (4.4)$$

$$\Gamma_{k_f,f}^n P_{k_f,f}^n g_{k_m,f}^n \leq \zeta_{k_m}^n, \quad \forall f, n \qquad (4.5)$$

$$\Gamma_{k_f,f}^n P_{k_f,f}^n g_{k_j,f}^n \leq \zeta_{k_j}^n, \quad \forall f, n \qquad (4.6)$$

$$\sum_{f \in c_l} \Gamma_{k_f,f}^n = 1, \quad \forall n \qquad (4.7)$$

where $\Gamma_{k_f,f}^n \in \{0, 1\}$ is an indicator if sub-channel n is allocated to the link between user k_f and femtocell f, $P_{k_f,f}^n \geq 0$ is the power assigned to the link between them.

In the optimization problem (4.2), the sum-rate maximization is subject to a data rate requirement R_f and total power budget $P_{f\max}$ for each femtocell f as indicated in (4.3) and (4.4), respectively. We have interference constraints for co-channel MUEs served by macrocell m and FUEs served by femtocells j in a neighboring cluster as given in (4.5) and (4.6), respectively. The reference user concept [14] is applied here, where the interference constraints are for co-channel MUEs and FUEs having the highest channel gain to the target femtocell f. The interference constraints in (4.5) and (4.6) can be further simplified to a single constraint by choosing one of them having the higher channel gain to the target femtocell f. In this way, the interference constraints can be rewritten as follows:

$$\Gamma_{k_f,f}^n P_{k_f,f}^n \leq \min \left(\frac{\zeta_{k_m}^n}{g_{k_m,f}^n}, \frac{\zeta_{k_j}^n}{g_{k_j,f}^n}\right) = \frac{\zeta_k^n}{g_{k,f}^n}, \quad \forall f, n. \qquad (4.8)$$

A user k now is either a co-channel MUE or an FUE whichever has higher channel gain to the target femtocell f. Finally, (4.7) is the exclusion constraint indicating that sub-channel n can be used in one femtocell only.

Since this problem is an MINLP, obtaining its optimal solution will be prohibitive. Therefore, several approaches will be proposed to solve it. One approach will be to reformulate the problem as a convex one by using the idea of time sharing. The other approaches will solve the problem in two phases:

- *Phase 1*: Given power allocation, we can perform sub-channel allocation. We assume the initial power $P_{k_f,f}^n$ on each sub-channel as the minimum of either $\frac{P_{f\max}}{N}$ or $\frac{\zeta_k^n}{g_{k,f}^n}$. The idea is to keep power as uniform as possible and at the same time not to violate the interference constraints. For Phase 1, several schemes with varying performance and complexity will be proposed.
- *Phase 2*: Given sub-channel allocation, we perform power allocation.

It is worth mentioning that, the **CH**s perform resource allocation in parallel. The FUEs in a cluster measure the interference levels in the previous time slot, and report them to their FAPs which, in turn, pass them to the **CH**. Based on those measurements, the **CH** performs resource allocation in the current time slot [15].

4.5 JOINT SUB-CHANNEL AND POWER ALLOCATION USING CONVEX REFORMULATION

A common approach in the literature is to relax the constraint that sub-channels can be used by one femtocell. Thus $\Gamma_{k_f,f}^n$ is reinterpreted as the sharing factor of femtocell f for sub-channel n. In addition, we define a new variable $s_{k_f,f}^n = p_{k_f,f}^n \Gamma_{k_f,f}^n$. Then $s_{k_f,f}^n$ becomes the actual transmitted power [16–18]. For each cluster c_l, problem in (4.2) can thus be reformulated as follows:

$$\max_{\Gamma_{k_f,f}^n, s_{k_f,f}^n} \sum_{f \in c_l} \sum_{n=1}^N \Gamma_{k_f,f}^n \Delta f \log_2 \left(1 + \frac{s_{k_f,f}^n \gamma_{k_f,f}^n}{\Gamma_{k_f,f}^n} \right) \tag{4.9}$$

subject to

$$\sum_{n=1}^N \Gamma_{k_f,f}^n \Delta f \log_2 \left(1 + \frac{s_{k_f,f}^n \gamma_{k_f,f}^n}{\Gamma_{k_f,f}^n} \right) \geq R_f, \quad \forall f \in c_l \tag{4.10}$$

$$\sum_{n=1}^N s_{k_f,f}^n \leq P_{f,max}, \quad \forall f \in c_l \tag{4.11}$$

$$s_{k_f,f}^n g_{k,f}^n \leq \xi_k^n, \quad \forall n, f \in c_l \tag{4.12}$$

$$\sum_{f \in c_l} \Gamma_{k_f,f}^n = 1, \quad \forall n \tag{4.13}$$

$$s_{k_f,f}^n \geq 0, \quad \forall n, f \in c_l \tag{4.14}$$

$$\Gamma_{k_f,f}^n \in (0, 1], \quad \forall n, f \in c_l \tag{4.15}$$

where $\gamma_{k_f,f}^n = \frac{g_{k_f,f}^n}{\sum_{j \notin c_l, j \in \mathbf{F}} s_{k_j,j}^n g_{k_f,j}^n + p_{k_m,m}^n g_{k_f,m}^n + N_o}$.

In (4.9), $\Gamma_{k_f,f}^n$ is not allowed to be zero since the objective function in (4.9) is not defined for $\Gamma_{k_f,f}^n = 0$. However, the objective function approaches close to zero when $\Gamma_{k_f,f}^n$ is arbitrarily small. Hence, the nature of the objective function remains the same [16]. A nice property of the optimization problem in (4.9) is that it is convex. The objective function is concave, inequality constraint in (4.10) is convex and all the remaining inequality and equality constraints are affine. Therefore, this problem can be efficiently solved by the interior point method [19]. It is worth mentioning that this approach gives an upper bound solution to (4.2).

4.6 SUB-CHANNEL ALLOCATION

In this section, knowing power allocation, we perform sub-channel allocation using various schemes of different performance and complexity.

4.6.1 Branch and Bound

For given power allocation, we have an ILP that can be optimally solved using the BnB technique. We have the following optimization problem:

$$\max_{\Gamma_{k_f,f}^n} Z_{BnB} = \sum_{f \in c_l} \sum_{n=1}^{N} \Gamma_{k_f,f}^n \Delta f \log_2 \left(1 + P_{k_f,f}^n \gamma_{k_f,f}^n\right) \qquad (4.16)$$

subject to

$$\sum_{n=1}^{N} \Gamma_{k_f,f}^n \Delta f \log_2 \left(1 + P_{k_f,f}^n \gamma_{k_f,f}^n\right) \geq R_f, \quad \forall f \qquad (4.17)$$

$$\sum_{f \in c_l} \Gamma_{k_f,f}^n = 1, \quad \forall n \qquad (4.18)$$

where $\Gamma_{k_f,f}^n \in \{0, 1\}$.

BnB is guaranteed to find the optimal sub-channel allocation but its complexity in the worst case is as high as that of exhaustive search which is $O(F^N)$.

4.6.2 Linear Programming

By relaxing the integrality constraint on $\Gamma_{k_f,f}^n$ to be taking any value $[0, 1]$, we have a Linear Program (LP) with an objective function value Z_{LP-TS} that can be solved using simplex or interior point method. In this way, $\Gamma_{k_f,f}^n$ can now be defined as a time sharing factor. It indicates the amount of time for which sub-channel n is allocated to user k_f in femtocell f. The solution obtained in this way is an upper bound solution to the optimal sub-channel allocation problem.

To retrieve the integrality property again for the LP solution, we assign sub-channel n to user k_f in femtocell f according to:

$$k_f(n) = \arg \max_f \Gamma_{k_f,f}^n, \quad \forall n \qquad (4.19)$$

where $k_f(n)$ denotes sub-carrier n allocated to user k_f in femtocell f. We denote the resulting objective function value as Z_{LP} and call this technique as LP with rounding. Using the interior point method, the complexity of sub-channel allocation using LP is $O(F^3 N^3)$.

4.6.3 Lagrangian Relaxation

The idea behind Lagrangian relaxation is that some optimization problems can be easily solved if some set of annoying constraints are removed (dualized). By doing so, we can solve an easier version of the original problem [20].

In our sub-channel allocation problem, if we dualize the data rate requirement constraints, we have the following optimization problem:

$$\max_{\Gamma_{k_f,f}^n} Z_{LG}(u_f) = \sum_{f \in c_l} \sum_{n=1}^{N} \Gamma_{k_f,f}^n \Delta B_f^n \left(1 + u_f\right) - \sum_{f \in c_l} u_f R_f \qquad (4.20)$$

subject to

$$\sum_{f \in c_l} \Gamma_{k_f,f}^n = 1, \quad \forall n \qquad (4.21)$$

where $\Gamma_{k_f,f}^n \in \{0, 1\}$, u_f is a non-negative Lagrange multiplier and

$$\Delta B_f^n = \Delta f \log_2 \left(1 + P_{k_f,f}^n \gamma_{k_f,f}^n\right). \qquad (4.22)$$

The function $Z_{LG}(u_f)$ is the Lagrange dual function.

This problem is an assignment problem that can be solved, for a given multiplier u_f, using a greedy approach, where for each sub-channel n, we allocate it to user k_f in femtocell f according to the following:

$$k_f(n) = \arg \max_f \Delta B_f^n \left(1 + u_f\right) - u_f R_f. \qquad (4.23)$$

The complexity of this greedy approach is $O(FN)$. Now, to obtain the optimal multiplier u_f, we have the following optimization problem:

$$\min_{u_f} Z_{LG}(u_f) \qquad (4.24)$$

subject to

$$u_f \geq 0. \qquad (4.25)$$

To obtain the values of the Lagrange multipliers u_f, we shall use the sub-gradient method. For a multiplier u_f we have

$$u_f^{(y+1)} = \left[u_f^{(y)} - t^{(y)}\left(\left(\sum_{n=1}^{N} \Gamma_{k_f,f}^{n,(y)} \Delta B_f^n\right) - R_f\right)\right]^+ \qquad (4.26)$$

where $[x]^+ = \max(x, 0)$, and $t^{(y)}$ is the step size obtained in the yth iteration using the following expression:

$$t^{(y)} = \frac{\lambda^{(y)}\left(Z_{LG}\left(u_f^{(y)}\right) - Z^*\right)}{\sum_{f \in c_l}\left(\left(\sum_{n=1}^{N} \Gamma_{k_f,f}^{n,(y)} \Delta B_f^n\right) - R_f\right)^2} \qquad (4.27)$$

where $\lambda^{(y)} \in [0, 2]$, and Z^* is the objective value of the best known feasible solution to (4.16). Iterations proceed until the multipliers converge or we reach a maximum number of iterations.

4.6.4   Heuristic Scheme

In this section, we propose a low complexity heuristic scheme for sub-channel allocation. If we reformulate the objective function and data rate constraint in (4.2) and (4.3) to $\Delta f \log_2 \left(1 + \Gamma_{k_f,f}^n P_{k_f,f}^n \gamma_{k_f,f}^n \right)$, we will have expressions that are equivalent to the original ones for binary values of $\Gamma_{k_f,f}^n$ [21]. If we further relax $\Gamma_{k_f,f}^n$ to take any value [0, 1], we have an NLP that is convex in $\Gamma_{k_f,f}^n$ [19].

Now, for the optimization problem (4.2), if we apply KKT conditions, we obtain the following formula for sub-channel allocation among femtocells:

$$
\Gamma_{k_f,f}^n = \left[\frac{\left(1 + \lambda_{1,f} \Delta f\right)}{\ln 2 \left(\lambda_{2,f} P_{k_f,f}^n + \lambda_{3,f}^n P_{k_f,f}^n g_{k,f}^n - \lambda_4^n \right)} - \frac{1}{P_{k_f,f}^n \gamma_{k_f,f}^n} \right]^+ \quad (4.28)
$$

where $\lambda_{1,f}$, $\lambda_{2,f}$, $\lambda_{3,f}^n$, and λ_4^n are Lagrange multipliers associated with data rate constraints, total power budget, interference constraints and exclusion constraints, respectively. Therefore, we can deduce that a sub-channel n is better be allocated to a user k_f in femtocell f having good SINR value $\gamma_{k_f,f}^n$ and having low channel gain $g_{k,f}^n$ to a reference user (hence, low interference to this co-channel user).

We can propose a heuristic by constructing the following ratio for each femtocell f on sub-channel n: $\frac{\gamma_{k_f,f}^n}{g_{k,f}^n}$. A sub-channel n is allocated to the link between user k_f and femtocell f based on the following criterion:

$$
k_f(n) = \arg \max_f \left(\frac{\gamma_{k_f,f}^n}{g_{k,f}^n} \right). \quad (4.29)
$$

We denote the resulting objective function value by Z_{HEUR}. The complexity of this approach is $O(FN)$. This approach, however, has lower complexity than that of the Lagrangian relaxation approach because it does not involve the iterations of the sub-gradient method.

4.6.5   Feasibility Guarantee Algorithm

A problem with the proposed sub-channel allocation schemes in Sections 4.6.2 (LP with rounding), 4.6.3 and 4.6.4 is that we can have some femtocells with no sub-channels allocated, leading to an infeasible solution to the original problem. Hence, to restore feasibility back, an algorithm called Feasibility Guarantee Algorithm (FGA) is proposed.

The idea of the algorithm is for the unsatisfied femtocells to choose the best sub-channel from the set of allocated sub-channels to the most satisfied femtocells in a cluster c_l. In the algorithm, we denote by N_f, the set of sub-channels allocated to femtocell f. Γ is the set of sub-channel allocation indicators for all femtocells $f \in c_l$. S and U are the set of satisfied femtocells and unsatisfied femtocells, respectively. Iterations are repeated until all unsatisfied femtocells are allocated a sub-channel. Based on the fact that the FUE is very close to its femtocell, and hence has a very good channel condition, the power allocation step can satisfy its data rate requirement.

Algorithm 4.1 FGA FOR c_l

Input: Set of Sub-Channel Allocation Γ in c_l
while $U \neq \emptyset$ **do**
    (a) Find Femtocell $f' \in S$ with maximum number of allocated sub-channels
    (b) Femtocell $f \in U$ chooses sub-channel n such that $n = \arg\max_{n \in N_{f'}} g_{k_f,f}^n$
    (c) Set $\Gamma_{k_f,f}^n = 1$, $\Gamma_{k_{f'},f'}^n = 0$ and update Γ, U, and S
end while

The complexity of this algorithm is $O(F^2 + FN)$.

4.7 POWER ALLOCATION

Given sub-channel allocation, we can perform power allocation in an optimal manner. We have the following optimization problem:

$$\max_{P_{k_f,f}^n} \sum_{f \in c_l} \sum_{n=1}^{N} \Gamma_{k_f,f}^n \Delta f \log_2 \left(1 + P_{k_f,f}^n \gamma_{k_f,f}^n\right) \tag{4.30}$$

subject to

$$\sum_{n=1}^{N} \Gamma_{k_f,f}^n \Delta f \log_2 \left(1 + P_{k_f,f}^n \gamma_{k_f,f}^n\right) \geq R_f, \quad \forall f \tag{4.31}$$

$$\sum_{n=1}^{N} \Gamma_{k_f,f}^n P_{k_f,f}^n \leq P_{f\max}, \quad \forall f \tag{4.32}$$

$$\Gamma_{k_f,f}^n P_{k_f,f}^n g_{k,f}^n \leq \zeta_k^n, \quad \forall f, n \tag{4.33}$$

where $P_{k_f,f}^n \geq 0$.

The optimization is a non-linear convex problem that can be efficiently solved by the interior point method [19].

4.8 PERFORMANCE EVALUATION

4.8.1 Parameters

We consider 10 femtocells deployed in an indoor area of dimensions 100 m $\times$ 100 m to constitute a dense environment and they are within the coverage area of a macrocell. The FUEs and MUEs exist indoor. Each femtocell has a single FUE and the MUEs are served by their macrocell. The communication links between BSs and UEs are affected by path-loss, shadowing, and Rayleigh fading. Indoor femto channel models for urban deployment are used [22]. For path-loss between macrocell m and UE in the indoor area, we have $PL(\text{dB}) = 15.3 + 40 \log_{10} R + L_{ow}$. For path-loss between femtocell and its FUE, we have $PL(\text{dB}) = 38.46 + 20 \log_{10} R$ and for path-loss between femtocell and another UE, we have $PL(\text{dB}) = \max\left(15.3 + 37.6 \log_{10} R, 38.46 + 20 \log_{10} R\right) + q L_{iw}$. For interference measurements, the worst-case initial condition is assumed for all

TABLE 4.1   Parameters

Parameter	Value
Carrier frequency	2 GHz
Number of femtocells	10
Maximum femtocell transmission power	20 mW
FUEs per femtocell	1
Number of sub-channels	10
Sub-channel bandwidth Δf	180 KHz
Noise power N_o	-174 dBm/Hz $+ 10\log_{10}(\Delta f)$
Macrocell radius	200 m
Standard deviation for log-normal shadowing in macrocell	6 dB
Standard deviation for log-normal shadowing in femtocell	4 dB
Outdoor wall loss L_{ow}	20 dB
Penetration loss of the inner walls qL_{iw}	15 dB

femtocells in the network, where all femtocells transmit with uniform power on all sub-channels. The simulation parameters are shown in Table 4.1.

We study the network performance in terms of average achieved data rate and average transmission power percentage for different cluster sizes. Beside the *distributed scheme* and the *centralized scheme*, we consider semi-distributed schemes with clusters of two (2 femtocells per cluster) and clusters of five (5 femtocells per cluster). Note that we only consider equal-sized clusters.

To assess the proposed sub-channel allocation schemes, the optimal solution is obtained by an exhaustive search. For each cluster, all possible combinations of sub-channel allocations are tried for all femtocells, and then the transmission power is allocated in an optimal fashion. The sub-channel allocation combined with the corresponding power allocation yielding the highest sum-rate of the femtocells is chosen as the optimal solution. The sub-channel allocation using the exhaustive search has a complexity of $O(F^N)$.

4.8.2   Numerical Results

Average achieved data rate versus interference threshold: We study the average achieved data rate for all femtocells versus the interference threshold for different cluster sizes. Different MUE densities are considered as well. Sub-channel allocation is performed using LP with time sharing.

In Figure 4.1(a), we study the average achieved data rate for all femtocells versus the interference threshold for the FUEs and the MUE. We have a single MUE. We observe that at loose interference thresholds, the distributed scheme has the highest average data rate. Though it has the highest amount of co-tier interference, each femtocell is allowed to use all the available sub-channels. As the interference constraints become tighter, the benefit of cooperation through clustering starts to appear with the cluster size of two offering the highest average data rate. At very

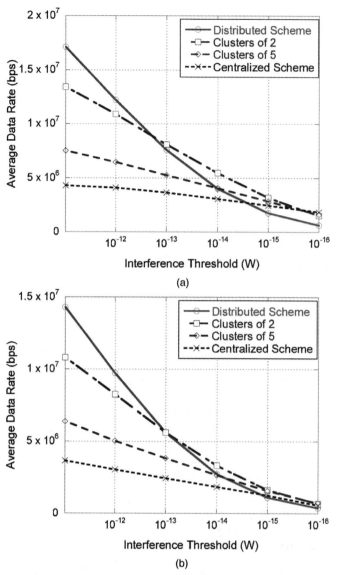

Figure 4.1   Average data rate of femtocells versus interference threshold ($R_f = 10^2$ bps): (a) single MUE case and (b) multiple MUE case.

tight interference thresholds, different clustering configurations and the centralized one have similar performance, which is better than that of the distributed scheme. For the centralized scheme, although the share of each femtocell in the transmission bandwidth is the lowest, it has eliminated interference among the femtocells.

Figure 4.1(b) shows similar results but with multiple MUEs. The achieved data rates for all of the schemes generally decrease due to the increased number of MUEs. From Figures 4.1(a) and (b), we can conclude that cooperation among femtocells is generally effective in improving the femtocell performance.

Average transmission power percentage versus interference threshold: We study the average percentage of femtocell transmission powers defined as $\frac{\sum_{f=1}^{F}\sum_{n=1}^{N}P_{k_f,f}^{n}}{P_{f\max}F} \times 100$ versus the interference threshold for different cluster sizes. Different MUE densities are considered as well. Sub-channel allocation is performed using LP with time sharing.

Figure 4.2(a) shows the percentage of average transmission power for all femtocells considering a single MUE. As the interference threshold becomes tighter, the

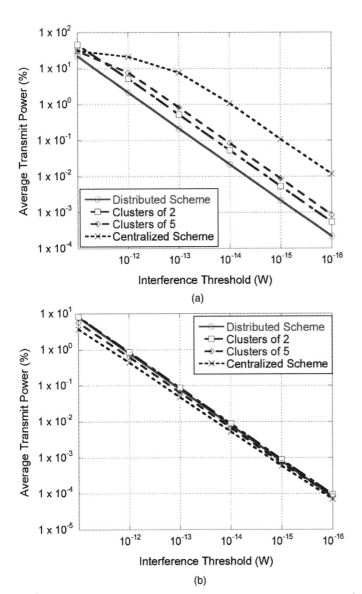

Figure 4.2   Femtocells average transmission power percentage versus interference threshold ($R_f = 10^2$ bps): (a) single MUE case and (b) multiple MUE case.

transmission power of all femtocells decreases. Since the centralized scheme has the MUE only to take care of, it is allowed to transmit at the maximum amount of power. On the other hand, the distributed scheme has the least allowed transmission power, since any femtocell has to take care of interference with all its neighbours.

Figure 4.2(b) shows the percentage of average transmission power for all femtocells considering multiple MUEs. Since we have a larger number of MUEs, all configurations have similar performance of decreased power with tighter interference thresholds. A generic femtocell in the distributed scheme and in the clustering scheme with cluster size of two has a higher number of sub-channels to use, and hence a higher opportunity to transmit more power.

Comparison among the proposed schemes: Figures 4.3 and 4.4 show the average achieved data rate versus the interference threshold for the clustering scheme with cluster size of two and five, respectively. The performances of the different schemes are shown for comparison. The optimal scheme is found for the cluster size of two only. From these two figures we can observe that the LP with time sharing and the convex approach have the highest average data rate since both give an upper bound to the optimal solution. Since the convex approach performs sub-channel and power allocation jointly, it has a higher data rate than that of the LP with time sharing approach. It is worth mentioning that both the schemes, however, solve a different problem from the original one. The BnB technique comes next. For the clustering scheme with cluster size of two, the BnB solution coincides with the optimal solution. Finally, we have the LP technique with rounding, Lagrangian Relaxation, and heuristic schemes, where they all have similar performance which is near optimal with the benefit of reduced complexity.

From the performance evaluation results, we have the following conclusion: in a dense deployment scenario with tight interference constraints, cooperation and

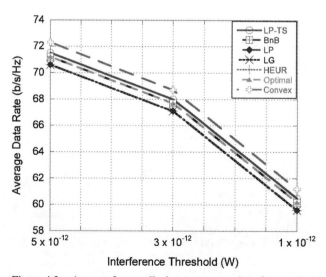

Figure 4.3   Average femtocells data rate versus interference threshold for cluster size of two ($R_f = 10^5$ bps).

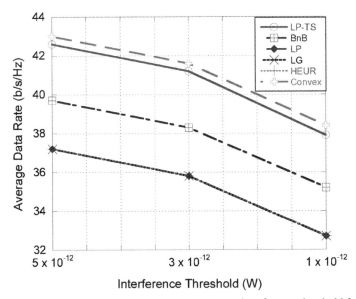

Figure 4.4   Average femtocells data rate versus interference threshold for cluster size of five ($R_f = 10^5$ bps).

coordination among femtocells is beneficial. The MINLP problem of sub-channel and power allocation in femtocell clusters can be solved with efficient schemes of reduced complexity and near optimal performance.

4.9   SUMMARY AND FUTURE RESEARCH DIRECTIONS

We have investigated the effect of clustering of densely deployed femtocells into cooperative groups on the system performance. At tight interference constraints, which will be used to protect co-channel FUEs and MUEs, clustering has been shown to be an efficient technique. The problem of sub-channel and power allocation in femtocell clusters turns out to be an MINLP. A reformulation has been done for the problem to transform the MINLP problem into a convex one which gives an upper bound to the optimal solution. Efficient suboptimal schemes offering near optimal solution and having lower complexity have been proposed as well.

Extension of this work can include formulating clustering as an optimization problem to obtain the optimal cluster size. Also, resource allocation for MUEs, can be considered to have a complete framework for a two-tier network. The effects of the channel gain uncertainties and backhaul constraints on the resource allocation in a clustered femtocell network can be investigated as well.

REFERENCES

1. N. Saquib, E. Hossain, L. B. Le, and D. I. Kim, "Interference management in OFDMA femtocell networks: Issues and approaches," *IEEE Wireless Communications*, vol. 19, pp. 86–95, June 2012.

2. F. Pantisano, M. Bennis, W. Saad, R. Verdone, and M. Latva-aho, "Coalition formation games for femtocell interference management: A recursive core approach," in *Proceedings of IEEE Wireless Communications and Networking Conference (WCNC)*, 2011, pp. 1161–1166, March 2011.

3. F. Pantisano, M. Bennis, R. Verdone, and M. Latva-aho, "Interference management in femtocell networks using distributed opportunistic cooperation," in *Proceedings of IEEE 73rd Vehicular Technology Conference (VTC Spring)*, pp. 1–5, May 2011.

4. F. Pantisano, M. Bennis, W. Saad, and M. Debbah, "Cooperative interference alignment in femtocell networks," in *Proceedings of IEEE Global Telecommunications Conference (Globecom)*, pp. 1–6, December 2011.

5. F. Pantisano, M. Bennis, W. Saad, M. Debbah, and M. Latva-aho, "Interference alignment for cooperative femtocell networks: A game- theoretic approach," *IEEE Transactions on Mobile Computing*, 2012.

6. G. Ning, Q. Yang, S. K. Kwak, and L. Hanzo, "Macro- and femtocell interference mitigation in OFDMA wireless systems," in *Proceedings of IEEE Global Telecommunications Conference (Globecom)*, December 2012.

7. S.-E. Wei, C.-H. Chang, Y.-E. Lin, H.-Y. Hsieh, and H.-J. Su, "Formulating and solving the femtocell deployment problem in two-tier heterogeneous networks," in *Proceedings of IEEE Int. Conference on Communications (ICC)*, pp. 5053–5058, June 2012.

8. A. Hatoum, R. Langar, N. Aitsaadi, and G. Pujolle, "Q-FCRA: Qos-based OFDMA femtocell resource allocation algorithm," in *Proceedings of 2012 IEEE International Conference on Communications (ICC)*, 2012.

9. W. Li, W. Zheng, W. Xiangming, and T. Su, "Dynamic clustering based sub-band allocation in dense femtocell environments," in *Proceedings of 2012 IEEE 75th Vehicular Technology Conference (VTC Spring)*, pp. 1–5, May 2012.

10. F. Pantisano, M. Bennis, R. Verdone, and M. Latvaaho, "Interference management in femtocell networks using distributed opportunistic cooperation," in *Proceedings of 2011 IEEE 73rd Vehicular Technology Conference (VTC Spring)*, pp. 1–5, May 2011.

11. X. Li, L. Qian, and D. Kataria, "Downlink power control in co-channel macrocell femtocell overlay," in *Proceedings of 43rd Annual Conference on Information Sciences and Systems, 2009 (CISS 2009)*, pp. 383–388, March 2009.

12. G. Cao, D. Yang, X. Ye, and X. Zhang, "A downlink joint power control and resource allocation scheme for co-channel macrocell-femtocell networks," in *Proceedings of 2011 IEEE Wireless Communications and Networking Conference (WCNC)*, pp. 281–286, March 2011.

13. J. Kim and D.-H. Cho, "A joint power and sub-channel allocation scheme maximizing system capacity in indoor dense mobile communication systems," *IEEE Transactions on Vehicular Technology*, vol. 59, pp. 4340–4353, November 2010.

14. K. Son, S. Lee, Y. Yi, and S. Chong, "REFIM: A practical interference management in heterogeneous wireless access networks," *IEEE Journal on Selected Areas in Communications*, vol. 29, pp. 1260–1272, June 2011.

15. F. Wang, M. Krunz, and S. Cui, "Price-based spectrum management in cognitive radio networks," *IEEE Journal of Selected Topics in Signal Processing*, vol. 2, pp. 74–87, February 2008.

16. Z. Shen, J. G. Andrews, and B. L. Evans, "Adaptive resource allocation in multiuser OFDM systems with proportional rate constraints," *IEEE Transactions on Wireless Communications*, vol. 4, pp. 2726–2737, November 2005.

17. M. Tao, Y.-C. Liang, and F. Zhang, "Resource allocation for delay differentiated traffic in multiuser OFDM systems," *IEEE Transactions on Wireless Communications*, vol. 7, no. 6, pp. 2190–2201, June 2008.

18. D. W. K. Ng and R. Schober, "Resource allocation and scheduling in multi-cell OFDMA systems with decode-and-forward relaying," *IEEE Transactions on Wireless Communications*, vol. 10, no. 7, pp. 2246–2258, July 2011.

19. S. Boyd and L. Vandenberghe, *Convex Optimization*. Cambridge University Press, New York, NY, USA, 2004.

20. M. Fisher, "An applications oriented guide to Lagrangian relaxation," Interfaces, vol. 15, no. 2, pp. 10–21, 1985.

21. D. Bharadia, G. Bansal, P. Kaligineedi, and V. K. Bhargava, "Relay and power allocation schemes for OFDM-based cognitive radio systems," *IEEE Transactions on Wireless Communications*, vol. 10, pp. 2812–2817, September 2011.

22. 3GPP, "Further advancements for E-UTRA physical layer aspects (Re-lease 9)," TR 36.814, 3rd Generation Partnership Project (3GPP), March 2010.

CHAPTER 5

RESOURCE ALLOCATION IN TWO-TIER NETWORKS USING FRACTIONAL FREQUENCY REUSE

5.1 INTRODUCTION

In OFDMA-based two-tier networks such as the macrocell–femtocell networks, co-tier/cross-tier and UL/DL interferences occur only when the aggressor (or the source of interference) and the victim use the same sub-channel. Therefore, it is essential to adopt an effective and robust interference management scheme that would mitigate the co-tier interference and reduce the CTI considerably in order to enhance the throughput of the overall network. As has been discussed in Chapter 2, different techniques such as cooperation among Macro evolved Node B (MeNB) and HeNBs and collaborative frequency scheduling [1], formation of groups of HeNBs and exchange of information (such as path-loss, geographical location, etc.) among neighboring HeNBs [2], power control approach [3, 4], and intelligent spectrum access [5] have been considered in the recent literature to reduce co-tier interference and CTI. However, in this chapter, we concentrate on an interference avoidance technique known as the FFR method (also advocated by Femtoforum in [6][1]), which requires minimal cooperation among BSs, has a less complex operational mechanism, and is well-suited for OFDMA-based LTE-Advanced systems.

The basic mechanism of FFR corresponds to partitioning the macrocell service area into spatial regions and each sub-region is assigned with different frequency sub-bands. Therefore, the cell edge-zone MUEs do not interfere with the center-zone MUEs, and with an efficient channel allocation method, the cell edge-zone MUEs may not interfere with neighboring cell edge-zone MUEs. As a result, the cell edge-zone MUEs receive an acceptable signal quality which subsequently reduces the outage probability and increases the network capacity. Note that this type of FFR

[1]Femtoforum has been renamed as Small Cell Forum: http://www.smallcellforum.org/.

Radio Resource Management in Multi-Tier Cellular Wireless Networks, First Edition.
Ekram Hossain, Long Bao Le, and Dusit Niyato
© 2014 John Wiley & Sons, Inc. Published 2014 by John Wiley & Sons, Inc.

scheme, when operating on a relatively large time-scale, is referred to as a static FFR scheme. In contrast, dynamic FFR schemes [7] can operate on short time scales and can be optimized for system utility with varying network dynamics. However, they are more complex and less scalable compared to the static schemes.

In this chapter, for OFDMA-based two-tier macrocell–femtocell networks, we evaluate three different static FFR schemes originally proposed for homogeneous networks, namely, *strict FFR*, *soft FFR*, and *sectored-FFR* (in particular, FFR-3) schemes. Also, a new static FFR scheme is proposed in this chapter, which is referred to as the *optimal static FFR (OSFFR)* scheme. We provide a broad comparison among all these different FFR schemes based on performance metrics such as outage probability, average network sum-rate, and spectral efficiency in two-tier macrocell–femtocell networks.

5.2 DIFFERENT FFR SCHEMES

5.2.1 Strict FFR Scheme

The basic mechanism here is to apply frequency reuse factor (FRF) of one to the center-zone MUEs and FRF of N to the edge-zone MUEs. The available frequency band is partitioned in such a way that, in a cluster of N cells, the center-zone MUEs in each macrocell are allocated with a common sub-band of frequencies while the rest of the frequencies are equally partitioned into sub-bands according to the FRF of the edge-zone and assigned separately to each cell edge-zone of the cluster. Therefore, a total number of $(N+1)$ sub-bands are required. Figure 5.1(a)(i) illustrates a cellular network with strict FFR deployment. Figure 5.1(a)(ii) illustrates a strict FFR deployment scenario with FRF of $N = 3$ to edge-zone MUEs. In Figure 5.1(a)(iii), the vertical bar represents the labeling of different sub-bands that are used by both MeNB(s) and HeNBs in the cluster of cell(s) in Figure 5.1(a)(ii).

In this scheme, the cell-edge MUEs in a macrocell (e.g., macrocell 1) are not interfered by any other MeNB in tier-1. This significantly reduces the inter-cell co-tier interference. Also, since the cell center-zone and edge-zone MUEs use different sub-bands, intra-cell co-tier interference for the MUEs is mitigated. To reduce intra-cell CTI, an HeNB located in the center-zone needs to choose a sub-channel from a sub-band that is assigned to the MUEs in the edge-zone. With $N = 3$, since only two sub-bands are allocated per cell in a cluster, the HeNB situated in the cell edge-zone has to select a sub-channel from the same sub-band as used by the MUEs in the center-zone (Figure 5.1(a)(ii)). For such an allocation, the CTI would be significant especially near the transition areas of center-zone and edge-zone in a macrocell. Under this frequency allocation scenario, the HeNBs are constantly interfered by the omnidirectional transmission from the MUEs on the same sub-channel even though the MUEs and the HeNBs use different sub-bands in both center-zone and edge-zone. Also, the co-tier interference between HeNBs may become severe especially in the edge-zone since all the neighboring cell edge-zone HeNBs use limited number of sub-channels from the same sub-band.

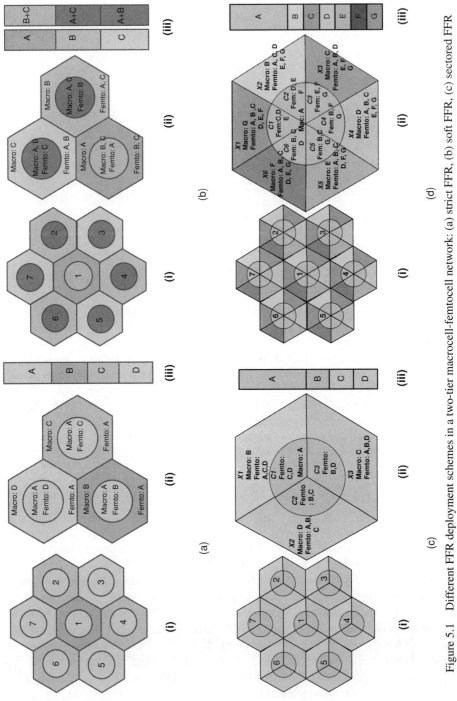

Figure 5.1 Different FFR deployment schemes in a two-tier macrocell-femtocell network: (a) strict FFR, (b) soft FFR, (c) sectored FFR (FFR–3) and (d) sectored FFR (FFR-6) scheme (© [2013] IEEE).

One of the important design parameters here is the radius of the center-zone of the macrocell. Using Monte-Carlo simulations, it was shown in [8] that, for uniformly distributed MUEs, if the cell center-zone radius (r_{center}) is 0.65 times the macrocell radius (R), then the average network throughput is maximized. We consider the same channel allocation and center-zone radius for strict FFR in a two-tier macrocell-femtocell network where in each cell of a cluster of size N, the total sub-channels allocated to center-zone MUEs is given by [8]: $K_{\text{center}} = \left\lceil K_{\text{band}} \left(\frac{r_{\text{center}}}{R} \right)^2 \right\rceil$, where K_{band} is the total number of available sub-channels in a macrocell. The total sub-channels allocated to the edge-zone MUEs is given by $K_{\text{edge}} = \lfloor (K_{\text{band}} - K_{\text{center}})/N \rfloor$.

5.2.2 Soft FFR Scheme

The soft FFR scheme uses a cell-partitioning technique similar to that of the strict FFR scheme. However, the center-zone MUEs of any cell are allowed to use the sub-bands of cell edge-zone MUEs of the neighboring cells within the cluster. For a cluster of N cells, the total number of available sub-channels in a cell is divided into N sub-bands with one sub-band assigned to each edge-zone. Figure 5.1(b)(i) depicts a cellular network with soft FFR deployment. Figure 5.1(b)(ii) illustrates the deployment of a soft FFR scheme with FRF of 3 to the edge-zone MUEs. The entire frequency is divided into sub-bands A, B, and C, and assigned to the cell edge-zone MUEs of macrocell 1, macrocell 2, and macrocell 7, respectively. Now, the center-zone MUEs of macrocell 1 are allowed to use sub-band B and sub-band C, that is, the sub-bands of cell edge-zone MUEs of macrocell 2 and macrocell 7, respectively. Therefore, soft FFR is more bandwidth-efficient than strict FFR.

In this scheme, both center-zone and edge-zone MUEs will experience interference from the tier-1 macrocells (Figure 5.1(b)(i)). A power control factor (ϵ) is therefore introduced for the edge-zone MUEs to reduce ICI. That is, if an MUE m is located in the center-zone, then the transmission power from the tagged MeNB is P_m^k on sub-channel k and if the MUE is located in the edge-zone, then the transmission power is, ϵP_m^k ($\epsilon > 1$). The optimal number of sub-channels allocated to center-zone MUEs is the same as that of the strict FFR case [8] and the total number of sub-channels allocated to edge-zone MUEs is given by

$$K_{\text{edge}} = \min \left(\lceil K_{\text{band}}/N \rceil, K_{\text{band}} - K_{\text{center}} \right). \qquad (5.1)$$

One of the major advantages of soft FFR is that it has better spectrum efficiency in comparison with strict FFR. Similar to strict FFR, an HeNB located in the center-zone may select the sub-band that is used by the MUEs in the edge-zone, and if the HeNB is located in the edge-zone, it chooses the sub-bands that are used by the MUEs in the center-zone (Figure 5.1(b)(ii)). Now, the HeNBs in the edge-zone have more options to choose a sub-channel, and therefore, the co-tier interference would be reduced. However, the CTI would be significant for users near the boundary of the center-zone and the edge-zone.

5.2.3 Sectored FFR (FFR-3) Scheme

The macrocell coverage area is partitioned into center-zone and edge-zone including three sectors per each zone (Figure 5.1(c)(i)). The entire frequency band is divided into two parts—one part is solely assigned to the center-zone (e.g., sub-band A in Figure 5.1(c)(ii)) and the other part is partitioned into three sub-bands (e.g., sub-bands B, C, and D) and assigned to the three edge-zones. An HeNB chooses a sub-band which is not used in the macrocell sub-area. When the HeNB is located in the center-zone, it also excludes the sub-band that is used by the MUEs in the edge-zone of the current sector [9].

As an example, when an HeNB is in edge-zone $X1$, it would only use sub-band A, C, or D and exclude sub-band B since sub-band B is used by the MUEs in region $X1$. Similarly, when an HeNB is in center-zone $C1$, it would avoid sub-band A which is used by the MUEs in the center-zone. It would also avoid sub-band B which is used by the MUEs in edge-zone $X1$, because the received signal power in sub-band B would be relatively strong for that HeNB which may create severe CTI [9]. Therefore, the HeNB in center-zone $C1$ would use sub-band C or D (Figure 5.1(c)(ii)). In this way, the intra-cell CTI will be minimized significantly. Due to sectoring, the inter-cell CTI would be reduced. For example, when a user is in region $X1$ in macrocell 1, CTI will be mainly from macrocell 2 and macrocell 7 rather than from all the six MeNBs in tier-1 of the network (Figure 5.1(c)(i)). As a result, the overall network sum-rate increases in comparison with strict FFR and soft FFR schemes.

The performance of a sectored FFR scheme such as the FFR-3 scheme can be improved by optimizing the edge-zone FRF, the center-zone radius, and the allocation of frequency resources in center-zone and edge-zone MUEs such that the overall network throughput is maximized. Therefore, similar to [10], an optimization problem can be formulated with the objective of maximizing the total network throughput subject to the minimum data rate requirement of MUEs in presence of HeNBs. By solving this optimization problem, we observe that the optimal edge-zone FRF for which the total network throughput is maximized is 6. The resulting FFR scheme is referred to as the optimal static fractional frequency reuse (OSFFR) scheme.

5.3 OPTIMAL STATIC FRACTIONAL FREQUENCY REUSE (OSFFR): AN IMPROVED FFR-BASED SCHEME

We consider the DL of a two-tier OFDMA-based heterogeneous wireless network consisting of an MeNB overlaid with several HeNBs. In this network model, each UE usually communicates on a specific sub-channel corresponding to the BS from which it receives the strongest signal strength, while the signals received from other BSs on the same sub-channel are considered as interference signals. We assume that an open access policy in employed within HeNBs where an MUE can connect to a nearby HeNB if the MUE experiences severe CTI or can achieve a higher throughput. We assume that there is a circular region corresponding to each HeNB where MUEs inside that region would request to connect to that HeNB (Figure 5.2). We assume that each sub-channel is assigned to at most one FUE or MUE. We consider

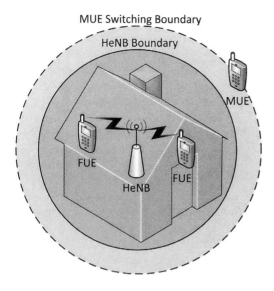

Figure 5.2   Switching boundary for MUEs.

sub-channel allocation in the network according to sector-based FFR. We assume that each macrocell has a total number of K sub-channels which are divided into $(N + 1)$ non-overlapping frequency sub-bands. Each sub-band consists of several sub-channels.

5.3.1   System Model and Assumptions

In the OSFFR scheme, the macrocell coverage is partitioned into center-zone and edge-zone with six sectors in each zone[2] (Figure 5.1(d)(i)). The center-zone MUEs (i.e., the UEs situated within the optimal center-zone radius of the cell) are allocated with sub-band A with the number of sub-channels in this sub-band obtained from the solution of an optimization problem (to be described in Section 5.3.3). The rest of the available sub-channels are divided into 6 sub-bands (B, C, D, E, F, and G) each of which is allocated to one of the edge-zone sectors. The allocation of different frequency sub-bands to the different areas in the cell is shown in Figure 5.1(d)(ii). Thus, in the OSFFR, FRF of one is applied in the center-zone, while FRF of six is applied to the edge-zone MUEs.

Let $K_m^{(c)}$ and $K_f^{(c)}$ be the total number of sub-channels allocated for MUEs and FUEs in the center-zone, respectively. Let $K_m^{(e)}$ and $K_f^{(e)}$ be the total number of sub-channels allocated for MUEs and FUEs in the edge-zone, respectively. The sub-channel allocation in each zone is done based on optimal values of the system design parameters (e.g., optimal FRF, center-zone radius, and proportion of total number of sub-channels allocated to center-zone/edge-zone). The information about the spatial allocation of sub-channels for the MUEs can then be broadcast so that each femtocell

---

[2]Note that the future generation cellular systems (e.g., 4G and 5G systems) may introduce 6-sector cell operation to mitigate ICI.

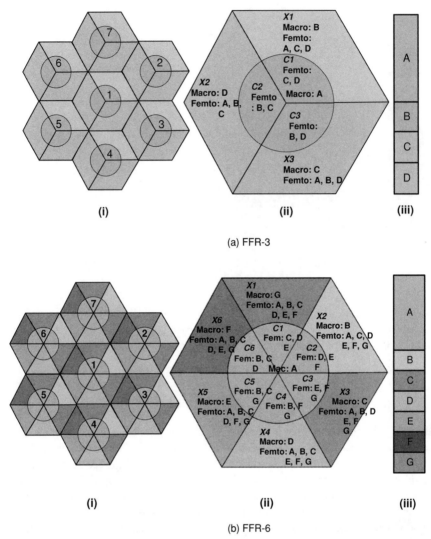

Figure 5.3   Sectored FFR schemes in multi-tier cellular networks: (a) FFR-3 and (b) FFR-6.

knows which sub-channels it can use. We consider that each UE is allocated one sub-channel to satisfy the minimum rate requirement.[3]

5.3.2   Channel Allocation

The sub-channel allocation for sector-based FFR is illustrated in Figure 5.3. For such sub-channel allocation an HeNB executes **Algorithm 5.1** to select the usable sub-channels in a distributed manner. In sector-based FFR, the available frequency band is divided into $N + 1$ sub-bands. Let $\mathbf{J}$ be the set of all available frequency sub-bands in

---

[3]Note that each sub-channel consists of several sub-carriers (i.e., 12) which may be increased to satisfy the minimum rate requirement.

Algorithm 5.1 OPERATIONAL ALGORITHM FOR SUB-CHANNEL ALLOCATION IN HeNB

Input: N, **J**
Output: $\mathbf{J_U}$ = Usable frequency sub-bands for HeNB

1: **for all** HeNB $f \in \mathbf{F_m}$ **do**
2:     $\mathbf{J_U} \leftarrow \mathbf{J}$ {Initialization}
3: **end for**
4: **for all** HeNB $f \in \mathbf{F_m}$ **do**
5:     $\mathbf{T} \leftarrow$ Set of RSSI values for all sub-bands
6:     $R_j^* \leftarrow$ Highest RSSI value
7:     **if** $(R_j^* == R_A)$ **then**
8:         $\mathbf{S}^* \leftarrow$ Set of $\lceil \frac{N+1}{2} \rceil$ sub-bands with comparatively high RSSI value
9:         $\mathbf{J_U} \leftarrow \mathbf{J_U} \setminus \mathbf{S}^*$ {Usable frequency sub-bands for HeNB} f in center-zone
10:     **else**
11:         $\mathbf{S}^* \leftarrow$ Set of one sub-band with the highest RSSI value
12:         $\mathbf{J}_U \leftarrow \mathbf{J}_U \setminus \mathbf{S}^*$ Usable frequency sub-bands for HeNB f in edge-zone
13:     **end if**
14: **end for**

---

an MeNB. For example, in FFR-3 (Figure 5.3(a)(iii)), $\mathbf{J} = \{A, B, C, D\}$, and in FFR-6 (Figure 5.3(b)(iii)), $\mathbf{J} = \{A, B, C, D, E, F, G\}$.[4] Let $\mathbf{J_U}$ be the set of usable frequency sub-bands for the HeNB f ($f \in \mathbf{F_m}$) which is set to $\mathbf{J}$ in the initialization phase of the HeNB. When HeNB f is turned on, it senses and collects the neighboring macrocell pilots signals. Hence it obtains the received signal strength indication (RSSI) [9] value (R_j, $j \in \mathbf{J}$) associated to each frequency sub-band from the pilot signals. Let $\mathbf{T}$ denote the set of RSSI values for all available frequency sub-bands in the macrocell, and R_j^* denote the highest RSSI value. If the RSSI value of sub-band A is the highest (i.e., $R_j^* = R_A$), then HeNB f is located in the center-zone. In this case, HeNB f forms $\mathbf{S}^*$, a set of $\lceil \frac{N+1}{2} \rceil$ sub-bands (including sub-band A) whose RSSI values are comparatively higher than those of other sub-bands. Now, $\mathbf{S}^*$ is excluded from $\mathbf{J_U}$, the set of usable frequency sub-bands for HeNB f located in any of the center-zones *C1–CN*. However, if the RSSI value of sub-band A is not the strongest, then HeNB f is located in one of the edge-zones *X1–XN*. Then, the set $\mathbf{S}^*$ would consist of only one frequency sub-band that has the strongest RSSI value (i.e., the frequency sub-band which is used by the macrocell in the edge-zone of the current sector). Thus, $\mathbf{S}^*$ is excluded from $\mathbf{J_U}$, the set of usable frequency sub-bands for HeNB f located in any of the edge-zones.

As an example, let us consider that an HeNB f is located in *C1* center-zone. Now, for both FFR-3 (Figure 5.3(a)(ii)) and FFR-6 (Figure 5.3(b)(ii)) schemes, the HeNB f will exclude sub-bands used by the MUEs in center-zone and edge-zone of the current sector since the RSSI values for these sub-bands will be high. However,

---

[4]Note that the sub-band notations are only used for symbolic purpose. Each sub-band consists of several sub-channels. However, the number of sub-channels available in each sub-band is obtained from the solution of an optimization problem.

due to reduced center-zone sub-area for FFR-6, the HeNB will also exclude two more sub-bands that are used by the MUEs in the edge-zone just adjacent to the current sector. Hence, the set S^* for FFR-3 would be: $S^* = \{A, B\}$, and for FFR-6 it would be: $S^* = \{A, B, G, F\}$. Now, the HeNB f would exclude S^* from J_U. Therefore, the set of usable frequency sub-bands for HeNB f located in center-zone $C1$ would be: $J_U = \{C, D\}$ and $J_U = \{C, D, E\}$ for FFR-3 and FFR-6, respectively. In an HeNB, we consider uniform transmission power on the sub-channels allocated to FUEs.

Note that, in Figure 5.1(d)(i), any MUE in the edge-zone would experience ICI mainly from one macrocell in tier-1. In other words, any MUE x located in edge-zone $X1$ of macrocell 1, which is allocated a sub-channel from sub-band G, will experience interference from only macrocell 4 if any MUE located in edge-zone $X1$ in this cell is using the same sub-channel. This reduces the ICI among MUEs substantially. In addition, since the center-zone MUEs do not share the spectrum with edge-zone MUEs, the intra-cell interference is mitigated. Furthermore, the entire macrocell adopts an FRF of 1. Under such deployment, when an HeNB is turned on, it senses the neighboring macrocell signals, executes **Algorithm 5.1** in a *distributed* manner and chooses sub-bands which are not used in the macrocell sub-area. Similar to [9], when the HeNB is located in the center-zone, it excludes the sub-band used in the center-zone and the sub-band which is used by macrocell in the edge-zone of current sector (Figure 5.1(d)(ii)). The HeNB additionally excludes two sub-bands which are used by macrocell in the edge-zones just adjacent to the current sector. Note that a low complexity and low cost implementation of HeNBs for such autonomous operation will be an important issue for successful deployment of this scheme.

As an example, when the HeNB is in edge-zone $X1$, it would use sub-band A, B, C, D, E, or F and exclude sub-band G since the sub-band G is used by the macrocell in region $X1$. Now, when the HeNB is located in center-zone $C1$, it would avoid sub-band A which is used by the macrocell in the center-zone. It would avoid sub-band G which is used by the macrocell in edge-zone $X1$, because the received signal power in sub-band G would be strong for that HeNB. In addition, it would exclude sub-band B and F (two sub-bands used by the macrocell in the edge-zones of the adjacent sectors of the current sector for that HeNB) since the received signal power of sub-band B and F would be relatively strong for that HeNB. Therefore, the HeNB located in center-zone $C1$ would use sub-band C, D, or E.

For the proposed scheme, with a reduced macrocell sub-area, an HeNB has more sub-bands to select from. Therefore, the co-tier interference is reduced significantly in comparison with the FFR-3 scheme. Also, the intra-cell CTI to an FUE may only result from the MUE in the same sector in the center-zone or near the transition regions of the edge-zones of the neighboring sectors within a cell. The inter-cell CTI would be only from one neighboring MeNB. For example, an FUE in region $X1$ will experience CTI mainly from the corresponding sector of macrocell 7 (Figure 5.1(d)(i)). In addition, an HeNB in the edge-zone would have six sub-bands to select from. This reduces the probability of intra-cell co-tier interference in comparison to other FFR schemes.

In the next section, we will discuss how to optimize the spatial channel allocation parameters for the proposed scheme.

5.3.3 Optimization of Spatial-Channel Allocation Parameters

The objective is to maximize the total network throughput of two-tier femtocell network, subject to minimum rate requirement for the UEs. The optimization parameters of sector-based FFR are as follows: (i) Percentage of the overall frequency spectrum of the system that should be attributed to center-zone and edge-zone of macrocell service area; (ii) the dimension of the center-zone and edge-zone, that is, the center-zone radius with respect to to the macrocell radius; and (iii) the FRF of the macrocell edge-zone.

The formulation of the optimization problem is presented as follows (see Table 5.1 for the notations used):

$$\max_{N, K_c, \Gamma^{k^{(s)}}_{x_c,m}, \Gamma^{k^{(s)}}_{x_e,m}} \sum_{m=1}^{M} \sum_{x_c \in \mathbf{X_c}} \sum_{k^{(s)} \in \mathbf{K^{(c)}_m}} \Gamma^{k^{(s)}}_{x_c,m} \cdot \Delta B \cdot \log_2 \left(1 + \alpha SINR^{k^{(s)}}_{x_c,m} \right) \quad (5.2)$$

$$+ \sum_{m=1}^{M} \sum_{x_e \in \mathbf{X_e}} \sum_{k^{(s)} \in \mathbf{K^{(e)}_m}} \Gamma^{k^{(s)}}_{x_e,m} \cdot \Delta B \cdot \log_2 \left(1 + \alpha SINR^{k^{(s)}}_{x_e,m} \right)$$

$$+ \sum_{f \in \mathbf{F_m}} \sum_{y_f \in \mathbf{Y_f}} \sum_{k^{(s)} \in \mathbf{J_U}} \Gamma^{k^{(s)}}_{y_f,f} \cdot \Delta B \cdot \log_2 \left(1 + \alpha SINR^{k^{(s)}}_{y_f,f} \right)$$

subject to
$$\Gamma^{k^{(s)}}_{x_c,m} \cdot \Delta B \cdot \log_2 \left(1 + \alpha SINR^{k^{(s)}}_{x_c,m} \right) \geq \Gamma^{k^{(s)}}_{x_c,m} \cdot C^{x_c,m}_{min}, \quad (5.3)$$

$$\forall m \in \mathbf{M}, x_c \in \mathbf{X_c}, k^{(s)} \in \mathbf{K^{(c)}_m}$$

$$\Gamma^{k^{(s)}}_{x_e,m} \cdot \Delta B \cdot \log_2 \left(1 + \alpha SINR^{k^{(s)}}_{x_e,m} \right) \geq \Gamma^{k^{(s)}}_{x_e,m} \cdot C^{x_e,m}_{min}, \quad (5.4)$$

$$\forall m \in \mathbf{M}, x_e \in \mathbf{X_e}, k^{(s)} \in \mathbf{K^{(e)}_m}$$

$$\Gamma^{k^{(s)}}_{y_f,f} \cdot \Delta B \cdot \log_2 \left(1 + \alpha SINR^{k^{(s)}}_{y_f,f} \right) \geq \Gamma^{k^{(s)}}_{y_f,f} \cdot C^{y_f,f}_{min}, \quad (5.5)$$

$$\forall f \in \mathbf{F_m}, y_f \in \mathbf{Y_f}, k^{(s)} \in \mathbf{J_U}$$

$$\Gamma^{k^{(s)}}_{x_c,m} \in \{0, 1\}, \quad \forall m \in \mathbf{M}, \forall x_c \in \mathbf{X_c} \quad (5.6)$$

$$\Gamma^{k^{(s)}}_{x_e,m} \in \{0, 1\}, \quad \forall m \in \mathbf{M}, \forall x_e \in \mathbf{X_e} \quad (5.7)$$

$$\Gamma^{k^{(s)}}_{y_f,f} \in \{0, 1\}, \quad \forall f \in \mathbf{F_A}, \forall y_f \in \mathbf{Y_f}. \quad (5.8)$$

The above optimization problem is a Mixed Integer Non-Linear Optimization Problem which may be solved based on the approaches presented in [9]. However, these approaches, for example, time sharing for sub-channel allocation are not suitable for modeling admission control of UEs (as will be discussed in details in Chapter 6). Hence, we use Monte Carlo simulations to obtain the best configuration (which we refer to as the optimal solution) for the design parameters of sector-based FFR.[5]

---

[5]Note that, based on the system requirements and configuration, the design parameters can be obtained off-line and stored in a look-up table.

TABLE 5.1   Summary of key notations

Notation	Definition
$\mathbf{M}, \mathbf{F}$	Set of all MeNBs/HeNBs
m, f	Serving MeNB/HeNB, $m \in \mathbf{M}$ and $f \in \mathbf{F}$
$\mathbf{M}'$	Set of MeNBs except the serving MeNB, $\mathbf{M}' \subset \mathbf{M}$
$\mathbf{F}'$	Set of neighboring HeNBs except the serving HeNB, $\mathbf{F}' \subset \mathbf{F}$
m', f'	Neighboring MeNB/HeNB, $m' \in \mathbf{M}'$ and $f' \in \mathbf{F}'$
$\mathbf{F_m}$	Set of all HeNBs in macrocell m
$\mathbf{X_m}, \mathbf{Y_f}$	Set of macro UEs (MUEs)/femto UEs (FUEs) associated with MeNB(m)/HeNB(f)
x_m, y_f	MUE/FUE served by MeNB(m)/HeNB(f), $x_m \in \mathbf{X_m}$ and $y_f \in \mathbf{Y_f}$
$\mathbf{K}$	Set of all sub-channels allocated in a macrocell, $\mathbf{K} = \{1, 2, \ldots, K\}$
$k^{(s)}$	A sub-channel under consideration, $k^{(s)} \in \mathbf{K}$
$P_m^{k^{(s)}}, P_f^{k^{(s)}}$	Transmission power from MeNB(m)/HeNB(f) on sub-channel $k^{(s)}$
$G_{x_m,m}^{k^{(s)}}, G_{y_f,f}^{k^{(s)}}$	Path-loss associated with sub-channel $k^{(s)}$ between MUE/FUE and MeNB/HeNB
$G_{x_m,f}^{k^{(s)}}, G_{y_f,m}^{k^{(s)}}$	Path-loss associated with sub-channel $k^{(s)}$ between MUE/FUE and HeNB/MeNB
$h_{x_m,m}^{k^{(s)}}, h_{y_f,f}^{k^{(s)}}$	Exponentially distributed channel power gain of unit mean associated with sub-channel $k^{(s)}$ between MUE/FUE and MeNB/HeNB
$h_{x_m,f}^{k^{(s)}}, h_{y_f,m}^{k^{(s)}}$	Exponentially distributed channel power gain of unit mean associated with sub-channel $k^{(s)}$ between MUE/FUE and HeNB/MeNB
$SINR_{x_m,m}^{k^{(s)}}, SINR_{y_f,f}^{k}$	SINR at MUE x_m/FUE y_f on sub-channel $k^{(s)}$
$C_{min}^{x_m,m}$	Minimum required rate for MUE x_m
$C_{min}^{y_f,f}$	Minimum required rate for FUE y_f
$\Gamma^{k^{(s)}}$	Indicator function and set to 1 if a sub-channel $k^{(s)}$ is assigned to a UE, otherwise it is set to 0
P_n	White noise power spectral density
ΔB	Bandwidth of a sub-channel
N	FFR frequency reuse factor of the edge-zone of MeNB (FRF Set, $\mathbf{N_m} = \{3, 6\}$, $N \in \mathbf{N_m}$)
Δr_m	Macrocell radius resolution (e.g., $\Delta r_m = 0.025$)
$\mathbf{K_m^{(c)}}, \mathbf{K_m^{(e)}}$	Set of sub-channels allocated in a macrocell (m) center-zone and edge-zone, respectively

In general, the signal-to-interference-plus-noise-ratio (SINR) for DL transmission to MUE x_m from MeNB m on sub-channel $k^{(s)}$ is given by[6]

$$SINR_{x_m,m}^{k^{(s)}} = \frac{P_m^{k^{(s)}} \, h_{x_m,m}^{k^{(s)}} \, G_{x_m,m}^{k^{(s)}}}{P_n \Delta B + \sum_{m' \in \mathbf{M}'} P_{m'}^{k^{(s)}} \, h_{x_m,m'}^{k^{(s)}} \, G_{x_m,m'}^{k^{(s)}} + \sum_{f \in \mathbf{F}} P_f^{k^{(s)}} \, h_{x_m,f}^{k^{(s)}} \, G_{x_m,f}^{k^{(s)}}} \quad (5.9)$$

where $P_m^{k^{(s)}}$ is the transmission power from MeNB m on sub-channel $k^{(s)}$. $h_{x_m,m}^{k^{(s)}}$ is the exponentially distributed channel fading power gain associated with sub-channel $k^{(s)}$.

[6]Here $x_m = x_c$ and $x_m = x_e$, if the MUE is located in center-zone and edge-zone, respectively.

$G^{k^{(s)}}_{x_m,m}$ is the path-loss associated with sub-channel $k^{(s)}$ between MUE x_m and MeNB m which is given as $G^{k^{(s)}}_{x_m,m} = 10^{-PL_{outdoor}/10}$, where $PL_{outdoor} = 28 + 35 \log_{10}(d)$ dB [10], where d is the Euclidean distance between a BS and a UE in meters. However, $G^{k^{(s)}}_{x_m,f}$ is affected by both indoor and outdoor path-loss. In this case, d would be the Euclidean distance between an HeNB f and the edge of the indoor wall in the direction of MUE x_m. This path-loss corresponds to indoor path-loss. After the wall, the path-loss will be based on the outdoor path-loss model. The indoor path-loss is modeled as follows [9]:

$$PL_{indoor} = \begin{cases} 38.5 + 20\log_{10}(d) + 7 \text{ dB}, & 0 < d \le 10 \\ 38.5 + 20\log_{10}(d) + 10 \text{ dB}, & 10 < d \le 20 \\ 38.5 + 20\log_{10}(d) + 15 \text{ dB}, & 20 < d \le 30. \end{cases} \qquad (5.10)$$

In (5.12), $\mathbf{M}'$ is the set of interfering MeNBs, which depends on the location of the MUEs and the specific sector-based FFR scheme deployed in the network.[7] $\mathbf{F}$ is the set of interfering HeNBs adjacent to the MUE x_m. Here, the adjacent HeNBs are defined as those HeNBs which are inside a circular area of radius $2r_f$ centred at the location of MUE x_m. Here, r_f is the transmission radius of an HeNB in meters. P_n represents noise power spectral density and ΔB represents bandwidth of a sub-channel. The practical capacity for an MUE x_m on sub-channel $k^{(s)}$ is then given by [10]: $C^{k^{(s)}}_{x_m,m} = \Delta B \cdot \log_2(1 + \alpha \text{SINR}^{k^{(s)}}_{x_m,m})$, where α is a constant defined by $\alpha = -1.5/\ln(5 \times BER)$ [9]. Here, BER represents the target BER (e.g., 10^{-6}) [9].

For an FUE y_f communicating with the HeNB f on sub-channel $k^{(s)}$, the DL SINR is given by

$$SINR^{k^{(s)}}_{y_f,f} = \frac{P^{k^{(s)}}_f \, h^{k^{(s)}}_{y_f,f} \, G^{k^{(s)}}_{y_f,f}}{P_n \Delta B + \sum_{m \in \mathbf{M}} P^{k^{(s)}}_m \, h^{k^{(s)}}_{y_f,m} \, G^{k^{(s)}}_{y_f,m} + \sum_{f' \in \mathbf{F}'} P^{k^{(s)}}_{f'} \, h^{k^{(s)}}_{y_f,f'} \, G^{k^{(s)}}_{y_f,f'}} \qquad (5.11)$$

where $\mathbf{F}'$ is the set of all interfering (or adjacent) HeNBs and $\mathbf{M}$ is the set of interfering MeNBs. Here, $G^{k^{(s)}}_{y_f,f}$ represents indoor channel gain for distance d between the FUE and its serving HeNB. On the other hand, $G^{k^{(s)}}_{y_f,m}$ corresponds to both indoor and outdoor channel gain. Since the interfering signal is coming from the MeNB, we include the channel fading power gain in the denominator. For co-tier interference at HeNB, we only assume indoor channel gain since the transmission radius of the HeNB is relatively small. Again, note that the interfering HeNBs are defined as those HeNBs which are within a circular area of radius $2r_f$ (e.g., $r_f = 30$ m) centred at FUE y_f. The practical capacity for an FUE y_f is given as $C^{k^{(s)}}_{y_f,f} = \Delta B \cdot \log_2(1 + \alpha \text{SINR}^{k^{(s)}}_{y_f,f})$.

For each sector-based FFR (i.e., FFR-3 and FFR-6), we obtain the best configuration consisting of the number of sub-channels allocated in the center-zone and edge-zone as well as the center-zone radius with respect to the macrocell radius.

---

[7]For example, in a 1-tier MeNB network model, if the MUE is located at the center-zone, then for both the FFR-3 and FFR-6 schemes, the interfering MeNBs will be MeNB2–MeNB7 (Figure 5.3). On the other hand, if the MUE is located in edge-zone then the set $\mathbf{M}'$ would consist two MeNBs and one MeNB for FFR-3 and FFR-6, respectively. For example, in FFR-6 based scheme (Figure 5.3(b)), if MUE is positioned at $X1$ edge-zone, then the only interfering MeNB will be MeNB4.

In this optimization problem, we assume that the MUEs and HeNBs are uniformly distributed in the macrocell service area. We assume one active FUE per HeNB. The algorithm corresponding to the solution methodology of the optimization problem for channel allocation and interference avoidance is presented in **Algorithm 5.2**. We use Monte-Carlo simulations to obtain the best configuration. We consider 1000 spatial realizations of network topology to obtain the design parameters for sector-based FFR. First, for each realization, we execute **Algorithm 5.2** 1000 times (capturing the channel variation) to obtain the best configuration for this particular spatial realization. This process is performed 1000 times, that is, by changing the network topology (capturing the spatial distribution of the MUEs, HeNBs, and FUEs) to obtain the best configuration for the proposed framework. The main steps for solving the optimization problem can be stated as follows:

- For each FFR and center-zone radius, we first classify the macro users as center-zone MUEs and edge-zone MUEs. Then, we divide the available frequency band into center-zone and edge-zone sub-band for center-zone and edge-zone MUEs, respectively. The number of sub-channels in center-zone and edge-zone sub-band are $K_m^{(c)}$ and $K_m^{(e)}$, respectively. However, $K_m^{(e)}$ is equally partitioned according to FRF of the edge-zone and allocated to the MUEs of the edge-zone sectors.

- Under the sub-channel allocation for MUEs, the HeNBs execute **Algorithm 5.1** (and as illustrated in Figure 5.3) to allocate sub-channels for FUEs.

- We calculate the maximum achievable capacity for each UE while satisfying its minimum rate and obtain the total throughput of the cell. Then, we obtain the optimal sub-channel allocation in center-zone and edge-zone that maximizes total throughput of the cell for each proportional center-zone radius (with respect to macrocell radius) and FFR scheme. Finally, we can obtain the best configuration consisting of optimal cell edge-zone FRF, percentage of the overall frequency spectrum allocated cell center-zone and edge-zone, and the radius of the cell center-zone radius with respect to the macrocell radius.

5.4 PERFORMANCE EVALUATION

5.4.1 Performance Metrics

We evaluate the performances of the different static FFR schemes in a two-tier macrocell–femtocell networking scenario by simulations (in MATLAB R2010a) in terms of outage probability, network throughput (or network sum-rate), and spectral efficiency.

Signal-to-interference-plus-noise-ratio (SINR): For DL transmission to MUE x_m from MeNB m on sub-channel k, $\text{SINR}_{x_m,m}^k$ is given by

$$\text{SINR}_{x_m,m}^k = \frac{P_m^k \, h_{x_m,m}^k \, G_{x_m,m}^k}{N_0 \Delta B + \sum_{m' \in \mathbf{M'}} P_{m'}^k \, h_{x_m,m'}^k \, G_{x_m,m'}^k + \sum_{f \in \mathbf{F}} P_f^k \, G_{x_m,f}^k}$$

Algorithm 5.2 ALGORITHM FOR CHANNEL ALLOCATION AND INTERFERENCE
COORDINATION

Input: $\mathbf{N}_m$, Δr_m, K, $C_{min}^{x_c,m}$, $C_{min}^{x_e,m}$, $C_{min}^{y_f,f}$, $\mathbf{F}_m$, $\mathbf{X}_m$, $\mathbf{Y}_f$, Target SNR at MUE.
Output: FFR Design Parameters

1: **for all** $N \in \mathbf{N}_m$ **do**
2:     **for** $r = 0.1$ **to** 1 **step** Δr **do**
3:        $\mathbf{X}_c^{(r)} = \{\}$ {The set center-zone MUEs when center-zone radius is r}
4:        $\mathbf{X}_e^{(r)} = \{\}$ {The set edge-zone MUEs when center-zone radius is r}
5:        **for all** $x_m \in \mathbf{X}_M$ **do**
6:           Calculate the Euclidean distance between x_m and the center of the MeNB.
7:           **if** the Euclidean distance is less than the radius r **then**
8:              $\mathbf{X}_c^{(r)} \leftarrow x_m$
9:           **else**
10:             $\mathbf{X}_e^{(r)} \leftarrow x_m$
11:           **end if**
12:        **end for**
13:        **for** $K_m^{(c),(r)} = 1$ **to** $K - 1$ **step** 1 **do**
14:           $Total\_throughput(K_m^{(c),(r)}) = 0$
15:           **for** $x_c^{(r)} \in \mathbf{X}_c^{(r)}$ **do**
16:              Calculate the SINR of $x_c^{(r)}$ where $k^{(s)} \in K_m^{(c),(r)}$
17:              Calculate the maximum achievable capacity of $x_c^{(r)}$, $C_{x_c,m}^{k^{(s)},(r)}$.
18:              **if** $C_{x_c,m}^{k^{(s)},(r)} > C_{min}^{x_c,m}$ **then**
19:                 $Total\_throughput(K_m^{(c),(r)}) = Total\_throughput(K_m^{(c),(r)}) + C_{x_c,m}^{k^{(s)},(r)}$
20:              **end if**
21:           **end for**
22:           Compute, $K_m^{(e),(r)} = K - K_m^{(c),(r)}$
23:           **for** $x_e^{(r)} \in \mathbf{X}_e^{(r)}$ **do**
24:              Calculate the SINR of $x_e^{(r)}$ where $k^{(s)} \in K_m^{(e),(r)}$
25:              Calculate the maximum achievable capacity of $x_e^{(r)}$, $C_{x_e,m}^{k^{(s)},(r)}$.
26:              **if** $C_{x_e,m}^{k^{(s)},(r)} > C_{min}^{x_e,m}$ **then**
27:                 $Total\_throughput(K_m^{(c),(r)}) = Total\_throughput(K_m^{(c),(r)}) + C_{x_e,m}^{k^{(s)},(r)}$
28:              **end if**
29:           **end for**
30:           **for all** HeNB $f \in \mathbf{F}_m$ **do**
31:              Execute **Algorithm** based on the sub-channel allocation for
                MUEs.
32:              **for all** $y_f^{(r)} \in \mathbf{Y}_f^{(r)}$ **do**
33:                 Calculate the SINR of $y_f^{(r)}$ where $k^{(s)} \in \mathbf{J}_U$
34:                 Calculate the maximum achievable capacity of $y_f^{(r)}$, $C_{y_f,f}^{k^{(s)},(r)}$.
35:                 **if** $C_{y_f,f}^{k^{(s)},(r)} > C_{min}^{y_f,f}$ **then**
36:                    $Total\_throughput(K_m^{(c),(r)}) = Total\_throughput(K_m^{(c),(r)}) + C_{y_f,f}^{k^{(s)},(r)}$
37:                 **end if**
38:              **end for**

39:      **end for**
40:      Obtain the center-zone sub-channels $K_m^{(c)}$ for which the total throughput is maximized for given r.
41:      **end for**
42:      Obtain the center-zone sub-channels $K_m^{(c)}$ and r for which the total throughput is maximized for given N.
43:      **end for**
44:      Obtain the center-zone sub-channels $K_m^{(c)}$, r, and N for which the total throughput is maximized.
45: **end for**

---

where P_m^k is the transmission power from MeNB m on sub-channel k, $h_{x_m,m}^k$ is the exponentially distributed channel fading power gain associated with sub-channel k, and $G_{x_m,m}^k$ is the path-loss associated with sub-channel k between MUE x_m and MeNB m which is given as $G_{x_m,m}^k = 10^{-PL_{outdoor}/10}$. This path-loss corresponds to outdoor path-loss and is modeled as $PL_{outdoor} = 28 + 35\log_{10}(d)$ dB, where d is the Euclidean distance between a BS and a user in meters. However, $G_{x_m,f}^k$ is affected by both indoor and outdoor path-loss. In this case, d would be the Euclidean distance between an HeNB f and the edge of the indoor wall in the direction of MUE x_m. After the wall, the path-loss will be based on an outdoor path-loss model.

Outage probability: We define the outage probability as the probability that a UE's instantaneous SINR on a given sub-channel k falls below the SINR threshold γ and given as $\mathbb{P}(\text{outage}) = \mathbb{P}\left(\text{SINR}_{x_m,m}^k < \gamma\right)$.

Sum-rate for FUEs in a macrocell: The maximum achievable capacity for an FUE y_f is given as $C_{y_f,f}^k = \Delta B \cdot \log_2(1 + \alpha\text{SINR}_{y_f,f}^k)$.

Average network sum-rate: The average network sum-rate, C_{avg} is

$$C_{avg} = \frac{\sum_{x_m \in \mathbf{X_m}} \sum_{k \in \mathbf{K}} \Gamma_{x_m,m}^k C_{x_m,m}^k}{MUE_{total}} + \frac{\sum_{f \in \mathbf{F_A}} \sum_{y_f \in \mathbf{Y_f}} \sum_{k \in \mathbf{K}} \Gamma_{y_f,f}^k C_{y_f,f}^k}{FUE_{total}} \quad (5.12)$$

where, in general, $\Gamma^k = 1$ when a sub-channel k is assigned to a UE. Otherwise, it is set to 0.

Spectral efficiency: We define the spectral efficiency (bits/s/Hz) in terms of average bits per second successfully received by a UE per unit spectrum. The spectral efficiency of transmission to MUE x_m on sub-channel k is given as $S_{x_m,m}^k = \log_2(1 + \alpha\text{SINR}_{x_m,m}^k)$ and that for FUE y_f is given as $S_{y_f,f}^k = \log_2(1 + \alpha\text{SINR}_{y_f,f}^k)$. The average network spectral efficiency, S is thus given by

$$S = \frac{\sum_{x_m \in \mathbf{X_m}} \sum_{k \in \mathbf{K}} \Gamma_{x_m,m}^k S_{x_m,m}^k}{K_{band}} + \frac{\sum_{f \in \mathbf{F_A}} \sum_{y_f \in \mathbf{Y_f}} \sum_{k \in \mathbf{K}} \Gamma_{y_f,f}^k S_{y_f,f}^k}{K_{band}}. \quad (5.13)$$

TABLE 5.2   Simulation parameters

Parameter	Value
Network size	1-tier (7 macrocells)
Radius of a macro cell	280 m
Radius of a femtocell	30 m
SNR at an MUE	10 dB
HeNB transmission power	20 mW
Number of MUEs in a macrocell	50
Maximum number of FUEs per femtocell	1
Channel bandwidth	10 MHz
Number of sub-channels	50
Sub-carrier spacing	15 kHz
White noise power spectral density	−174 dBm/Hz
Power control factor, ϵ	4

5.4.2   Simulation Parameters

The simulation parameters are shown in Table 5.2. The network is composed of 7 macrocells, and the HeNBs (i.e., femtocells) are randomly deployed over the macrocells. The number of HeNBs is varied up to 40 in one macrocell coverage area. We assume that the HeNBs operate in closed access mode (i.e., only registered FUEs will be able to access the HeNBs). The MUEs are uniformly distributed in the network. The MUEs and FUEs are randomly allocated with available sub-channels from the designated frequency bands corresponding to each sub-area for each scheme. We assume a "snap-shot" model, where all the network parameters (in Table 5.2) remain constant during a simulation run.

5.4.3   Simulation Results

Figure 5.4 shows variations in the network throughput with the fraction of the center-zone radius (with respect to the macrocell radius) for various cell edge-zone FRF. For each cell edge-zone FRF and fraction of the center-zone radius, the optimal number of sub-channels for center-zone MUEs is obtained by enumeration for which the network throughput is maximized.

For the OSFFR scheme, the total network throughput (or spectral efficiency (b/s/Hz) of the network) is maximized if center-zone radius is 54% of the total macro-cell radius (Figure 5.4) and 36% of the total frequency resources are allocated for the center-zone MUEs (i.e., sub-band A). For FFR-3, the optimal center-zone radius is 61% of the macrocell radius and the optimal frequency resources for the center-zone MUEs is 48% of the whole frequency band. Table 5.3 shows the design parameters obtained for different sector-based FFR schemes as the minimum rate requirements and link-level transmission power vary within the network. The optimal values for the OSFFR and FFR-3 are used to obtain the performance evaluation results given below.

Figure 5.5(a) shows the variations in outage probability with SINR threshold for different FFR schemes (without HeNBs and with 40 HeNBs per macrocell to

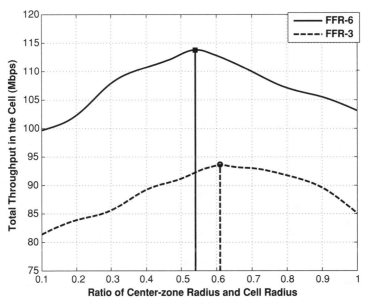

Figure 5.4 Network throughput versus fraction of the center-zone radius (with respect to the macrocell radius) for different FFR factor for the cell edge-zones (© [2013] IEEE).

demonstrate how the outage probability deteriorates in presence of large number of HeNBs). Note that the strict FFR scheme exhibits slightly better outage performance when the SINR targets are low. This is due to the fact that, in strict FFR, the edge-zone MUEs of the center MeNB (i.e., the MeNB under observation) are not interfered by any other MeNBs of the first tier of the network. When the SINR threshold increases (e.g., >11.5 dB), the outage probability of strict FFR scheme (in presence of HeNBs)

TABLE 5.3 Design parameters for sector-based FFR schemes

Sector-based FFR	Target SNR at MUE (dB)	Minimum Rate Requirement for UEs (kbps)	Center-zone Radius Allocation (%)	Center-zone Frequency Allocation (%)
FFR-3	15	100	61	48
		200	55	44
		275	53	42
	20	100	64	53
		200	59	47
		275	57	46
FFR-6	15	100	54	36
		200	48	33
		275	45	31
	20	100	56	43
		200	51	40
		275	48	38

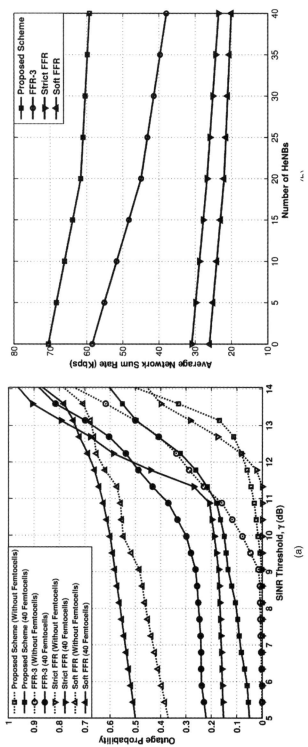

Figure 5.5 (a) Outage probability of MUEs with SNR $= 10$ dB for different FFR schemes as the SINR threshold varies, and (b) Average network sum rate of MUEs for different FFR schemes (© [2013] IEEE).

119

becomes higher than that of the proposed scheme and reaches close to that of the soft FFR scheme.

For the proposed scheme, ICI to the edge-zone MUEs is caused by only 1 MeNB, whereas for FFR-3 and soft FFR schemes, ICI is caused by 2 and 6 MeNBs, respectively. As a result, the outage probability is higher for these two FFR schemes. In comparison with the other FFR schemes, in the proposed scheme, the usable sub-bands for femtocells are increased in edge-zone and center-zone of a cell. As a result, the probability that two neighboring HeNBs would use the same sub-band and the same sub-channel is far more reduced as compared to the other FFR techniques. Therefore, the inter-HeNB interference is significantly reduced. Also, with the proposed scheme, due to the increased number of sub-bands for HeNBs in both center-zone and edge-zone, the number of sub-channels for FUEs per unit area increases. This results in a smaller probability of causing CTI with the MUEs in comparison with the other FFR schemes. As a result, the outage probability is comparatively low for the MUEs.

Figure 5.5(b) shows the variations in average network sum-rate as the number of HeNBs varies within the cell. We observe that the average network sum-rate for the proposed scheme is higher than that for each of the other FFR schemes. Again, this is due to reduced co-tier interference and CTI and hence higher SINR offered by the proposed scheme. Also, with the proposed scheme, the usable number of sub-channels per unit area increases when compared with the other FFR schemes, and consequently the spectral efficiency increases. Figure 5.6 shows variations in the spectral efficiency of the network as the number of HeNB varies. Note that, for the proposed scheme, with only 25 HeNBs per macrocell service area, the target spectral efficiency for LTE-A systems (i.e., 30 b/s/Hz [11]) is well satisfied. Also, from Figure 5.6, we observe that, for the edge-zone UEs, the average gains in spectral efficiency for the proposed scheme are 27%, 41%, and 49%, when compared with FFR-3, strict FFR, and soft FFR schemes, respectively. With the proposed scheme, for the UEs both in center-zone and edge-zone, the average gains in spectral efficiency are 23%, 43%, and 51%, when compared with FFR-3, strict FFR, and soft FFR schemes, respectively.

5.5   SUMMARY AND FUTURE RESEARCH DIRECTIONS

FFR is a simple and effective mechanism for interference management in OFDMA-based multi-tier networks. We have presented a broad comparison among four different FFR schemes, namely, strict FFR, soft FFR, FFR-3, and optimal static FFR (OSFFR) schemes, for two-tier networks. Simulation results have shown that the proposed OSFFR scheme offers a superior performance compared with the three other state-of-the-art FFR schemes. The FFR schemes described here correspond to partitioning and allocation of spectrum into different spatial regions of the macrocell service area in a static manner. Such static allocations may not be optimal under dynamic traffic load variation (e.g., due to the mobility of the UEs) and may increase the blocking probability. Note that an open access mode can reduce this blocking probability resulting from static resource partitioning. Optimal FFR schemes in the

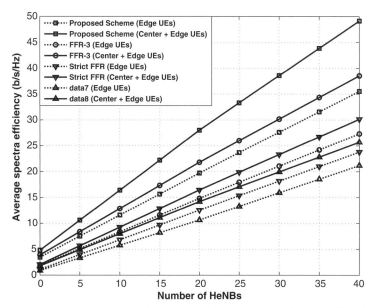

Figure 5.6   Average spectral efficiency of the network for different FFR schemes.

presence of mass deployment of HeNBs that satisfy the data rates for UEs as well the target blocking probabilities need to be developed. In this context, self-organizing and autonomous FFR frameworks will be desirable from the scalability point of view. In addition, dynamic power control methods can be developed to use in conjunction with FFR schemes to improve the capacity of multi-tier cellular networks. Such hybrid schemes based on resource partitioning through FFR as well as power control are currently being considered for LTE-Advanced systems.

REFERENCES

1. M. E. Sahin, I. Guvenc, M.-R. Jeong, and H. Arslan, "Handling CCI and ICI in OFDMA femtocell networks through frequency scheduling," *IEEE Transactions on Consumer Electronics*, vol. 55, no. 4, pp. 1936–1944, November 2009.
2. H. Li, X. Xu, D. Hu, X. Qu, X. Tao, and P. Zhang, "Graph method based clustering strategy for femtocell interference management and spectrum efficiency improvement," in *Proceedings of IEEE 6th International Conference on Wireless Communications Networking and Mobile Computing (WiCOM)*, pp. 1–5, 23–25 September 2010.
3. S. Park, W. Seo, Y. Kim, S. Lim, and D. Hong, "Beam subset selection strategy for interference reduction in two-tier femtocell networks," *IEEE Transactions on Wireless Communications*, vol. 9, no. 11, pp. 3440–3449, November 2010.
4. 3GPP R1-106052, "Per Cluster Based Opportunistic Power Control," 3GPP RAN1 Meeting, Jacksonville, USA, November 2010.
5. L. Zhang, L. Yang, and T. Yang, "Cognitive interference management for LTE-A femtocells with distributed carrier selection," in *Proceedings of IEEE 72nd Vehicular Technology Conference (VTC 2010-Fall)*, pp. 1–5, 6–9 September 2010.

6. www.femtoforum.org
7. A. Imran, M. A. Imran, and R. Tafazolli, "A novel self-organizing framework for adaptive frequency reuse and deployment in future cellular networks, in *Proceedings of IEEE 21st International Symposium on Personal Indoor and Mobile Radio Communications (PIMRC)*, 2010, pp. 2354–2359.
8. T. Novlan, J. G. Andrews, I. Sohn, R. K. Ganti, and A. Ghosh, "Comparison of fractional frequency reuse approaches in the OFDMA cellular downlink," in *Proceedings of IEEE Global Telecommunications Conference (Globecom)*, pp. 1–5, 6–10 December 2010.
9. M. Tao, Y. C. Liang, and F. Zhang, "Resource Allocation for Delay Differentiated Traffic in Multiuser OFDM Systems" IEEE Transactions on Wireless Communications, vol. 7, no. 6, pp. 2190–2201, June 2008.
10. L. Poongup, L. Taeyoung, J. Jangkeun, and S. Jitae, "Interference management in LTE femtocell systems using fractional frequency reuse," in *Proceedings of 12th International Conference on Advanced Communication Technology*, vol. 2, pp. 1047–1051, 7–10 February 2010.
11. M. Rumney, "Introducing LTE Advanced," Agilent Technologies, 22 May 2011.

CHAPTER 6

CALL ADMISSION CONTROL IN FRACTIONAL FREQUENCY REUSE-BASED TWO-TIER NETWORKS

In a cellular network, CAC is responsible for admission or rejection of an incoming request from a UE into the network based on some predefined criteria while taking the network load conditions and QoS requirements of both incoming and existing users into account. To enable spectrum sharing between macrocells and small cells in OFDMA-based two-tier networks, one of the three access modes, namely, open, closed and hybrid access, can be employed. In the open access mode, the MUEs are allowed to connect to either their own macrocell base stations (MBSs) or SBSs. In contrast, in the closed access mode, in a two-tier macrocell–femtocell network, for example, only certain users (subscribers) belonging to the so-called Closed Subscriber Group (CSG) are allowed to connect to each FBS. For a typical hybrid access scheme, limited spectrum access at each small cell is granted for MUEs, which wish to establish connections. An efficient CAC policy is required in this case to coordinate spectrum sharing and admission control for both types of users, which should strike a balance between achieving high spectrum utilization and protecting QoS requirements for small cell users. Note that in this case both the connection-level QoS measures at the network layer (e.g., connection blocking and connection dropping probabilities) and the packet-level QoS measures (e.g., packet error rate and packet throughput) at the MAC layer need to be considered. Such spectrum sharing and admission control mechanisms for OFDMA-based two-tier networks based on cross-layer design need to be developed. By using these methods, the traffic loads among macrocells and small cells can be dynamically balanced to improve the network capacity significantly.

6.1 RELATED WORK

There have been a lot of work on the CAC problem for single-tier (or homogeneous) cellular wireless networks. In [1], a mobility-based CAC method is presented where

Radio Resource Management in Multi-Tier Cellular Wireless Networks, First Edition.
Ekram Hossain, Long Bao Le, and Dusit Niyato
© 2014 John Wiley & Sons, Inc. Published 2014 by John Wiley & Sons, Inc.

various types of priority-based handoffs are investigated. Queueing-based analytical results are presented in [2] for different average call holding times for new and handoff calls for homogeneous networks.

There have been a few work on teletraffic modeling of hierarchical (or multi-tier) wireless networks. One of the advantages of layered network is that, the different layers may provide an alternate route for the admitted calls or handoff calls that are blocked due to congestion in a particular layer. For example, in an OFDMA-based two-tier macrocell–femtocell network, the ongoing traffic of the MUEs can overflow to neighboring femtocells (which are operating in either open access or hybrid access mode) when the macrocell is congested. Again, since the capacity of a femtocell is limited, blocked calls from femtocells can overflow to the macrocell [3]. Modeling the teletraffic performance in heterogeneous networks is challenging since it needs to consider different aspects such as, how to model new call and handoff calls at the different layers in a hierarchical layered architecture, how to model user mobility and traffic at the different layers, and how to model the radio network access performance [3, 4].

In [5], a hierarchically overlaid layout network is considered where microcells cover the high teletraffic areas and macrocells cover low teletraffic areas and provide overflow channels for overlaid microcells. In order to reduce the forced termination of calls in progress, handoff calls are given priority to access the channels in both the microcell and macrocell levels. The basic structure of the proposed hierarchical overflow system is that the microcells receive input streams of new and handoff calls, whereas macrocells receive input streams of new and handoff calls as well as overflow traffic components from the neighboring microcells, that is, the system shall operate in such a manner that a call served at a given hierarchical level will not request handoff to a cell that is lower in the hierarchy. A mathematical framework is presented for hierarchical overlaid macrocell network using multi-dimensional birth–death processes to characterize system states and evaluate the teletraffic performance of the cellular network that considers call overflow from one hierarchical level to the next as well as the issue of resource availability for handoff calls. The overflow traffic is modeled as Poisson traffic and the mathematical framework analyzes call blocking, handoff failure, and forced termination probabilities derived from the state probabilities. The simulation results show that as more channels are allocated to the overlaid macrocell, the blocking probability is improved under low and moderate offered traffic. However, under high offered traffic, blocking probability in microcells increases. Also, it is observed that the cutoff priority for handoff calls can lower the forced termination probability but increases blocking probability in general. However, the proposed framework requires a huge state space, particularly when the number of channels in each macrocell becomes large.

To reduce the complexity, in [6], an analytical model for the performance evaluation of a hierarchical cellular system is presented where more realistic assumptions are considered that relate to the distribution of call time and the cell residence time. In this model, the call time is characterized by a hyper-Erlang distribution where the Laplace transform (LT) of the channel occupancy time distribution for each type of call (i.e., new call, handoff call, and overflow call) is derived as a function of the LT of the cell residence time distribution (i.e., general distribution). The channel occupancy time is modeled as a renewal process that captures the inherent overflow from a lower layer of cells to a higher layer of cells and quantifies the channel occupancy time of overflow calls in a tractable manner. However, the overflow from a macrocell layer to a microcell layer is not mentioned in the paper.

Some models in the literature capture the non-Poisson characteristics of the overflow traffic. In these models, the overflow traffic is usually characterized by the mean and variance of the offered overflow traffic intensity. The mean of a specific class of overflow traffic can be directly obtained as the equivalent blocked traffic of this class in its overflowed network group [3]. The ratio of variance to mean is defined as the *peakedness*, which indicates the bursty nature of the overflow traffic.

In a hierarchical network where all different tiers have identical statistical characteristics (e.g., OFDMA-based two-tier macrocell–femtocell networks), a specific class of overflow traffic from a low tier to a higher tier holds identical statistical moments in these two tiers. Heterogeneous networks also give rise to different statistical characteristics of input traffic, mobility model, and service time distribution at different network tiers. The statistical moments of the overflow traffic from one network tier to another tier are thus required to be modeled with general distribution and constitute a major technical challenge in layered teletraffic modeling. In addition to this problem, the speed-sensitive handoff and traffic overflow between layered cellular systems (e.g., macrocell overlaid with femtocells) are required to be incorporated in the model since cells of different sizes in a multi-tiered structure provide multiple service coverages of UEs of various mobility classes [3, 7]. To tackle this problem, in [7], a speed-sensitive cell selection, cell re-selection, and handoff mechanism is proposed that assigns the UEs to the appropriate cell layer according to their speeds. Speed estimation can be based on the recent cell dwell time or the past dwell times. The basic mechanism of the proposed method is that, in a two-layer cell architecture, the lower cell layer comprises microcells primarily providing coverage for slow-moving UEs, while the higher cell layer consists of macrocells serving primarily fast-moving UEs. The authors provide a network-controlled speed-sensitive handoff control algorithm.

In [4], an analytical method that incorporates multiple overflow routes (i.e., correspond to the hierarchical multi-tier network) and non-Poisson behavior of the overflow traffic and at the same time handover and overflow of handover calls, is presented. In this model, the network performance is evaluated based on calculating the probability of call failure using a Markov chain model. In [3], an analytical framework is presented that attempts to solve the problem related with multiservice performance modeling for hierarchical network. The reason is that the previously mentioned multi-dimensional Markov chain model that aims to analyze the multiservice loss performance is somewhat intractable and requires complex computation. Here the authors consider the statistical heterogeneity in traffic and mobility models at different tiers when determining the statistical mean and variance of the inter-tier overflow traffic.

For multi-tier networks, in [8], the CAC problem is formulated as a semi-Markov decision process (SMDP) model and structural results on optimal cost function are presented for a network model where a WiMAX cell is overlaid with WLANs. A very recent work, presented in [9] addresses the joint resource allocation and admission control problem for OFDMA-based femtocell networks, where a power adaptation algorithm is proposed to adapt the transmission power of the HeNBs according to network dynamics.

The work presented in [8] does not consider all types of handovers (e.g., WLAN–WLAN) and does not take into account the link-level QoS requirements of the users. Also, no guideline for the selection of the cost value function is presented in [8]. The work presented in [9] considers an SMDP-based CAC, but the user mobility is not taken into account.

In this chapter, we will present a CAC method for a two-tier macrocell–femtocell network which uses a sector-based FFR for spatial channel allocation for macrocells and femtocells (see Chapter 5). The FFR parameters are optimized as discussed in Chapter 5. In this method, we consider all types of calls in a two-tier network along with random mobility of the UEs. Note that the existing work in the traditional setting (macro–microcell setting) usually assume that blocked calls from lower tiers (small cells) will be overflowed to higher tiers, and they do not explicitly capture interference in the network. We develop a CAC model that captures both co-tier interference and CTI. Also, the design of an overflow policy in our setting is different since calls from the macro-tier will actually be overflowed to the lower tier (femto tier).

6.2 CALL ADMISSION CONTROL MODEL

We assume that each of the MeNBs and HeNBs has an independent CAC unit that is capable of measuring the external call arrival rates. The admission control decision in each MeNB or HeNB is performed in a decentralized manner. We assume that an HeNB has the information about the total number of occupied sub-channels at an MeNB (and vice-versa). However, for an MeNB to make a call admission decision, we assume that it considers the average of the total number of sub-channels occupied by FUEs in HeNBs in each sector. The exchange of information may be done via backhaul or over-the-air. We assume that, at a given time instance, two adjacent HeNBs do not use the same sub-band and sub-channel.[1] We assume that calls arrive to the BSs according to Poisson process and the call holding times are exponentially distributed. This assumption is essential in order to formulate the CAC problem based on the SMDP-model. We assume a random mobility pattern for the MUEs and FUEs. Therefore, the arrival processes for handoff calls (e.g., handoff from MeNB to HeNB, HeNB to HeNB, or HeNB to MeNB) do not follow Poisson processes. Hence, we consider "Hayward's Approximation" to approximate the overflowed non-Poisson traffic by Poisson traffic.

We assume that each HeNB has a fixed transmission region where MUEs inside the region would request to establish connection with the HeNB. The HeNB admits an MUE based on an admission control policy (which depends on the number of available channels) and also if the MUE achieves the minimum rate requirement

---

[1]In Chapter 5, in Figure 5.3 we have seen that an HeNB in center-zone and edge-zone has $\left\lceil \frac{N+1}{2} \right\rceil$ and N usable sub-bands, respectively. Hence, two neighboring HeNBs using the same sub-band and same sub-channel at a given time instance may have very low probability.

TABLE 6.1   Different call types in sector-based FFR

Call Type	Call Type
i. New calls to MeNB per sector:	v. Intra-sector macro–macro handoff:
1. New calls at center-zone	1. Handoff calls from edge-zone to center-zone
2. New calls at edge-zone	2. Handoff calls from center-zone to edge-zone
ii. New calls to HeNB per sector:	vi. Intra-sector femto–femto handoff:
1. New calls at center-zone	1. Handoff calls from edge-zone to center-zone
2. New calls at edge-zone	2. Handoff calls from center-zone to edge-zone
iii. Inter-sector macro–macro handoff:	vii. Inter-sector macro–femto and femto–macro handoff:
1. Handoff calls from edge-zone to edge-zone	1. Handoff calls from edge-zone to edge-zone
2. Handoff calls from center-zone to center-zone	2. Handoff calls from center-zone to center-zone
3. Handoff calls from edge-zone to center-zone	3. Handoff calls from edge-zone to center-zone
4. Handoff calls from center-zone to edge-zone	4. Handoff calls from center-zone to edge-zone
iv. Inter-sector femto–femto handoff:	viii. Intra-sector macro–femto and femto–macro handoff:
1. Handoff calls from edge-zone to edge-zone	1. Handoff calls from edge-zone to center-zone
2. Handoff calls from center-zone to center-zone	2. Handoff calls from center-zone to edge-zone
3. Handoff calls from edge-zone to center-zone	
4. Handoff calls from center-zone to edge-zone	

(or higher) when connected to the HeNB. We assume that the path-loss information is available at the BSs. We assume that the average transmission rate of a UE depends on the Euclidean distance between the UE and the BS.

The call arrival and handoff processes in a two-tier sector-based FFR are listed in Table 6.1. Due to a large number of call types, it would be very complicated to keep track of all the call arrival processes in the macrocell and would involve high computational complexity to obtain the CAC policy (e.g., by using a value iteration algorithm (VIA) for an SMDP-based model [9]). However, due to fixed channel allocation in sector-based FFR, we can obtain the admission control policy for any sector and use the same policy in all the sectors of the macrocell. Furthermore, in each sector, due to static frequency allocation in center-zone and edge-zone, we may obtain the policy for each zone/region separately. This will reduce the computational complexity of the VIA significantly.

Figure 6.1 illustrates the call arrival and departure per sector of an FFR-6 based two-tier network. The basic call arrival/handoff processes along with their

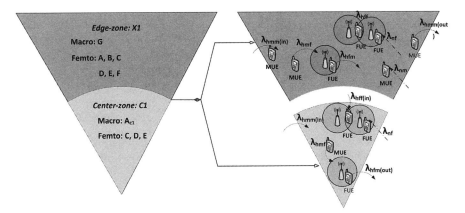

Figure 6.1   Arrival and departure of different call types in a sector for FFR-6.

TABLE 6.2 Arrival rate and rejection costs for different call types

Call Type	Arrival Rate	Call Rejection Cost	Call Type Description
1	λ_{nm}	R_{nm}	New calls to MeNB
2	λ_{nf}	R_{nf}	New calls to HeNB
3	λ_{hmm}	R_{hmm}	Handoff calls from MeNB to MeNB (intra-sector or inter-sector)
4	λ_{hmf}	R_{hmf}	Handoff calls from MeNB to HeNB
5	λ_{hfm}	R_{hfm}	Handoff calls from HeNB to MeNB
6	λ_{hff}	R_{hff}	Handoff calls from HeNB to HeNB

descriptions, arrival rates, and call rejection costs are presented in Table 6.2. The average call holding times for MUE and FUE are given by $1/\mu_m$ and $1/\mu_f$, respectively.

6.3 CALL ADMISSION CONTROL POLICY FOR FFR-BASED MULTI-TIER CELLULAR NETWORKS

6.3.1 Problem Formulation

The CAC policy in two-tier macrocell-femtocell networks corresponds to a decision theoretic optimization problem [10–12] that comprises long-term call admission decisions on whether to admit or reject the arriving calls in MeNB and HeNBs. The CAC decision is usually based on the available sub-channels at the BSs, the distribution and rate of the call arrival process and call rejection cost or acceptance reward associated with the admission decision. Our objective is to obtain a stationary admission policy for a multi-tier network that minimizes the total expected cost for new and handoff calls resulting from the admission control decision. Since the arrival and departure of different call types have Markov property, we can formulate this CAC problem as an SMDP [10]. We start the analysis by obtaining the CAC policy for a finite-horizon problem and later we will show that the structural results (i.e., the convexity and monotonically non-decreasing property of the cost function) also hold for an infinite-horizon CAC problem. We obtain the CAC policy based on a VIA. The obtained policy is a threshold-based policy and has a two-dimensional structure. We refer to this policy as a *macro–femto threshold policy* and show that this two-dimensional threshold-based policy can be considered as an optimal solution for a static FFR-based multi-tier newtork.

The SMDP-model comprises of the following five major components: decision epochs, state space, action space, expected cost value function, and state transition probabilities.

Decision Epochs: The decision epochs are considered at the occurrence of different call type events, that is, new call, horizontal handoff call, and vertical handoff call arrivals. However, no decisions are taken at the departure of the calls. Fictitious decision epochs corresponding to the departure events are considered to keep track of the departures. The fictitious decision epochs do not change the Markovian nature of the decision process since the call holding time for the UEs is exponentially distributed, thus having memoryless property [10]. Inclusion of fictitious decision epochs simplifies the formulation of the SMDP-model as it ensures that the system state only changes to adjacent states. As a result, the transition probability matrix has many zero terms which makes the VIA computationally less complex [8, 10]. We introduce a fictitious call event of type "0" along with other six call arrival types as shown in Table 6.2.

State space: The state space consists of all possible combinations of occupied sub-channels for macrocell, femtocell and type of calls, within a particular zone of a sector. Note that this state space will be different for center-zone and edge-zone as the numbers of available sub-channels for MeNB and HeNB are different in center-zone and edge-zone. Consider state variable $s = (i, j, k)$, where, i is the number of occupied macrocell sub-channels, j is the number of occupied femtocell sub-channels, and k is the call type. Let K_m^{max} and K_f^{max} denote, respectively, the maximum number of sub-channels allocated for MUEs and FUEs for a specific zone of a sector (i.e., for center-zone of any sector, $K_m^{max} = K_m^{(c)}/N$ and $K_f^{max} = K_f^{(c)}$). Define the state space, $S = \{s : s = (i, j, k), \forall i, j, k\}$, where $i \in \{0, 1, 2, \ldots, K_m^{max}\}$, $j \in \{0, 1, 2, \ldots, K_f^{max}\}$ and $k \in \{0, 1, 2, 3, 4, 5, 6\}$. Let s_{max} be the total number of states in the state space S.

Action space: At any state $s = (i, j, k)$, for every call arrival event, either new or handoff (vertical or horizontal) call, two actions are possible: accept (denoted by 1) or reject (denoted by 0), that is,

$$a(s) = \begin{cases} 1, & \text{if the call is admitted} \\ 0, & \text{if the call is rejected.} \end{cases}$$

The action state space can be defined as $A = \{a : a \in \{0, 1\}^{s_{max}}\}$.

Expected cost value function: The expected cost value function $c_s(a)$ corresponds to the expected cost incurred until the next decision epoch if action $a(s)$ is taken in the present state $s \in S$. The one-step expected cost value function is given by

$$c_s(a) = \begin{cases} R_{nm}, & s = (i, j, 1) \text{ and } a = 0 \\ R_{nf}, & s = (i, j, 2) \text{ and } a = 0 \\ R_{hmm}, & s = (i, j, 3) \text{ and } a = 0 \\ R_{hfm}, & s = (i, j, 4) \text{ and } a = 0 \\ R_{hmf}, & s = (i, j, 5) \text{ and } a = 0 \\ R_{hff}, & s = (i, j, 6) \text{ and } a = 0 \\ 0, & \text{otherwise.} \end{cases}$$

The rejection cost values can be based on the upper bound of the blocking/dropping probability constraints and the priority of the call types. For example, let us consider that the blocking/dropping probability constraints of different call types for a two-tier macrocell–femtocell network are given by $\wp_{hmf} \leq \wp_{hff} \leq \wp_{hfm} \leq \wp_{hmm} \leq \wp_{nm}(=\wp_{nf})$. This signifies that the system prefers macro-to-femto handoff to femto-to-femto handoff and so forth. Thus, the rejection cost value associated with different call types can be set as follows: $R_{hmf} \geq R_{hff} \geq R_{hfm} \geq R_{hmm} \geq R_{nm}(=R_{nf})$.

State transition probabilities: Let $P_{st}(a)$ be the state transition probability from state $s \in S$ to state $t \in S$ at the next decision epoch of the system if action $a(s)$ is chosen. Now, considering the current state $s = (i, j, k)$, if action $a(s)$ is taken when $k \in \{1, 3\}$, the state transition probability is given by

$$
P_{st}(a = 1) = \begin{cases}
(i+1)(\lambda_{hmm} + \mu_m) \cdot \tau_s(a = 1), & t = (i, j, 0) \\
j(\lambda_{hff} + \mu_f) \cdot \tau_s(a = 1), & t = (i+1, j-1, 0) \\
\lambda_{nm} \cdot \tau_s(a = 1), & t = (i+1, j, 1) \\
\lambda_{nf} \cdot \tau_s(a = 1), & t = (i+1, j, 2) \\
\lambda_{hmm} \cdot \tau_s(a = 1), & t = (i+1, j, 3) \\
(i+1)\lambda_{hmf} \cdot \tau_s(a = 1), & t = (i+1, j, 4) \\
j\lambda_{hfm} \cdot \tau_s(a = 1), & t = (i+1, j, 5) \\
\lambda_{hff} \cdot \tau_s(a = 1), & t = (i+1, j, 6)
\end{cases}
$$

where $\tau_s(a = 1) = \tau_{mf}(i + 1, j)$, and

$$
P_{st}(a = 0) = \begin{cases}
i(\lambda_{hmm} + \mu_m) \cdot \tau_s(a = 0), & t = (i-1, j, 0) \\
j(\lambda_{hff} + \mu_f) \cdot \tau_s(a = 0), & t = (i, j-1, 0) \\
\lambda_{nm} \cdot \tau_s(a = 0), & t = (i, j, 1) \\
\lambda_{nf} \cdot \tau_s(a = 0), & t = (i, j, 2) \\
\lambda_{hmm} \cdot \tau_s(a = 0), & t = (i, j, 3) \\
i\lambda_{hmf} \cdot \tau_s(a = 0), & t = (i, j, 4) \\
j\lambda_{hfm} \cdot \tau_s(a = 0), & t = (i, j, 5) \\
\lambda_{hff} \cdot \tau_s(a = 0), & t = (i, j, 6)
\end{cases}
$$

where $\tau_s(a = 0) = \tau_{mf}(i, j)$. Here, $1/\tau_{mf}(i, j)$ is the rate of going out of the state $s = (i, j, k)$, and $\tau_{mf}(i, j)$ is given by

$$
\tau_{mf}(i, j) = [\lambda_{nm} + \lambda_{nf} + \lambda_{hmm} + i\lambda_{hmm} + \lambda_{hff} + j\lambda_{hff} + i\lambda_{hmf} + j\lambda_{hfm} + i\mu_m + j\mu_f]^{-1}.
$$

The state transition probabilities for $k \in \{2, 4, 5, 6\}$ are tabulated in the **Appendix**.

6.3.2 Optimal CAC Policy

We use a VIA [8] based on Bellman iterative equation (shown in **Algorithm 6.1**) to obtain the stationary admission policy for the SMDP-based CAC problem for two-tier macrocell-femtocell networks. In each iteration n of the VIA, we obtain a policy $\delta_n(s)$ for state $s \in S$ that minimizes the total cost function $U_n(s)$. Based on a *stopping rule*[2] and accuracy factor ϵ, we obtain the stationary policy and the optimal cost function. This CAC policy is obtained considering that the network dynamics remains constant.

---

[2]Let B_n denote the difference between the upper bound (H_n) and the lower bound (L_n) of $U_n(t) - U_{n-1}(t)$ at nth iteration where $t \in \mathbf{S}$. The stopping rule is defined as: $B_n \leq \epsilon L_n$ [10].

In [13], it is shown that for homogeneous network, the optimal CAC policy has a one-dimensional threshold structure and is independent of system states. However, since our system model comprises of MeNB and HeNBs, we may claim that the stationary policy has a two dimensional macro–femto threshold structure. In order to justify our claim, we consider an approach similar to that in [8,13]. First, we prove that, for infinite-horizon CAC problem the cost function is monotonically non-decreasing and convex. Then we show that the two-dimensional macro–femto threshold policy is an optimal solution to the CAC problem for a two-tier macrocell–femtocell network.

We start our analysis with a finite-horizon cost function. For an n-stage problem, let $U_n(i, j)$ refer to the optimal cost function for any call type k at the beginning of a decision epoch. Using uniformization technique [8, 13], we can recursively write $U_{n+1}(i, j)$ as

$$
\begin{aligned}
U_{n+1}(i, j) \\
&= \lambda_{nm} \tau_{mf}(K_m^{max}, K_f^{max}) \left[\min \left(U_n(i, j) + R_{nm}, U_n(i + 1, j)\right)\right] \\
&\quad + \lambda_{nf} \tau_{mf}(K_m^{max}, K_f^{max}) \left[\min \left(U_n(i, j) + R_{nf}, U_n(i, j + 1)\right)\right] \\
&\quad + \lambda_{hmm} \tau_{mf}(K_m^{max}, K_f^{max}) \left[\min \left(U_n(i, j) + R_{hmm}, U_n(i + 1, j)\right)\right] \\
&\quad + i \lambda_{hmf} \tau_{mf}(K_m^{max}, K_f^{max}) \left[\min \left(U_n(i - 1, j) + R_{hmf}, U_n(i - 1, j + 1)\right)\right] \\
&\quad + j \lambda_{hfm} \tau_{mf}(K_m^{max}, K_f^{max}) \left[\min \left(U_n(i, j - 1) + R_{hfm}, U_n(i + 1, j - 1)\right)\right] \\
&\quad + \lambda_{hff} \tau_{mf}(K_m^{max}, K_f^{max}) \left[\min \left(U_n(i, j) + R_{hff}, U_n(i, j + 1)\right)\right] \\
&\quad + i \mu_m \tau_{mf}(K_m^{max}, K_f^{max}) U_n(i - 1, j) + j \mu_f \tau_{mf}(K_m^{max}, K_f^{max}) U_n(i, j - 1) \\
&\quad + i \lambda_{hmm} \tau_{mf}(K_m^{max}, K_f^{max}) U(i - 1, j) + j \lambda_{hff} \tau_{mf}(K_m^{max}, K_f^{max}) U_n(i, j - 1) \\
&\quad + \left(1 - \frac{\tau_{mf}(K_m^{max}, K_f^{max})}{\tau_{mf}(i, j)}\right) U_n(i, j)
\end{aligned}
\tag{6.1}
$$

where the boundary conditions are given by

$$
U_n(-1, j) = 0 \text{ and } U_n(K_m^{max} + 1, j) = \infty, \text{ where } j \in \{0, 1, \ldots, K_f^{max}\}
$$
$$
U_n(i, -1) = 0 \text{ and } U_n(K_f^{max} + 1, j) = \infty, \text{ where } i \in \{0, 1, \ldots, K_m^{max}\}.
$$

Here, $\tau_{mf}(K_m^{max}, K_f^{max})$ is the uniformization parameter [8, 10]. In (6.1), the first two terms correspond to new call arrival in MeNB and HeNB, respectively. The third and fifth terms correspond to horizontal handoffs involving MeNB and HeNB, respectively. The fourth and sixth terms correspond to vertical handoffs to HeNB and MeNB, respectively. The seventh to tenth terms correspond to departure events. The last term is due to the uniformization technique that corresponds to the probability of being in the same state. Let $s = (i, j, k)$ be the current state of the system at the beginning of the decision epoch and a call type $\bar{k}$ arrives at the system. From (6.1), it can be seen that the call is admitted only if $U_n(\bar{i}, \bar{j}, \bar{k}) - U_n(i, j, k) \leq R_{\bar{k}}$, where $t = (\bar{i}, \bar{j}, \bar{k})$ is the next state of the system and $R_{\bar{k}}$ is the rejection cost associated with the call type $\bar{k}$.

It is shown in [10] that for expected cost function problems with finite state and action spaces, the obtained admission policy is stationary. As $U_n(i, j)$ has a provision to remain in the same state due to uniformization technique, Theorem (6.6.2) in [10] holds. In other words, given a specific call type, for an irreducible and aperiodic

Markov decision process, the difference between the upper bound (H_n) and the lower bound (L_n) of $U_{n+1}(i, j) - U_n(i, j)$ converges to the optimal average cost per unit time as $n \to \infty$, [8, 10]. This can be achieved by adjusting the *stopping rule* of the VIA by setting the accuracy factor to a small value. Hence, we may conclude that, the structural results obtained for finite-horizon $U_n(s)$ correspond to the infinite-horizon per unit time cost function. Therefore, if we can prove that the cost function of the finite-horizon problem is convex and monotonically non-decreasing (**Lemma 6.1**), then it can be concluded that the cost function has the same properties for two-tier infinite-horizon CAC problem.

Lemma 6.1.   In a two-tier network, for a specific call type, the cost function is monotonically non-decreasing and convex in i (or j) for every fixed j (or i).

In order to prove Lemma 6.1, let us define two difference operators for $U_n(i, j)$ for specific call type k.

$$\Delta U_n^i(i, j) = U_n(i, j) - U_n(i - 1, j), \quad i = 1, 2, \ldots, K_m^{max}$$
$$\Delta U_n^j(i, j) = U_n(i, j) - U_n(i, j - 1), \quad j = 1, 2, \ldots, K_f^{max}.$$

Here, $\Delta U_n^i(i, j)$ and $\Delta U_n^j(i, j)$ correspond to a sequence for every fixed j and i, respectively. Now, we need to show that, the sequences of $\Delta U_n^i(i, j)$ and $\Delta U_n^j(i, j)$ are (i) non-decreasing and (ii) convex in i and j, respectively. For proof, we refer to **Appendix** at the end of this chapter.

From **Lemma 1**, we can conclude that, for a given call type k, the sequence of $\Delta U_n^i(i, j)$ is increasing in i for every fixed j. Hence, there exists an $\tilde{i}$ for every fixed j and call type k such that $\Delta U_n^i(\tilde{i} + 1, j) > R_k$ and $\Delta U_n^i(\tilde{i}, j) \leq R_k$. In other words, for a given call type, for every fixed j, there exists an $\tilde{i}$, which is the threshold for admitting the call. Hence, for a given call type, we have a two-dimensional optimal macro and femto threshold based set, where the number of occupied channels in HeNB and MeNB (respectively) are taken into consideration to decide on the admission of a type of call. Executing the VIA, we can obtain these macro and femto threshold sets which are denoted by $\mathbf{T_m}$ and $\mathbf{T_f}$, respectively. **Algorithm 6.1** is executed independently in the BSs to decide on the admission or rejection of a call. The algorithm corresponds to a cross-layer framework where a call is only admitted at a BS depending on the availability of a sub-channel and the minimum rate requirement. In the next section, we will provide the performance evaluation results for the proposed CAC framework.

6.4   PERFORMANCE EVALUATION

To evaluate the performance of the CAC method, we use an event-driven simulation using MATLAB R2010a to obtain the average arrival rates using Hayward's approximation. Next, we use the average arrival rates to obtain the CAC policy based on VIA using MATLAB R2010a. Finally, the admission control policy is used in the

---

Algorithm 6.1 VALUE ITERATION ALGORITHM

Input: P_{st}, $c_s(a)$, $\tau_s(a)$, **S**
Output: δ_n = Optimal CAC decision
$\mathbf{T_m}$ = Set of optimal macro thresholds
$\mathbf{T_f}$ = Set of optimal femto thresholds

1: **Initialization :**
   Select $U_0(s)$ s.t. $0 \leq U_0(s) \leq \min_{a(s)} \{c_s(a)/\tau_s(a)\}$, $\forall s \in \mathbf{S}$
   $\epsilon > 0$ {Accuracy factor}
   $\mathbf{T_m}(j, k) = \{\}$
   $\mathbf{T_f}(i, k) = \{\}$
   $n = 0$
   $\delta_n(s) = 1, \forall s \in \mathbf{S}$
2: $n = n + 1$
3: **Compute** $\forall s \in \mathbf{S}$

$$U_n(s) = \min_{a(s) \in A} \left[\frac{c_s(a)}{\tau_s(a)} + \frac{\tau_{mf}(K_m^{max}, K_f^{max})}{\tau_s(a)} \sum_{t \in S} P_{st} U_{n-1}(t) \right.$$
$$\left. + \left(1 - \frac{\tau_{mf}(K_m^{max}, K_f^{max})}{\tau_s(a)} \right) U_{n-1}(s) \right]$$

$$\delta_n(s) = \arg \min_{a(s) \in A} \left[\frac{c_s(a)}{\tau_s(a)} + \frac{\tau_{mf}(K_m^{max}, K_f^{max})}{\tau_s(a)} \sum_{t \in S} P_{st} U_{n-1}(t) \right.$$
$$\left. + \left(1 - \frac{\tau_{mf}(K_m^{max}, K_f^{max})}{\tau_s(a)} \right) U_{n-1}(s) \right]$$

4: **Compute**
   $T_n^m(j, k) = \arg \max_i \{\delta_n(s) = 1\}$, $\forall j$ and $k \in \{1, 3, 5\}$ {MeNB threshold}
   $T_n^f(i, k) = \arg \max_j \{\delta_n(s) = 1\}$, $\forall i$ and $k \in \{2, 4, 6\}$ {HeNB threshold}
5: **Compute**
   $L_n = \min_{t \in S} [U_n(t) - U_{n-1}(t)]$
   $H_n = \max_{t \in S} [U_n(t) - U_{n-1}(t)]$
6: $B_n = H_n - L_n$
7: **if** $B_n \leq \epsilon L_n$ {Stopping rule}
   then
8:     δ_n is optimal.
9:     $\mathbf{T_m}(j, k) \leftarrow T_n^m(j, k)$
10:    $\mathbf{T_f}(i, k) \leftarrow T_n^f(i, k)$
11:    **Stop**
12: **else**
13:    **Go to step 2**
14: **end if**

Algorithm 6.2 MACRO–FEMTO THRESHOLD-BASED CAC ALGORITHM

Input: T_m = Set of optimal macro thresholds
T_f = Set of optimal femto thresholds
A call of type k arrives at the BS
Output: Call Admission Decision

  1: **if** a call arrives at MeNB **then**
  2:    **if** $(i < T_m(j, k))$ **then**
  3:       Calculate the average rate of the UE based on path-loss
  4:       **if** expected rate > minimum rate requirement **then**
  5:          Accept the call
  6:       **else**
  7:          Reject the call
  8:       **end if**
  9:    **else**
 10:       Reject the call
 11:    **end if**
 12: **end if**
 13: **if** a call arrives at HeNB **then**
 14:    **if** $(j < T_f(i, k))$ **then**
 15:       Calculate the average rate of the UE based on path-loss
 16:       **if** expected rate > minimum rate requirement **then**
 17:          Accept the call
 18:       **else**
 19:          Reject the call
 20:       **end if**
 21:    **else**
 22:       Reject the call
 23:    **end if**
 24: **end if**

---

event-driven simulation. In the simulations, we consider that each UE is allocated one sub-channel to fulfill the minimum rate requirement. We consider the FFR-6 for the CAC problem. FFR-6 corresponds to high gain in throughput and less outage probability for edge-zone UEs in comparison to FFR-3 (see Chapter 5). In addition, since the system state space for FFR-6 is smaller than that of FFR-3, the complexity of the CAC problem and the running time of VIA are reduced. Figure 6.2 depicts a snap-shot of the simulation environment (i.e., network topology and UE location) during the event-driven simulation. The HeNBs are uniformly distributed within each sector. The locations of the MeNBs and HeNBs generating new calls are also uniformly distributed. The new call arrivals at the MeNB and HeNB are assumed to follow Poisson processes.

During the simulation runs, the UEs move randomly at different velocity and thus the horizontal and vertical handovers are superposition of Poisson traffic and non-Poisson overflow traffic. However, to obtain the optimal macro–femto threshold

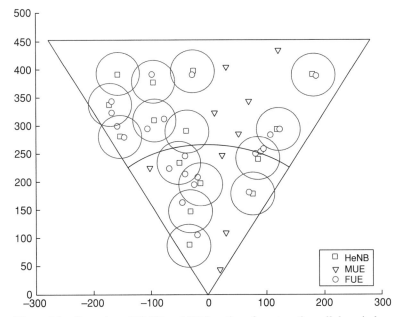

Figure 6.2   Snap shot of HeNB and UE locations for a two-tier cellular wireless network.

policy using an SMDP model, the call arrival rates have to be Poisson. Therefore, we have to approximate the offered load of different call types by Poisson processes. Hayward's approximation [14, 15] is one of the acceptable methodologies to approximate non-Poisson traffic by Poisson traffic. Using this approximation, the non-Poisson traffic with intensity ρ and peakedness A for a single sub-channel, can be approximated to Poisson traffic with intensity $\frac{\rho}{A}$. Peakedness is defined as the ratio of the variance of the number of active calls to its mean. For Poisson traffic the peakedness is $A = 1$.

6.4.1   Optimal CAC Policy

We investigate the effect of call arrival rate and the values corresponding to normalized cost factors on the optimal CAC policy for different call types in a two-tier macrocell–femtocell network.[3] As an example, we show the results for call type 2 (i.e., new femto call). Note that the priorities of the call types are set according to the system requirements, and the normalized cost factor is set according to blocking/dropping probability of different call types. In our model, we set the highest priority to the handoff calls from macro-to-femto, then femto-to-femto handoff and so forth.

Figure 6.3 illustrates the policy obtained from the VIA for normalized cost factor *set-1* (Table 6.3) where the arrival rates for all call types are set to 0.3 calls/min. This figure also corresponds to the macro–femto threshold policy for a new femto

---

[3]We consider the minimum rate at the UEs to be 100 kbps and the target SNR at the MUEs to be 15 dB, unless otherwise specified. The simulation results are obtained according to the design parameters presented in Table 5.3.

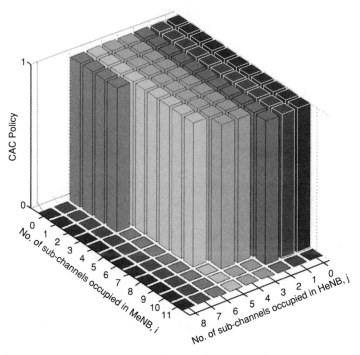

Figure 6.3   CAC policy for new femto call for normalized cost factor *set-1*: arrival rate 0.3 calls/min for all call types.

TABLE 6.3   Simulation parameters

Parameter	Value
Number of HeNBs per sector	15
New call arrival rate (Poisson) to MeNB	0.01–0.8 calls/min
New call arrival rate (Poisson) to HeNB	0.01–0.8 calls/min
Inter-region MeNB–MeNB handover rate (Poisson)	0.01–0.8 calls/min
Inter-region HeNB–HeNB handover rate (Poisson)	0.01–0.15 calls/min
Inter-region MeNB–HeNB handover rate (Poisson)	0.01–0.15 calls/min
Inter-region HeNB–MeNB handover rate (Poisson)	0.01–0.15 calls/min
Average service rate for MUE	0.25/min
Average service rate for FUE	0.25/min
HeNB access mode	Open access mode
Maximum number of FUEs per HeNB	8
UE mobility	0–3 m/s
Cost-weighting factor *set-1*, $R_{hmf} : R_{hff} : R_{hfm} : R_{hmm} : R_{nf} : R_{nm}$	5:4:3:2:1:1
Cost-weighting factor *set-2*, $R_{hmf} : R_{hff} : R_{hfm} : R_{hmm} : R_{nf} : R_{nm}$	15:10:6:3:1:1
Accuracy factor, ϵ	10^{-3}
Event-driven simulation run time	1 hour

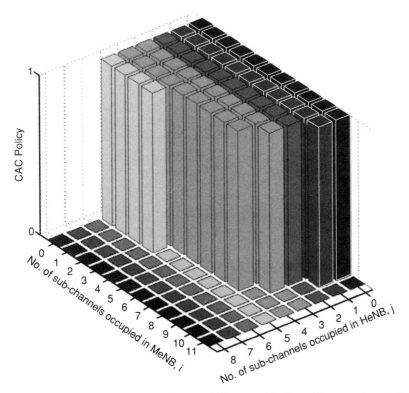

Figure 6.4   CAC policy for new femto call for normalized cost factor *set-1*: arrival rate of macro-to-femto handover calls is increased to 0.7 calls/min.

call. For example, a new femto call will be blocked when the numbers of occupied sub-channels in MeNB and HeNB are 4 and 6, respectively. Figure 6.4 illustrates the obtained policy for new femto calls as we only increase the arrival rate of handoff calls from macro-to-femto (i.e., call type 4 λ_{hmf}) to 0.7 calls/min. We can see that since the priority of handoff calls from macro-to-femto is set to be the highest, more new femto calls are blocked, resulting in higher blocking probability.

Figure 6.5 depicts the policy obtained for new femto calls for the normalized cost factor *set-2* where the arrival rates for all call types are set to 0.3 calls/min. As we compare Figures 6.3 and 6.5, we can notice that more new femto calls are blocked as the number of occupied sub-channels in the HeNB increases. This is due to the fact that in the normalized cost factor *set-2*, the rejection cost for macro-to-femto handoff is proportionately higher than that in *set-1*. As the arrival rate of macro-to-femto handoffs increases to 0.7 calls/min (Figure 6.6), we observe that fewer new femto calls are admitted into the HeNB.

6.4.2   Call Blocking/Dropping Probability Performance

The call-level QoS performance of the FFR-6 based two-tier network is illustrated in Figures 6.7–6.10. The CAC policy obtained from the VIA is used in the event-driven

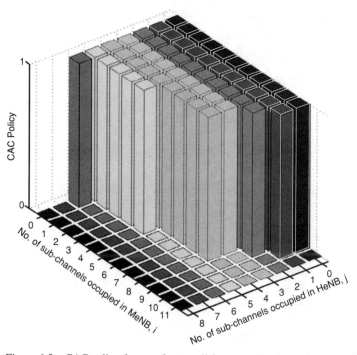

Figure 6.5   CAC policy for new femto call for normalized cost factor *set-2*: arrival rate 0.3 calls/min for all call types.

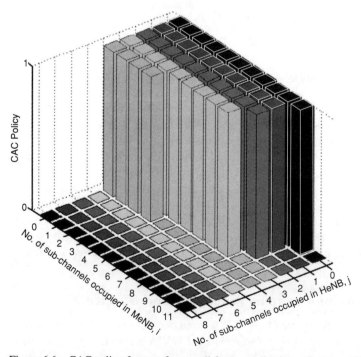

Figure 6.6   CAC policy for new femto call for normalized cost factor *set-2*: arrival rate of macro-to-femto handover calls increased to 0.7 calls/min.

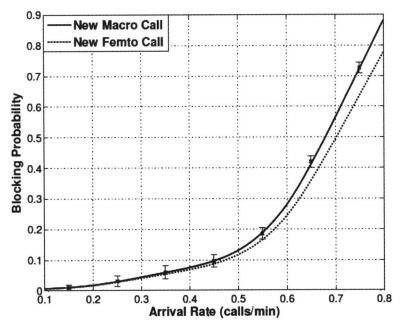

Figure 6.7   New call blocking probability for normalized cost factor *set-1*.

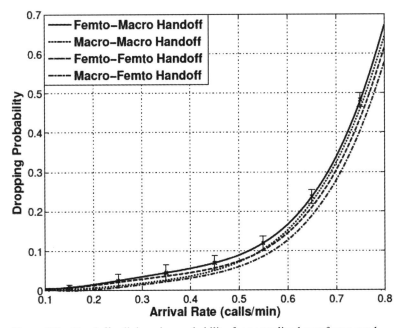

Figure 6.8   Handoff call dropping probability for normalized cost factor *set-1*.

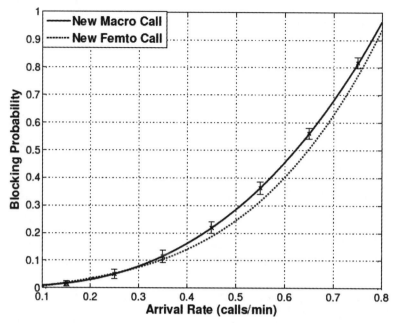

Figure 6.9   New call blocking probability for normalized cost factor *set-2*.

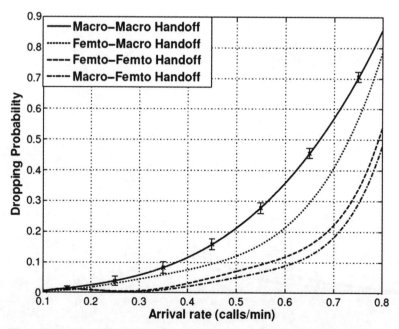

Figure 6.10   Handoff call dropping probability for normalized cost factor *set-2*.

simulation to obtain these performance results. Figures 6.7 and 6.8 correspond to new call (macro and femto) blocking probability and handoff call (vertical and horizontal handoffs for MeNB and HeNB) dropping probability, respectively, for normalized cost factor *set-1*. It is notable that in general the handoff (both vertical and horizontal) calls have better QoS performance in the system at high arrival rates. This is due to the fact that more handoff calls are admitted into the network in comparison to new calls since the cost factor values for handoff call are higher in comparison to those for new calls. Specifically, the macro-to-femto handoff calls have better QoS performance in comparison to other call types. However, since the values of the cost factors are relatively close to each other, we see that the call blocking and dropping probability curves are relatively close to each other. In addition, the curves show a pattern in QoS performance that reflects the priority of the call types that can be associated with the system requirements and call blocking/dropping probability constraints. With the cost factor *set-1*, the system has call blocking/dropping probability ≤ 0.1 for all types of calls with arrival rate up to 0.45 calls/min.[4] It is also notable from Figures 6.7 and 6.8 that, for the proposed framework, the maximum deviation of blocking/dropping probability for new macro and femto-to-macro calls is $\pm 2.11\%$ and $\pm 2.03\%$, respectively.

Figures 6.9 and 6.10 show, respectively, new call blocking probability and handoff call dropping probability for cost factor *set-2*. Now, as the proportional values of the cost factors for handoff calls are relatively higher in comparison to those in *set-1*, more new calls are blocked at high mobility resulting in higher call blocking probability. However, we notice similar pattern and trend for the new call blocking probability (see Figures 6.7 and 6.9). This is due to the fact that in both the cost factor sets, the new calls for macro and femto have the same proportional values. On the other hand, the cost factor values for handoff calls in *set-2* are proportionally higher in comparison to those in *set-1*. As a result, the call dropping probability curves (Figure 6.10) are not close to each other as they are in Figure 6.8. Therefore, unlike the previous case, the threshold arrival rates are different for each call type to ensure that the call blocking/dropping probability requirement is satisfied within the system. For example, if we consider that the system requires call blocking/dropping probability ≤ 0.1, then the threshold arrival rates for new calls (both macro and femto) and macro-to-femto are approximately 0.33 and 0.6 calls/min, respectively.[5] It is also notable from Figures 6.9–6.10 that, for the proposed framework, the maximum deviation of blocking/dropping probability for new macro and macro-to-macro calls is $\pm 2.25\%$ and $\pm 1.86\%$, respectively.

Thus we can see that, the system QoS performance depends on the priority of the call type and the values assigned to the normalized cost factors. If the system requires a steady QoS for all types of calls, then we may choose the cost factor values as in *set-1*. If the system requires to maintain a strict QoS performance for specific call type, the cost factor values can be chosen similar to those in *set-2*.

---

[4]We refer this setup as configuration-1.

[5]We refer this setup as configuration-2.

6.4.3 Comparison of Call-Level QoS

Figures 6.11–6.13 depict the comparison between the proposed framework and a reference model in terms of blocking/dropping probability of MUEs considering the target SNR at MUEs and minimum rate requirement as 15 dB and 275 kbps, respectively. In these figures "ch. unavail" refer to the case where call blocking is due to the unavailability of channels during the CAC process. For the reference model, we consider a non-optimal configuration where the radius of cell center-zone is half of the macrocell radius and the number of allocated sub-channels is in proportion to the areas of center-zone and edge-zone. Since the FFR design parameters for the reference model are not optimal, we notice an increase in the call blocking/dropping probability for MUEs. For example, from Figure 6.11, we can notice that, the average increase in new macro call blocking probability for the reference model is approximately 26.54%.

Figures 6.14–6.16 depict the comparison between the proposed framework and the reference model in terms of blocking/dropping probability of the MUEs considering the target SNR at the MUEs and the minimum rate requirement as 20 dB and 275 kbps, respectively. In this case, we also notice from Figure 6.14 that, the average increase in new macro call blocking probability for the reference model is approximately 29.91%. Also, since the FFR design parameters for the reference model are not optimized, the average increase in new macro call blocking probability is also very high.

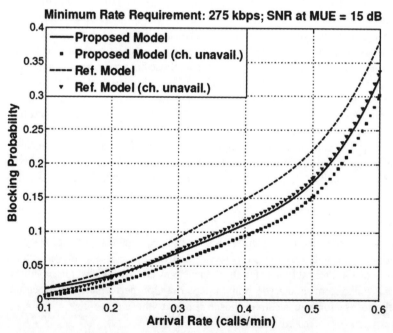

Figure 6.11 Comparison between the proposed model and the reference model in terms of new macro call blocking probability for normalized cost factor *set-1* and target SNR at MUE = 15 dB.

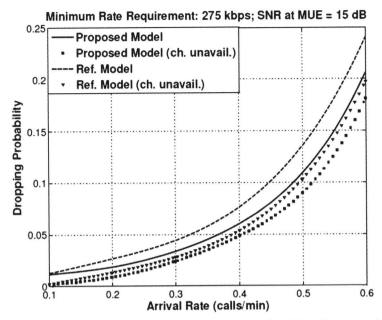

Figure 6.12   Comparison between the proposed model and the reference model in terms of macro-to-macro handoff call dropping probability for normalized cost factor *set-1* and target SNR at MUE = 15 dB.

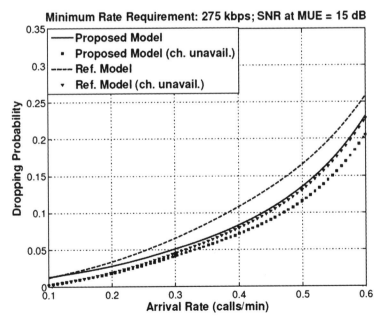

Figure 6.13   Comparison between the proposed model and the reference model in terms of femto-to-macro handoff call dropping probability for normalized cost factor *set-1* and target SNR at MUE = 15 dB.

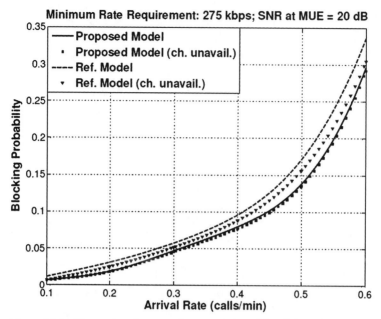

Figure 6.14 Comparison between the proposed model and the reference model in terms of new macro call blocking probability for normalized cost factor *set-1* and target SNR at MUE = 20 dB.

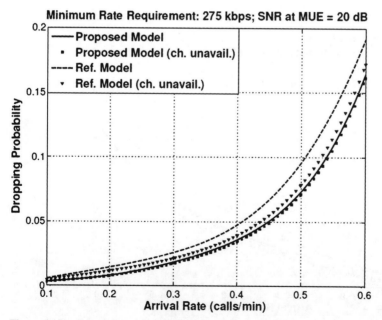

Figure 6.15 Comparison between the proposed model and the reference model in terms of macro-to-macro handoff call dropping probability for normalized cost factor *set-1* and target SNR at MUE = 20 dB.

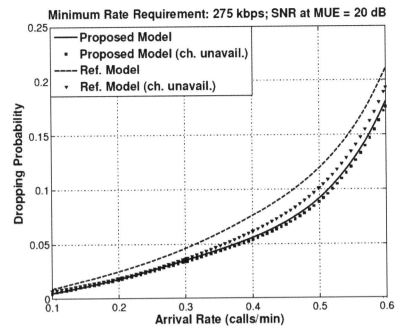

Figure 6.16   Comparison between the proposed model and the reference model in terms of femto-to-macro handoff call dropping probability for normalized cost factor *set-1* and target SNR at MUE = 20 dB.

6.5   SUMMARY AND FUTURE RESEARCH DIRECTIONS

For FFR-6, we have presented a decision-theoretic framework to model and solve the CAC problem in a two-tier macrocell-femtocell network. The problem has been formulated as an SMDP and a VIA has been used to obtain the optimal macro–femto threshold policy. In the process of obtaining the admission policy, we have shown that the cost function is monotonically non-decreasing and convex, and hence the structural results obtained holds for an infinite-horizon CAC problem to ensure long-term QoS performance. To facilitate the application of the SMDP framework under user mobility, we have used "Hayward's Approximation" to approximate the non-Poisson traffic by Poisson traffic. We have shown how the mobility and cost factor values of call types affect the admission policy obtained from the VIA. The normalized cost factors and the FFR design parameters can be chosen according to call blocking/dropping probability thresholds (for different call types) as determined by the system requirements.

In the context of handoff management and admission control, it is important to develop realistic models for input traffic and overflow (or handoff) at the different tiers of the network in presence of user location, mobility, and different service classes with prioritization among them, as well as an access mode of the small cells. A multi-tier network model will give rise to different statistical characteristics of input traffic, mobility model, and service time distribution at different network tiers [16, 17]. Tractable performance analysis models for handoff management and admission control need to be developed considering this statistical heterogeneity at different tiers along with the physical-layer models for spectrum sharing.

The overflow policy may be designed differently if there are more than two tiers in the network. When there are multiple (>2) tiers, the resource allocation among tiers will also impact the admission control design since it determines how many channels (e.g., RBs) are available at each tier to support new and handoff calls from different tiers.

The design of CAC policies for multi-tier networks should also consider the multiple radio access technology (multi-RAT) scenarios where an MUE or an FUE is also equipped with 3G and/or Wi-Fi radio in addition to the OFDMA radio and a seamless handoff is possible among the different RATs. In this context, microeconomic and game-theoretic models can be developed to model the RAT selection as well as the BS selection problem in multi-tier networks.

For efficient CAC and handoff management, cross-layer methods need to be developed with the following considerations. Since the interference structures in the UL and DL in the multi-tier networks are very different, constraints at the physical layer will depend on whether the arriving/handoff calls need to establish communication in both UL and DL or in only one direction as well as the access mode in use. For DL communications, the MUEs may not be able to connect to a nearby SBS in the closed access mode, or the MUEs could be highly mobile which require them to be connected with the MBSs. Then, to avoid excessive interference from nearby SBSs, some channels should not be used by the nearby SBSs, which are reserved for MUEs. Therefore, the problem of joint selection of network tier and resource allocation will need to be investigated for both UL and DL communications.

APPENDIX

State Transition Probability

The state transition probabilities from state $s \in S$ to state $t \in S$ at the next decision epoch of the system for action $a(s)$ when $k \in \{2, 4, 5, 6\}$ are as follows:

- For $k = 2, 6$

$$P_{st}(a = 1) = \begin{cases} i(\lambda_{hmm} + \mu_m) \cdot \tau_s, & t = (i - 1, j + 1, 0) \\ (j + 1)(\lambda_{hff} + \mu_f) \cdot \tau_s, & t = (i, j, 0) \\ \lambda_{nm} \cdot \tau_s, & t = (i, j + 1, 1) \\ \lambda_{nf} \cdot \tau_s, & t = (i, j + 1, 2) \\ \lambda_{hmm} \cdot \tau_s, & t = (i, j + 1, 3) \\ i\lambda_{hmf} \cdot \tau_s, & t = (i, j + 1, 4) \\ (j + 1)\lambda_{hfm} \cdot \tau_s, & t = (i, j + 1, 5) \\ \lambda_{hff} \cdot \tau_s, & t = (i, j + 1, 6) \end{cases}$$

where $\tau_s(a = 1) = \tau_{mf}(i, j + 1)$.

$$P_{st}(a = 0) = \begin{cases} i(\lambda_{hmm} + \mu_m) \cdot \tau_s, & t = (i - 1, j, 0) \\ j(\lambda_{hff} + \mu_f) \cdot \tau_s, & t = (i, j - 1, 0) \\ \lambda_{nm} \cdot \tau_s, & t = (i, j, 1) \\ \lambda_{nf} \cdot \tau_s, & t = (i, j, 2) \\ \lambda_{hmm} \cdot \tau_s, & t = (i, j, 3) \\ i\lambda_{hmf} \cdot \tau_s, & t = (i, j, 4) \\ j\lambda_{hfm} \cdot \tau_s, & t = (i, j, 5) \\ \lambda_{hff} \cdot \tau_s, & t = (i, j, 6) \end{cases}$$

where $\tau_s(a = 0) = \tau_{mf}(i, j)$.

- For $k = 4$

$$P_{st}(a = 1) = \begin{cases} (i - 1)(\lambda_{hmm} + \mu_m) \cdot \tau_s, & t = (i - 2, j + 1, 0) \\ (j + 1)(\lambda_{hff} + \mu_f) \cdot \tau_s, & t = (i - 1, j, 0) \\ \lambda_{nm} \cdot \tau_s, & t = (i - 1, j + 1, 1) \\ \lambda_{nf} \cdot \tau_s, & t = (i - 1, j + 1, 2) \\ \lambda_{hmm} \cdot \tau_s, & t = (i - 1, j + 1, 3) \\ (i - 1)\lambda_{hmf} \cdot \tau_s, & t = (i - 1, j + 1, 4) \\ (j + 1)\lambda_{hfm} \cdot \tau_s, & t = (i - 1, j + 1, 5) \\ \lambda_{hff} \cdot \tau_s, & t = (i - 1, j + 1, 6) \end{cases}$$

where $\tau_s(a = 1) = \tau_{mf}(i - 1, j + 1)$, and

$$P_{st}(a = 0) = \begin{cases} (i - 1)(\lambda_{hmm} + \mu_m) \cdot \tau_s, & t = (i - 2, j, 0) \\ j(\lambda_{hff} + \mu_f) \cdot \tau_s, & t = (i - 1, j - 1, 0) \\ \lambda_{nm} \cdot \tau_s, & t = (i - 1, j, 1) \\ \lambda_{nf} \cdot \tau_s, & t = (i - 1, j, 2) \\ \lambda_{hmm} \cdot \tau_s, & t = (i - 1, j, 3) \\ (i - 1)\lambda_{hmf} \cdot \tau_s, & t = (i - 1, j, 4) \\ j\lambda_{hfm} \cdot \tau_s, & t = (i - 1, j, 5) \\ \lambda_{hff} \cdot \tau_s, & t = (i - 1, j, 6) \end{cases}$$

where $\tau_s(a = 0) = \tau_{mf}(i - 1, j)$.

- For $k = 5$

$$P_{st}(a = 1) = \begin{cases} (i+1)(\lambda_{hmm} + \mu_m) \cdot \tau_s, & t = (i, j-1, 0) \\ (j-1)(\lambda_{hff} + \mu_f) \cdot \tau_s, & t = (i+1, j-2, 0) \\ \lambda_{nm} \cdot \tau_s, & t = (i+1, j-1, 1) \\ \lambda_{nf} \cdot \tau_s, & t = (i+1, j-1, 2) \\ \lambda_{hmm} \cdot \tau_s, & t = (i+1, j-1, 3) \\ (i+1)\lambda_{hmf} \cdot \tau_s, & t = (i+1, j-1, 4) \\ (j-1)\lambda_{hfm} \cdot \tau_s, & t = (i+1, j-1, 5) \\ \lambda_{hff} \cdot \tau_s, & t = (i+1, j-1, 6) \end{cases}$$

where $\tau_s(a = 1) = \tau_{mf}(i+1, j-1)$, and

$$P_{st}(a = 0) = \begin{cases} i(\lambda_{hmm} + \mu_m) \cdot \tau_s, & t = (i-1, j-1, 0) \\ (j-1)(\lambda_{hff} + \mu_f) \cdot \tau_s, & t = (i, j-2, 0) \\ \lambda_{nm} \cdot \tau_s, & t = (i, j-1, 1) \\ \lambda_{nf} \cdot \tau_s, & t = (i, j-1, 2) \\ \lambda_{hmm} \cdot \tau_s, & t = (i, j-1, 3) \\ i\lambda_{hmf} \cdot \tau_s, & t = (i, j-1, 4) \\ (j-1)\lambda_{hfm} \cdot \tau_s, & t = (i, j-1, 5) \\ \lambda_{hff} \cdot \tau_s, & t = (i, j-1, 6) \end{cases}$$

where $\tau_s(a = 0) = \tau_{mf}(i, j-1)$.

Proof of Monotonically Non-Decreasing Property of Cost Function

We use a mathematical induction method on n to show that the cost function is monotonically non-decreasing and convex in i (or j) for every fixed j (or i). In other words, we consider $U_n(i, j)$ to be monotonically non-decreasing and convex in i (or j) for every fixed j (or i). Now, we need to show that $U_{n+1}(i, j)$ is monotonically non-decreasing and convex in i (or j) for every fixed j (or i).

In this proof we consider the center-zone of a specific sector and hence the parameters presented are associated with the center-zone specification. We consider the hypothesis that $U_n(i, j)$ is monotonically non-decreasing in i for every fixed j, that is, $\Delta U_n^i(i, j) \geq 0$. Now we need to show that, $\Delta U_{(n+1)}^i(i, j) \geq 0$.

$$\Delta U_{(n+1)}^i(i, j) = U_{(n+1)}(i, j) - U_{(n+1)}(i-1, j)$$

$$= \lambda_{nm} \tau_{mf}(K_m^{(c)}, K_f^{(c)})\Phi_1(i, j) + \lambda_{nf} \tau_{mf}(K_m^{(c)}, K_f^{(c)})\Phi_2(i, j)$$

$$+ \lambda_{hmm} \tau_{mf}(K_m^{(c)}, K_f^{(c)})\Phi_3(i, j) + \lambda_{hmf} \tau_{mf}(K_m^{(c)}, K_f^{(c)})\Phi_4(i, j)$$

$$+ \lambda_{hfm} \tau_{mf}(K_m^{(c)}, K_f^{(c)})\Phi_5(i, j) + \lambda_{hff} \tau_{mf}(K_m^{(c)}, K_f^{(c)})\Phi_6(i, j)$$

$$+ \mu_m \tau_{mf}(K_m^{(c)}, K_f^{(c)})\Phi_7(i, j) + \mu_f \tau_{mf}(K_m^{(c)}, K_f^{(c)})\Phi_8(i, j)$$

$$+ \lambda_{hmm} \tau_{mf}(K_m^{(c)}, K_f^{(c)})\Phi_9(i, j) + \lambda_{hff} \tau_{mf}(K_m^{(c)}, K_f^{(c)})\Phi_{10}(i, j)$$

$$+ \tau_{mf}(K_m^{(c)}, K_f^{(c)})\Phi_{11}(i, j) \tag{6.2}$$

where

$$\Phi_1(i, j) = \min\left(U_n\,(i,\,j) + R_{nm},\, U_n\,(i+1,\,j)\right)$$
$$- \min\left(U_n\,(i-1,\,j) + R_{nm},\, U_n\,(i,\,j)\right)$$

$$\Phi_2(i, j) = \min\left(U_n\,(i,\,j) + R_{nf},\, U_n\,(i,\,j+1)\right)$$
$$- \min\left(U_n\,(i-1,\,j) + R_{nf},\, U_n\,(i-1,\,j+1)\right)$$

$$\Phi_3(i, j) = \min\left(U_n\,(i,\,j) + R_{hmm},\, U_n\,(i+1,\,j)\right)$$
$$- \min\left(U_n\,(i-1,\,j) + R_{hmm},\, U_n\,(i,\,j)\right)$$

$$\Phi_4(i, j) = i\,\min\left(U_n\,(i-1,\,j) + R_{hmf},\, U_n\,(i-1,\,j+1)\right)$$
$$-(i-1)\min\left(U_n\,(i-2,\,j) + R_{hmf},\, U_n\,(i-2,\,j+1)\right)$$

$$\Phi_5(i, j) = j[\min\left(U_n\,(i,\,j-1) + R_{hfm},\, U_n\,(i+1,\,j-1)\right)]$$
$$-j[\min\left(U_n\,(i-1,\,j-1) + R_{hfm},\, U_n\,(i,\,j-1)\right)]$$

$$\Phi_6(i, j) = \min\left(U_n\,(i,\,j) + R_{hff},\, U_n\,(i,\,j+1)\right)$$
$$- \min\left(U_n\,(i-1,\,j) + R_{hff},\, U_n\,(i-1,\,j+1)\right)$$

$$\Phi_7(i, j) = i\,U_n(i-1,\,j) - (i-1)U_n(i-2,\,j)$$

$$\Phi_8(i, j) = j\,(U_n(i,\,j-1) - U_n(i-1,\,j-1))$$

$$\Phi_9(i, j) = i\,U_n(i-1,\,j) - (i-1)U_n(i-2,\,j)$$

$$\Phi_{10}(i, j) = j\,(U_n(i,\,j-1) - U_n(i-1,\,j-1))$$

$$\Phi_{11}(i, j) = \left(\frac{1}{\tau_{mf}(K_m^{(c)},\,K_f^{(c)})} - \frac{1}{\tau_{mf}(i,\,j)}\right) U_n(i,\,j)$$
$$- \left(\frac{1}{\tau_{mf}(K_m^{(c)},\,K_f^{(c)})} - \frac{1}{\tau_{mf}(i-1,\,j)}\right) U_n(i-1,\,j).$$

In order to show that $\Delta U_{(n+1)}^i(i,\,j) \geq 0$, we consider part-by-part analysis of 6.2. In this analysis, first we note that for a specific call type, for every fixed j, the following initial conditions apply:

$$\left.\begin{array}{l} U_0(i,\,j) = 0, \quad \forall (i,\,j) \leq (K_m^{(c)},\,K_f^{(c)}) \\ U_0(K_m^{(c)}+1,\,j) = \infty, \quad \forall j. \end{array}\right\} \tag{6.3}$$

Now, let us consider $\Phi_1(i,\,j)$ of 6.2:

$$\Phi_1(i, j) = \min\left(U_n\,(i,\,j) + R_{nm},\, U_n\,(i+1,\,j)\right)$$
$$- \min\left(U_n\,(i-1,\,j) + R_{nm},\, U_n\,(i,\,j)\right)$$
$$= \min(R_{nm},\, U_n(i+1,\,j) - U_n(i,\,j)) + U_n(i,\,j)$$
$$- \min(R_{nm},\, U_n(i,\,j) - U_n(i-1,\,j)) - U_n(i-1,\,j)$$
$$= \{\min(R_{nm},\, U_n(i+1,\,j) - U_n(i,\,j))$$
$$- \min(R_{nm},\, U_n(i,\,j) - U_n(i-1,\,j))\}$$
$$+ \{U_n(i,\,j) - U_n(i-1,\,j)\}. \tag{6.4}$$

Since we are using the method of induction on variable n, the hypothesis that $U_n(i, j)$ is monotonically non-decreasing in i (or j) for every fixed j (or i) holds true. Therefore, for a given call type for every fixed j, $U_n(i, j) - U_n(i - 1, j) = \Delta U_n^i(i, j) \geq 0$. Concurrently, it also signifies that $\min(R_{nm}, U_n(i + 1, j) - U_n(i, j)) \geq 0$ and $\min(R_{nm}, U_n(i, j) - U_n(i - 1, j)) \geq 0$. However, the term, $\min(R_{nm}, U_n(i + 1, j) - U_n(i, j)) - \min(R_{nm}, U_n(i, j) - U_n(i - 1, j)) \geq 0$, since $U_n(i + 1, j) - U_n(i, j) \geq 0$. Therefore, we can conclude $\Phi_1(i, j) \geq 0$. Similarly, it can be proved that, $\Phi_3(i, j)$, $\Phi_5(i, j) \geq 0$. Now, let us consider $\Phi_2(i, j)$:

$$
\begin{aligned}
&\Phi_2(i, j) \\
&= \min\left(U_n(i, j) + R_{nf}, U_n(i, j + 1)\right) \\
&\quad - \min\left(U_n(i - 1, j) + R_{nf}, U_n(i - 1, j + 1)\right) \\
&= \underbrace{\min(R_{nf}, \Delta U_n^j(i, j + 1)) - \min(R_{nf}, U_n^j(i - 1, j + 1))}_{\geq 0} + \underbrace{\Delta U_n^i(i, j))}_{\geq 0} \geq 0.
\end{aligned}
$$

Similarly, it can be proved that $\Phi_6(i, j) \geq 0$. Since $U_n(i, j - 1) - U_n(i - 1, j - 1) = \Delta U_n^i(i, j - 1) \geq 0$, it can be readily shown that, $\Phi_8(i, j)$, $\Phi_{10}(i, j) \geq 0$. In order to ensure that the summation of the rest of the terms in 6.2 is non-negative, we first expand $\Phi_{11}(i, j)$ as follows:

$$
\begin{aligned}
\Phi_{11}(i, j) &= \left(\frac{1}{\tau_{mf}(K_m^{(c)}, K_f^{(c)})} - \frac{1}{\tau_{mf}(i, j)}\right) U_n(i, j) \\
&\quad - \left(\frac{1}{\tau_{mf}(K_m^{(c)}, K_f^{(c)})} - \frac{1}{\tau_{mf}(i - 1, j)}\right) U_n(i - 1, j) \\
&= \lambda_{hmm}[(K_m^{(c)} - i) U_n(i, j) - (K_m^{(c)} - i + 1) U_n(i - 1, j)] \\
&\quad + \lambda_{hff}[(K_f^{(c)} - j) U_n(i, j) - (K_f^{(c)} - j) U_n(i - 1, j)] \\
&\quad + \lambda_{hmf}[(K_m^{(c)} - i) U_n(i, j) - (K_m^{(c)} - i + 1) U_n(i - 1, j)] \\
&\quad + \lambda_{hfm}[(K_f^{(c)} - j) U_n(i, j) - (K_f^{(c)} - j) U_n(i - 1, j)] \\
&\quad + \mu_m[(K_m^{(c)} - i) U_n(i, j) - (K_m^{(c)} - i + 1) U_n(i - 1, j)] \\
&\quad + \mu_f[(K_f^{(c)} - j) U_n(i, j) - (K_f^{(c)} - j) U_n(i - 1, j)] \\
&= \lambda_{hmm} \Psi_1(i, j) + \lambda_{hff} \Psi_2(i, j) + \lambda_{hmf} \Psi_3(i, j) \\
&\quad + \lambda_{hfm} \Psi_4(i, j) + \mu_m \Psi_5(i, j) + \mu_f \Psi_6(i, j).
\end{aligned}
\tag{6.5}
$$

Let us consider $\Psi_2(i, j)$:

$$
\begin{aligned}
\Psi_2(i, j) &= [(K_f^{(c)} - j) U_n(i, j) - (K_f^{(c)} - j) U_n(i - 1, j)] \\
&= (K_f^{(c)} - j)\{U_n(i, j) - U_n(i - 1, j)\} \\
&= (K_f^{(c)} - j) \underbrace{\Delta U_n^i(i, j)}_{\geq 0} \geq 0.
\end{aligned}
$$

Similarly, it can be proved that $\Psi_4(i, j)$, $\Psi_6(i, j) \geq 0$. Now, consider the summation of $\Phi_9(i, j)$ of 6.2 and $\Psi_1(i, j)$ of 6.5,

$$
\begin{aligned}
\Phi_9(i, j) + \Psi_1(i, j) &= i\, U_n(i - 1, j) - (i - 1)U_n(i - 2, j) \\
&\quad + [(K_m^{(c)} - i)\, U_n(i, j) - (K_m^{(c)} - i + 1)\, U_n(i - 1, j)] \\
&= (K_m^{(c)} - i)\, \underbrace{\Delta U_n^i(i, j)}_{\geq 0} + (i - 1)(U_n(i - 1, j) - U_n(i - 2, j)) \\
&\geq (i - 1)\, \underbrace{\Delta U_n^i(i - 1, j)}_{\geq 0} \geq 0.
\end{aligned}
$$

Similarly, it can be proved that $\Phi_7(i, j) + \Psi_5(i, j) \geq 0$. Finally, consider the summation of $\Phi_4(i, j)$ of 6.2 and $\Psi_3(i, j)$ of 6.5:

$$
\begin{aligned}
&\Phi_4(i, j) + \Psi_3(i, j) \\
&= i \min(R_{hmf}, \Delta U_n^j(i - 1, j + 1)) - (i - 1)\min(R_{hmf}, \Delta U_n^j(i - 2, j + 1)) \\
&\quad + (K_m^{(c)} - i)\, \underbrace{\Delta U_n^i(i, j)}_{\geq 0} + (i - 1)(U_n(i - 1, j) - U_n(i - 2, j)) \\
&\geq i \min(R_{hmf}, \Delta U_n^j(i - 1, j + 1)) - (i - 1)\min(R_{hmf}, \Delta U_n^j(i - 2, j + 1)) \\
&\quad + (i - 1)\, \underbrace{\Delta U_n^i(i - 1, j)}_{\geq 0} \\
&\geq i \left\{ \min(R_{hmf}, \Delta U_n^j(i - 1, j + 1)) - \min(R_{hmf}, \Delta U_n^j(i - 2, j + 1)) \right\} \\
&\quad + \min(R_{hmf}, \underbrace{\Delta U_n^j(i - 2, j + 1)}_{\geq 0}) \\
&\geq i \underbrace{\left\{ \min(R_{hmf}, \Delta U_n^j(i - 1, j + 1)) - \min(R_{hmf}, \Delta U_n^j(i - 2, j + 1)) \right\}}_{\geq 0,\ \text{since } U_n(i, j) \text{ is convex in } i \text{ (or } j) \text{ for every fixed } j \text{ (or } i)} \geq 0.
\end{aligned}
$$

Therefore, it can be concluded that $\Delta U_{(n+1)}^i(i, j) = U_{(n+1)}(i, j) - U_{(n+1)}(i - 1, j) \geq 0$. Similarly, it can be shown that $\Delta U_{(n+1)}^j(i, j) = U_{(n+1)}(i, j) - U_{(n+1)}(i, j - 1) \geq 0$. Hence, for a given call type the cost function is monotonically non-decreasing in i (or j) for every fixed j (or i).

Proof of Convexity of Cost Function

In order to prove that for a given call type, $U_{n+1}(i, j)$ is convex in i for every fixed j, we need to show that $\Delta U_{n+1}^i(i + 1, j) \geq \Delta U_{n+1}^i(i, j)$, that is, $\Delta U_{n+1}^i(i + 1, j) - \Delta U_{n+1}^i(i, j) \geq 0$, where the initial conditions in (6.3) hold. Now,

$$
\begin{aligned}
&\Delta U_{n+1}^i(i + 1, j) - \Delta U_{n+1}^i(i, j) \\
&= \lambda_{nm}\tau_{mf}(K_m^{(c)}, K_f^{(c)})\Xi_1(i + 1, j) + \lambda_{nf}\tau_{mf}(K_m^{(c)}, K_f^{(c)})\Xi_2(i + 1, j) \\
&\quad + \lambda_{hmm}\tau_{mf}(K_m^{(c)}, K_f^{(c)})\Xi_3(i + 1, j) + \lambda_{hmf}\tau_{mf}(K_m^{(c)}, K_f^{(c)})\Xi_4(i + 1, j) \\
&\quad + \lambda_{hfm}\tau_{mf}(K_m^{(c)}, K_f^{(c)})\Xi_5(i + 1, j) + \lambda_{hff}\tau_{mf}(K_m^{(c)}, K_f^{(c)})\Xi_6(i + 1, j) \quad\quad (6.6) \\
&\quad + \mu_m\tau_{mf}(K_m^{(c)}, K_f^{(c)})\Xi_7(i + 1, j) + \mu_f\tau_{mf}(K_m^{(c)}, K_f^{(c)})\Xi_8(i + 1, j) \\
&\quad + \lambda_{hmm}\tau_{mf}(K_m^{(c)}, K_f^{(c)})\Xi_9(i + 1, j) + \lambda_{hff}\tau_{mf}(K_m^{(c)}, K_f^{(c)})\Xi_{10}(i + 1, j) \\
&\quad + \tau_{mf}(K_m^{(c)}, K_f^{(c)})\Xi_{11}(i + 1, j)
\end{aligned}
$$

where $\Xi_l(i + 1, j) = \Phi_l(i + 1, j) - \Phi_l(i, j), l = 1, 2, 3, \ldots, 11$.

Now, we consider part-by-part analysis of (6.6) to prove the convexity of $U_{n+1}(i, j)$ in i (or j) for every fixed j(or i) for a specific call type k.

$$
\begin{aligned}
\Xi_1&(i + 1, j)\\
= &\left\{\min(R_{nm}, \Delta U_n^i(i + 2, j)) - \min(R_{nm}, \Delta U_n^i(i + 1, j))\right\}\\
- &\left\{\min(R_{nm}, \Delta U_n^i(i + 1, j)) - \min(R_{nm}, \Delta U_n^i(i, j))\right\}\\
+ &\quad\underbrace{\left\{\Delta U_n^i(i + 1, j) - \Delta U_n^i(i, j)\right\}}_{\geq 0,\ \text{since } U_n(i, j) \text{ is convex in } i \text{ (or } j) \text{ for every fixed } j \text{ (or } i)}\\
\geq &\left\{\min(R_{nm}, \Delta U_n^i(i + 2, j)) - \min(R_{nm}, \Delta U_n^i(i + 1, j))\right\}\\
- &\left\{\min(R_{nm}, \Delta U_n^i(i + 1, j)) - \min(R_{nm}, \Delta U_n^i(i, j))\right\}.
\end{aligned}
$$

Now, with the hypothesis that $U_n(i, j)$ is convex in i (or j) for every fixed j (or i), we can justify that, $\min(R_{nm}, \Delta U_n^i(i + 2, j)) - \min(R_{nm}, \Delta U_n^i(i + 1, j)) \geq 0$ and $\min(R_{nm}, \Delta U_n^i(i + 1, j)) - \min(R_{nm}, \Delta U_n^i(i, j)) \geq 0$. In addition, since $\Delta U_n^i(i + 2, j) - \Delta U_n^i(i + 1, j) \geq 0$, we can conclude that, $\Xi_1(i + 1, j) \geq 0$. Similarly, it can be proved that, each of the terms $\Xi_2(i + 1, j)$, $\Xi_3(i + 1, j)$, $\Xi_5(i + 1, j)$, $\Xi_6(i + 1, j)$, $\Xi_8(i + 1, j)$, $\Xi_{10}(i + 1, j)$ is ≥ 0.

Now, let us consider expanding $\Xi_{11}(i + 1, j)$ as follows:

$$
\begin{aligned}
\Xi_{11}&(i + 1, j)\\
= &\lambda_{hmm}[(K_m^{(c)} - i - 1)\, U_n(i + 1, j) - (K_m^{(c)} - i)\, U_n(i, j)]\\
- &\lambda_{hmm}[(K_m^{(c)} - i)\, U_n(i, j) + (K_m^{(c)} - i + 1)\, U_n(i - 1, j)]\\
+ &\lambda_{hff}[(K_f^{(c)} - j)\, U_n(i + 1, j) - (K_f^{(c)} - j)\, U_n(i, j)]\\
- &\lambda_{hff}[(K_f^{(c)} - j)\, U_n(i, j) + (K_f^{(c)} - j)\, U_n(i - 1, j)]\\
+ &\lambda_{hmf}[(K_m^{(c)} - i - 1)\, U_n(i + 1, j) - (K_m^{(c)} - i)\, U_n(i, j)\\
- &\lambda_{hmf}[(K_m^{(c)} - i)\, U_n(i, j) + (K_m^{(c)} - i + 1)\, U_n(i - 1, j)]\\
+ &\lambda_{hfm}[(K_f^{(c)} - j)\, U_n(i + 1, j) - (K_f^{(c)} - j)\, U_n(i, j)\\
- &\lambda_{hfm}[(K_f^{(c)} - j)\, U_n(i, j) + (K_f^{(c)} - j)\, U_n(i - 1, j)]\\
+ &\mu_m[(K_m^{(c)} - i - 1)\, U_n(i + 1, j) - (K_m^{(c)} - i)\, U_n(i, j)]\\
- &\mu_m[(K_m^{(c)} - i)\, U_n(i, j) + (K_m^{(c)} - i + 1)\, U_n(i - 1, j)]\\
+ &\mu_f[(K_f^{(c)} - j)\, U_n(i + 1, j) - (K_f^{(c)} - j)\, U_n(i, j)\\
- &\mu_f[(K_f^{(c)} - j)\, U_n(i, j) + (K_f^{(c)} - j)\, U_n(i - 1, j)]\\
= &\lambda_{hmm}\, Z_1(i + 1, j) + \lambda_{hff}\, Z_2(i + 1, j) + \lambda_{hmf}\, Z_3(i + 1, j)\\
+ &\lambda_{hfm}\, Z_4(i + 1, j) + \mu_m\, Z_5(i + 1, j) + \mu_f\, Z_6(i + 1, j).
\end{aligned}
$$

First, let us consider $Z_2(i + 1, j)$ as follows:

$$
\begin{aligned}
Z_2&(i + 1, j)\\
= &(K_f^{(c)} - j)\, U_n(i + 1, j) - (K_f^{(c)} - j)\, U_n(i, j) - (K_f^{(c)} - j)\, U_n(i, j)\\
+ &(K_f^{(c)} - j)\, U_n(i - 1, j)\\
= &(K_f^{(c)} - j)\underbrace{\left\{\Delta U_n^i(i + 1, j) - \Delta U_n^i(i, j)\right\}}_{\geq 0} \geq 0.
\end{aligned}
$$

APPENDIX 153

Similarly, it can be proved that the terms $Z_4(i+1, j)$, $Z_6(i+1, j)$ is ≥ 0. Now, consider the summation of $\Xi_9(i+1, j)$ in (6.6) and $Z_1(i+1, j)$ in (6.7) as follows:

$$\Xi_9(i+1, j) + Z_1(i+1, j)$$
$$= (i-1)\{U_n(i, j) - U_n(i-1, j)\} - (i-1)\{U_n(i-1, j) - U_n(i-2, j)\}$$
$$+2 U_n(i, j) - 2 U_n(i-1, j) + (K_m^{(c)} - i - 1)\{U_n(i+1, j) - U_n(i, j)\}$$
$$-(K_m^{(c)} - i - 1)\{U_n(i, j) - U_n(i-1, j)\}$$
$$-2 U_n(i, j) + 2 U_n(i-1, j)$$
$$= (i-1)\underbrace{\{\Delta U_n^i(i, j) - \Delta U_n^i(i-1, j)\}}_{\geq 0}$$
$$+(K_m^{(c)} - i - 1)\underbrace{\{\Delta U_n^i(i+1, j) - \Delta U_n^i(i, j)\}}_{\geq 0} \geq 0.$$

Finally, let us consider the summation of $\Xi_4(i+1, j)$ in (6.6) and $Z_3(i+1, j)$ in (6.7) as follows:

$$\Xi_4(i+1, j) + Z_3(i+1, j)$$
$$= (i-1)\min(R_{hmf}, \Delta U_n^j(i, j+1)) - (i-1)\min(R_{hmf}, \Delta U_n^j(i-1, j+1))$$
$$+(i-1)\Delta U_n^i(i, j) - (i-1)\min(R_{hmf}, \Delta U_n^j(i-1, j+1))$$
$$-(i-1)\min(R_{hmf}, \Delta U_n^j(i-2, j+1)) - (i-1)\Delta U_n^i(i-1, j)$$
$$+2\min(R_{hmf}, \Delta U_n^j(i, j+1)) + 2 U_n(i, j) - 2\min(R_{hmf}, \Delta U_n^j(i-1, j+1))$$
$$-2 U_n(i-1, j) + (K_m^{(c)} - i - 1)\underbrace{\{\Delta U_n^i(i+1, j) - \Delta U_n^i(i, j)\}}_{\geq 0} - 2 U_n(i, j)$$
$$+2 U_n(i-1, j)$$
$$\geq (i-1)\left\{\min(R_{hmf}, \Delta U_n^j(i, j+1)) - \min(R_{hmf}, \Delta U_n^j(i-1, j+1))\right\}$$
$$+(i-1)\underbrace{\{\Delta U_n^i(i, j) - \Delta U_n^i(i-1, j)\}}_{\geq 0}$$
$$-(i-1)\left\{\min(R_{hmf}, \Delta U_n^j(i-1, j+1)) - \min(R_{hmf}, \Delta U_n^j(i-2, j+1))\right\}$$
$$+2\underbrace{\{\min(R_{hmf}, \Delta U_n^j(i, j+1)) - \min(R_{hmf}, \Delta U_n^j(i-1, j+1))\}}_{\geq 0,\ \text{since } U_n(i, j)\ \text{is convex in } i\ (\text{or } j)\ \text{for every fixed } j\ (\text{or } i)}$$
$$\geq (i-1)\Big[\left\{\min(R_{hmf}, \Delta U_n^j(i, j+1)) - \min(R_{hmf}, \Delta U_n^j(i-1, j+1))\right\}$$
$$- \left\{\min(R_{hmf}, \Delta U_n^j(i-1, j+1)) - \min(R_{hmf}, \Delta U_n^j(i-2, j+1))\right\}\Big]$$
$$\geq 0.$$

Since $U_n(i, j)$ is convex in i (or j) for every fixed j (or i), $\min(R_{hmf}, \Delta U_n^j(i, j+1)) - \min(R_{hmf}, \Delta U_n^j(i-1, j+1)) \geq 0$, and $\min(R_{hmf}, \Delta U_n^j(i-1, j+1)) - \min(R_{hmf}, \Delta U_n^j(i-2, j+1)) \geq 0$. In addition, since $\Delta U_n^i(i, j+1) > \Delta U_n^i(i-1, j+1)$, the summation of both the terms is greater than or equal to 0. Hence, $\Delta U_{n+1}^i(i+1, j) \geq \Delta U_{n+1}^i(i, j)$. Similarly, it can be shown that $\Delta U_{n+1}^j(i, j+1) \geq \Delta U_{n+1}^j(i, j)$. Hence, the cost function is convex.

REFERENCES

1. J. Hou and Y. Fang, "Mobility-based call admission control schemes for wireless mobile networks," *Wireless Communications and Mobile Computing*, vol. 1, no. 3, pp. 269–282, July–September 2001.
2. Y. Fang and Y. Zhang, "Call admission control schemes and performance analysis in wireless mobile networks," *IEEE Transactions on Vehicular Technology*, vol. 51, no. 2, pp. 371–382, March 2002.
3. Q. Huang, K. T. Ko, and V. B. Iversen, "Performance modeling for heterogeneous wireless networks with multiservice overflow traffic," *IEEE Globecom 2009*, pp. 1–7, November 30–December 4, 2009.
4. P. Fitzpatrick, C. S. Lee, and B. Warfield, "Teletraffic performance of mobile radio networks with hierarchical cells and overflow," *IEEE Journal on Selected Areas in Communications*, vol. 15, pp. 1549–1557, October 1997.
5. S. S. Rappaport and L. R. Hu, "Microcellular communications systems with hierarchical macro overlays: Traffic performance models and analysis," *Proceedings of the IEEE*, vol. 82, pp. 1383–1397, September 1994.
6. K. Yeo and C. Jun, "Modeling and analysis of hierarchical cellular networks with general distributions of call and cell residence times," *IEEE Communications Magazine*, vol. 51, no. 6, pp. 1361–1374, November 2002.
7. B. Jabbari and W. F. Fuhrmann, "Teletraffic modeling and analysis of flexible hierarchical cellular networks with speed-sensitive handoff strategy," *IEEE IEEE Journal on Selected Areas in Communications*, vol. 15, pp. 1539–1548, October 1997.
8. A. Farbod and B. Liang, "Efficient structured policies for admission control in heterogeneous wireless networks," *ACM/Springer Mobile Networks and Applications*, vol. 12, no. 5, pp. 309–323, December 2007.
9. L. B. Le, D. Niyato, E. Hossain, D. I. Kim, and D. T. Hoang, "QoS-aware and energy-efficient resource management in OFDMA femtocells," *IEEE Transactions on Wireless Communications*, vol. 12, no. 1, pp. 180–194, January 2013.
10. H. C. Tijms, *A First Course in Stochastic Models*. John Wiley and Sons Ltd., 2003.
11. M. Puterman, *Markov Decision Processes: Discrete Stochastic Dynamic Programming*. John Wiley & Sons, Inc. New York, NY, USA, 1994.
12. S. M. Ross, *Applied Probability Models with Optimization Applications*. Dover Publications, 1992.
13. R. Ramjee, D. Towsley, and R. Nagarajan, "On optimal call admission control in cellular networks," *Wireless Networks*, vol. 3, no. 1, pp. 29–41, 1997.
14. A. Kumar, D. Manjunath, and J. Kuri, *Communication Networking: An Analytical Approach*. Morgan Kaufmann Publishers, May 2004.
15. A. A. Fredericks, "Congestion in blocking systems: A simple approximation technique," *Bell System Technical Journal*, vol. 59, no. 6, pp. 805–827, 1980.
16. K. Yeo and C. Jun, "Modeling and analysis of hierarchical cellular networks with general distributions of call and cell residence times," *IEEE Transactions on Vehicular Technology*, vol. 51, no. 6, pp. 1361–1374, November 2002.
17. Q. Huang, K.-T. Ko, and V. B. Iversen, "Approximation of loss calculation for hierarchical networks with multiservice overflows," *IEEE Transactions on Communications*, vol. 56, no. 3, pp. 466–473, March 2008.

CHAPTER 7

GAME THEORETIC APPROACHES FOR RESOURCE MANAGEMENT IN MULTI-TIER NETWORKS

Game theory is a set of mathematical tools used to analyze the interactions among independent rational entities. Recently, game theory has found many applications in wireless communications and networking [1]. Game theory can be used in the decision making process of multiple rational entities, so that the wireless systems can operate efficiently and meet the users' QoS requirements. In a multi-tier cellular wireless network such as a two-tier macrocell–femtocell network, there are many decision problems involving multiple entities (e.g., macro base stations, femto access points, macro and femto users). These entities can be rational and self-interested to maximize their own benefits (or minimize their own costs).

In this chapter, we discuss the applications of game theory for resource management in two-tier macrocell–femtocell networks. Game theory is a suitable tool to model the interactions among different entities in these networks. We provide an introduction to game theory and discuss different types of game models. Then, we provide a review of the game theoretic models for resource allocation in two-tier networks. We first discuss the game formulations for power control and sub-channel allocation in an orthogonal frequency-division multiple access (OFDMA)-based two-tier (macrocell–femtocell) network. Then, we review the game formulations for resource management with pricing. Based on economic models, the pricing schemes can be used to reduce interference and enhance the efficiency of spectrum sharing. Then, we present the game theoretic models for access control. Users in a two-tier macrocell–femtocell network can choose to join either a femtocell or a macrocell based on their utility. Also, the femtocell and macrocell can control the data transmission of the users. Different game models can be developed for access control in a two-tier network. To this end, we outline some future research directions.

7.1 INTRODUCTION TO GAME THEORY

Game theory, developed as one of the fields in mathematics, has many applications in economics and social sciences. Recently, game theory has been applied to solve

Radio Resource Management in Multi-Tier Cellular Wireless Networks, First Edition.
Ekram Hossain, Long Bao Le, and Dusit Niyato
© 2014 John Wiley & Sons, Inc. Published 2014 by John Wiley & Sons, Inc.

various issues in wireless communication networks including small cell networks such as femtocell networks. In this section, we will provide an introduction to game theory. To give a background for understanding and using game theory in small cell networks, we will first discuss the motivations of using game theory. Then, we will briefly review different types of game models before discussing the design considerations to use game theory to model and analyze the resource management problems in small cell networks.

7.1.1   Motivations of Using Game Theory

The motivations of using game theory for the resource management in multi-tier and small cell networks can be summarized as follows:

- *Self-interested players:* In small cell networks, there are multiple entities which may not belong to or be controlled by a single provider or controller. For example, in two-tier macrocell–femtocell networks, multiple femto access points (i.e., SBSs) can be deployed in the service area of a macrocell. The femto access points can be operated by different owners which could also be different from the provider of the macro base station. Likewise, the macro users and femto users may subscribe to the different providers, and they have different subscription plans. In this situation, the entities in the femtocell network are independent and make decisions on resource management in such a way that they achieve the highest benefit (e.g., optimal performance) or the lowest cost. This is referred to as the self-interest or rational behavior. These entities can be considered as players in a game, in which their strategies are chosen based on the individual payoff rather than the system-wide objective.

- *Distributed environment:* Distributed architecture will be common for multi-tier networks such as femtocell networks. This is due to the limited communication capability of backhaul among macro base stations and femto access points. Therefore, without complete information and full control, distributed resource management algorithms for the small cell networks will be highly desirable. Game theory can be used to design and develop distributed resource management algorithms with minimum information exchange and control. A game theory framework allows different entities to collect the network state and wireless channel status to make a decision locally and intelligently. This can be referred to as the self-organizing capability. Distributed algorithms can be designed to achieve the optimal or near optimal solution. In addition, the distributed algorithms will be highly scalable to accommodate a large number of femto access points and users.

Game theory provides a rich set of mathematical and analytical tools to investigate the behaviors (e.g., convergence) and solutions (e.g., equilibrium) of game-theoretic resource management algorithms in multi-tier and small cell networks providing the opportunity to improve users' satisfaction and system's efficiency.

7.1.2   Types of Games

In general, a game model is defined by a set of players, a set of actions and strategies, and payoffs. The players are the decision makers who plan their strategies and accordingly choose actions to optimize their payoffs. The payoff, which represents the motivation of a player, is generally defined as a utility function. A brief introduction to the different types of game models typically used to model and analyze radio resource management (RRM) problems is provided in the following.

- *Noncooperative game:* In a noncooperative game, the players could have fully or partially conflicting interests in the decision making process. In particular, an action of one player (i.e., strategy) will impact the payoffs of other players. For example, in the transmission power control game [2], femto users (i.e., players) choose transmission power (i.e., strategy) to transmit data. The payoff of the femto user is defined as the utility in terms of transmission rate. The transmission rate is derived as a function of the signal-to-interference-plus-noise ratio (SINR). Therefore, the utility is a function of transmission powers of all femto users in the network. If one femto user increases its transmission power, this user can achieve a higher transmission rate due to higher SINR. However, other femto users will suffer from stronger interference from the transmitted signal of this femto user. This situation can be modeled as a noncooperative game since the femto users will choose their transmission powers to maximize their utilities, without concerning about the interference caused to other users. Given a set of players, a set of strategies, and payoff functions, the solution of the noncooperative game in terms of an equilibrium can be obtained. In the literature, there are a few common equilibrium solution concepts for the noncooperative game. The most widely used solution concept is a Nash equilibrium. The Nash equilibrium ensures that none of the players can increase his payoff by unilaterally changing his strategy. Other solution concepts, such as correlated equilibrium [3] and dominant-strategy equilibrium [4], can also be used.

- *Cooperative game:* A cooperative game can be applied to the situation where players in a group (i.e., coalition) can enforce cooperation among each other to achieve a common objective of the group. A cooperative game can be either a bargaining game or a coalitional game. While the bargaining game is based on a bargaining process among the players, the coalitional game focuses on players acting as groups. The coalitional game can be classified into two categories: canonical game and coalition formation game [5]. In the canonical coalitional game, all players aim to form and stabilize a grand coalition (i.e., coalition of all players in the game). A common approach is to divide the value of coalitions (i.e., payoff of the coalition) among the players such that none of the players has any incentive to leave the grand coalition. The most commonly used solution is a *core* which ensures that no other coalition exists such that any player can achieve a higher payoff. However, the core could be empty, and if exists, it is usually not unique. Therefore, alternative solution concepts such as Shapley value [6] can be applied. In the coalition formation game, the players are rational in forming coalitions. A common approach is to allow players to join and split from an existing coalition based on their received payoffs. The common solution of the coalition formation game is the Nash stable coalition, which is similar to the Nash equilibrium. That is, the players inside or outside the Nash stable coalition cannot improve their individual payoffs by unilaterally splitting or joining the coalition, respectively.

- *Static and dynamic games:* A game can be either static or dynamic. In a static game, all players choose their strategies and perform actions at the same time. The players do not have information or knowledge of the strategies to be used by other players. Also, the strategy and the outcome are determined regardless of time. On the other hand, in a dynamic game, there is an implication of time in choosing the strategy and determining the outcome. A dynamic game can be a repeated game, in which the players play the static game multiple times. In this case, the players may have the knowledge about the strategies used by the other players and can adjust their strategies to be used in the current and future time based on the obtained knowledge. Alternatively, a dynamic game can be a differential game. In the differential game, an optimal control method is applied in which the payoff is a function of not only the time but also the state of the system.

- *Hierarchical Game:* In a hierarchical game, also widely known as a Stackelberg game, some players (i.e., leaders) can make decisions before other players (i.e., followers). In this case, the followers can observe the strategies used by the leaders and the followers optimize their strategies accordingly (i.e., to maximize the payoffs). Based on the knowledge that the followers will act according to the strategies of the leaders, the leaders can choose the strategies that maximize their payoffs. In general, the payoffs of the leaders are not less than those in a non-hierarchical game. This is called the first move advantage. A hierarchical game can be used for resource management in femtocell networks. The typical leaders are the macrocell entities such as a macro base station or a macro user. The followers are the femtocell entities such as the femto access points or femto users. The solution is defined in terms of a Stackelberg equilibrium which is a strategy profile such that the leaders maximize their payoffs, while the followers achieve the highest possible payoffs (i.e., best response) given the strategies of the leaders. A common approach to obtain the Stackelberg equilibrium is to use backward induction. In the backward induction method, the optimization models are formulated for the leaders and the followers. The optimization model for the followers is solved first. Next, the optimal solution of the followers is used to obtain the optimal solution of the leaders.

Note that the detail of each type of game can be found in the standard game theory books [7, 8]. In addition, the details of game models formulated for solving RRM problems in wireless communications and networking can be found in [1].

7.1.3   Noncooperative Game

A noncooperative game in strategic or normal form is a triplet defined as follows:
$G = (\mathcal{N}, (S_i)_{i \in \mathcal{N}}, (u_i)_{i \in \mathcal{N}})$, where

- $\mathcal{N}$ is the finite set of players.
- S_i is the set of strategies of player $i \in \mathcal{N}$.
- u_i is the payoff function of player $i \in \mathcal{N}$, which is defined as a function of $S = S_1 \times \cdots S_i \times \cdots \times S_N$ for $N = |\mathcal{N}|$ where $|\mathcal{N}|$ is the Cardinality of the set $|\mathcal{N}|$.

In the strategic game, a strategy profile is defined as $\mathbf{s} = (\mathbf{s}_i, \mathbf{s}_{-i}) \in S$, where $\mathbf{s}_i \in S_i$ and $\mathbf{s}_{-i} \in S_{-i} = \prod_{j=1; j \neq i}^{N} S_j$ are the vectors of strategies of player i and all other players, respectively. The player in the strategic game chooses the strategy to optimize her payoff function. If the players deterministically choose strategies (i.e., the strategies are chosen with probability of one), then it is called pure strategy. On the other hand, if the players choose their strategies based on probability distributions, then it is called a mixed strategy.

A strategy is dominant for player i if

$$u_i(\mathbf{s}_i, \mathbf{s}_{-i}) \geq u_i(\mathbf{s}_i', \mathbf{s}_{-i}) \tag{7.1}$$

for all $\mathbf{s}_i' \in S_i$ and for all $\mathbf{s}_{-i} \in S_{-i}$. In other words, the dominant strategy is the best strategy of that player. In particular, the dominant strategy will yield the highest payoff for the player regardless of the strategies of other players. The dominant strategy is useful to define the solution of a strategic game, since if the dominant strategy exists for the player, the rational player has no incentive and will not deviate from the dominant strategy. As a result, if every player has the dominant strategy, then all players will choose their dominant strategies rationally. The solution concept derived from the dominant strategy is called dominant-strategy equilibrium. The equilibrium is the strategy profile, which is composed of the dominant strategies of all players.

Conversely, we can define the strictly dominated strategy. $\mathbf{s}_i'$ is the strictly dominated strategy of player i by a strategy $\mathbf{s}_i$ if

$$u_i(\mathbf{s}_i, \mathbf{s}_{-i}) > u_i(\mathbf{s}_i', \mathbf{s}_{-i}) \tag{7.2}$$

for all $\mathbf{s}_{-i} \in S_{-i}$. In other words, if the strategy is strictly dominated, there will be another better strategy (i.e., yielding higher payoff) regardless of the strategies of other players. A rational player will avoid choosing all the strictly dominated strategies. The elimination of a strictly dominated strategy is called *iterated strict dominance*, which could help in reducing the strategy set to be chosen toward the equilibrium solution. One important property of the iterated strict dominance is that the order of strategy elimination does not affect the outcome of a game. Therefore, regardless of which strictly dominated strategy is removed from the game first, the final remaining strategy set will be the same.

A similar concept is weakly dominated strategy. The strategy $\mathbf{s}_i' \in S_i$ of player i is said to be weakly dominated by strategy $\mathbf{s}_i \in S_i$ if

$$u_i(\mathbf{s}_i, \mathbf{s}_{-i}) \geq u_i(\mathbf{s}_i', \mathbf{s}_{-i}) \tag{7.3}$$

for all $\mathbf{s}_{-i} \in \mathcal{S}_{-i}$. Again, the concept of weakly dominated strategy can be used to reduce a strategy set of a player, by eliminating the strategy which will result in a lower payoff. Similarly, this concept is called *iterated weak dominance*. However, it is important to highlight that unlike iterated strict dominance, the iterated weak dominance may not lead to a unique final outcome. In other words, the order of strategy elimination will affect the final outcome.

Although the concepts of dominant and dominated strategies are useful, in some game, such strategies may not exist and hence the corresponding equilibrium cannot be obtained. The most widely used solution concept for a noncooperative game is called the Nash equilibrium. The intuition of the Nash equilibrium solution is that none of the players has an incentive to deviate from the equilibrium if other players keep their strategies unchanged. The Nash equilibrium strategy profile $(\mathbf{s}_i^*, \mathbf{s}_{-i}^*)$ is defined as follows:

$$u_i(\mathbf{s}_i^*, \mathbf{s}_{-i}^*) \geq u_i(\mathbf{s}_i, \mathbf{s}_{-i}^*) \tag{7.4}$$

for all $\mathbf{s}_i \in \mathcal{S}_i$ and for all $i \in \mathcal{N}$. At the Nash equilibrium, no player can unilaterally change her strategy, given that other players' strategies are fixed. If $u_i(\mathbf{s}_i^*, \mathbf{s}_{-i}^*) > u_i(\mathbf{s}_i, \mathbf{s}_{-i}^*)$, then the Nash equilibrium is said to be strict. The Nash equilibrium has a nice property that once the game reaches the equilibrium, if the players are rational, the strategy profile will not change, implying that the game has reached the stable point.

The typical approach to solve for the Nash equilibrium of a noncooperative game is through the use of a best response function. The best response function for player i can be defined as $b_i(\mathbf{s}_{-i})$, which is a set of strategies for that player such that

$$b_i(\mathbf{s}_{-i}) = \{\mathbf{s}_i \in \mathcal{S}_i | u_i(\mathbf{s}_i, \mathbf{s}_{-i}) \geq u_i(\mathbf{s}_i', \mathbf{s}_{-i}), \forall \mathbf{s}_i' \in \mathcal{S}_i\}. \tag{7.5}$$

In other words, the best response function is defined by $b_i(\mathbf{s}_{-i}) = \max_{\mathbf{s}_i} u_i(\mathbf{s}_i, \mathbf{s}_{-i})$. The best response ensures that the player will achieve the highest payoff given the fixed strategies of other players. Then, the Nash equilibrium can be defined based on the best response function. In this case, the strategy profile $\mathbf{s}^* \in \mathcal{S}$ is a Nash equilibrium if and only if every player's strategy is a best response to the other players' strategies, that is,

$$\mathbf{s}_i^* \in b_i(\mathbf{s}_{-i}^*) \tag{7.6}$$

for all players i. To solve for the Nash equilibrium, first the best response functions will need to be obtained, for example, by using standard optimization methods. Then, the strategy profile which satisfies the condition of a Nash equilibrium is identified.

There are some issues related to the Nash equilibrium. Firstly, with a pure strategy, the noncooperative game may possess zero, one, or many Nash equilibria. Secondly, the Nash equilibrium may not be the optimal solution. In other words, the Nash equilibrium may not yield the highest payoff for all players in the game. It is important to analyze the property of a Nash equilibrium in a game. In the following, the sketches of the widely used proofs for existence and uniqueness of the Nash equilibrium in a noncooperative game are given.

- *Existence:* The most commonly used approach for proving the existence of a Nash equilibrium is through showing that the game is a concave game [9]. A concave game is defined as a game in which player i chooses strategy $\mathbf{s}_i$ so that $\mathbf{s} \in \mathcal{S}$ where $\mathcal{S}$ is a closed bounded convex set. Also, the payoff function $u_i(\mathbf{s})$ of player i should be continuous and twice differentiable in $\mathbf{s} \in \mathcal{S}$ and concave in $\mathbf{s}_i \in \mathcal{S}_i$. The concavity of a payoff function can be proved through checking the Hessian matrix. In particular, if the Hessian matrix is negative definite, then the payoff function is concave.

- *Uniqueness:* The uniqueness of a Nash equilibrium in a noncooperative game can be proved by showing that the best response function is a standard function. Let $\mathbf{b}(\mathbf{s}) = (b_1(\mathbf{s}), \ldots, b_N(\mathbf{s}))$. The standard function possesses three properties: (i) positivity, where $\mathbf{b}(\mathbf{s}) > 0$, (ii) monotonicity, where if $\mathbf{s} \geq \mathbf{s}'$, then $\mathbf{b}(\mathbf{s}) \geq \mathbf{b}(\mathbf{s}')$, and (iii) scalability, where for all $\alpha > 1$, $\alpha\mathbf{b}(\mathbf{s}) > \mathbf{b}(\alpha\mathbf{s})$. Note that the vector inequality $\mathbf{s} > \mathbf{s}'$ is a strict inequality in all elements. Alternatively, the uniqueness of the Nash equilibrium can be verified through that the game is diagonally strictly concave on its strategy space [10]. Another approach is to prove that a noncooperative game is a potential game [11]. In the potential game, any changes in the payoff function of any player are also correspondingly reflected in a potential function due to a unilateral deviation by the player, that is, $\Delta u_i = u_i(\mathbf{s}_i, \mathbf{s}_{-i}) - u_i(\mathbf{s}_i', \mathbf{s}_{-i}) = \Delta P = P(\mathbf{s}_i, \mathbf{s}_{-i}) - P(\mathbf{s}_i', \mathbf{s}_{-i})$, where $P(\cdot)$ is a potential function.

It is also important to analyze the efficiency or optimality of a Nash equilibrium of a noncooperative game. One important measure of efficiency is the Pareto optimality. The strategy profile $\mathbf{s} \in \mathcal{S}$ is Pareto superior to another strategy profile $\mathbf{s}'$ if, for every player $i \in \mathcal{N}$,

$$u_i(\mathbf{s}_i, \mathbf{s}_{-i}) \geq u_i(\mathbf{s}_i', \mathbf{s}_{-i}') \tag{7.7}$$

with at least one player i' where

$$u_{i'}(\mathbf{s}_{i'}, \mathbf{s}_{-i'}) \geq u_{i'}(\mathbf{s}_{i'}', \mathbf{s}_{-i'}'). \tag{7.8}$$

Then, the strategy profile $\mathbf{s}^* \in \mathcal{S}$ is Pareto optimal if there exists no other strategy profile that is Pareto superior to $\mathbf{s}^*$. In other words, if a strategy profile is Pareto optimal, there is no other strategy profile that results in a higher payoff for one player without lowering the payoff of at least one other player. With the Pareto optimality, every player can gain the highest payoff without hurting other players. Therefore, it is desirable to find the Nash equilibrium which is also Pareto optimal.

An alternative efficiency measure of a Nash equilibrium is the *price of anarchy*. The price of anarchy is defined as a ratio of the maximum social welfare to the social welfare achieved at the worst-case equilibrium. The maximum social welfare or the total utility (or sum of payoffs of all the players in a game) is achieved by a centralized or a genie-aided solution. The social welfare at the worst-case Nash equilibrium is defined as follows:

$$u_{\text{NE}} = \min_{\mathbf{s} \in \mathcal{S}^{\text{NE}}} \sum_{i \in \mathcal{N}} u_i(\mathbf{s}) \tag{7.9}$$

where $\mathcal{S}^{\text{NE}}$ is the Nash equilibrium strategy space. The maximum social welfare from the centralized solution can be obtained from

$$u_{\text{MS}} = \max_{s \in \mathcal{S}} \sum_{i \in \mathcal{N}} u_i(\mathbf{s}). \tag{7.10}$$

Then, the price of anarchy is defined as follows:

$$\eta = \frac{u_{\text{MS}}}{u_{\text{NE}}}. \tag{7.11}$$

7.1.4 Cooperative Game

While a noncooperative game models the rational and self-interest behavior of the individual players, a cooperative game focuses on the players cooperating and acting as a group. In the cooperative game, the players are allowed to make agreements among themselves. The agreements will influence the strategies to be used by the players and their payoffs. The cooperative game can be divided into two categories: bargaining game and coalitional game. The bargaining game analyzes the bargaining process among players, while the coalitional game focuses on the formation of coalitions.

Bargaining game: Bargaining refers to a situation where two or more players can mutually benefit from reaching a certain agreement, mostly in terms of resource sharing. Although in the bargaining game literature, there are different approaches and solution concepts proposed, the most commonly used model is the Nash bargaining solution. The Nash bargaining solution considers the bargaining outcome at the final agreement, but not intermediate bargaining process. For a two player bargaining game (i.e., $\mathcal{N} = \{1, 2\}$), the payoff space (i.e., set of all possible payoffs that two players can achieve) can be defined as follows:

$$\mathcal{U} = \{(u_1(s_1), u_2(s_2)) | \mathbf{s} = (s_1, s_2) \in \mathcal{S}\}. \tag{7.12}$$

In addition, the disagreement point is defined as the payoffs that the players will receive if the agreement of the bargaining game cannot be reached. The disagreement point is defined as $\mathbf{d} = (d_1, d_2)$. In this case, $\mathcal{U}$ is a convex and compact set. There exists some $\mathbf{u} \in \mathcal{U}$ such that $\mathbf{u} > \mathbf{d}$. The Nash bargaining solution (u_1^*, u_2^*) is obtained from

$$(u_1^*, u_2^*) = \max_{u_1, u_2}(u_1 - d_1)(u_2 - d_2) \tag{7.13}$$

for $(u_1, u_2) \in \mathcal{U}$ and $(u_1, u_2) \geq (d_1, d_2)$. Let a bargaining problem be denoted by $(\mathcal{U}, \mathbf{d})$. Then, the bargaining solution is a function $\mathbf{f}$ that determines a unique outcome $\mathbf{f}(\mathcal{U}, \mathbf{d}) \in \mathcal{U}$ for every bargaining problem $(\mathcal{U}, \mathbf{d})$. In other words, $f_i(\mathcal{U}, \mathbf{d})$ is the payoff of player i in the bargaining outcome, where $f_i(\mathcal{U}, \mathbf{d})$ is an element of $\mathbf{f}(\mathcal{U}, \mathbf{d})$. The Nash bargaining solution satisfies the following properties.

- *Pareto optimality:* The bargaining solution $\mathbf{f}(\mathcal{U}, \mathbf{d})$ is Pareto optimal if there does not exist $\mathbf{u} = (u_1, u_2) \in \mathcal{U}$ such that $\mathbf{u} \geq \mathbf{f}(\mathcal{U}, \mathbf{d})$ and $u_i > f_i(\mathcal{U}, \mathbf{d})$ for some i.

- *Symmetry:* If the bargaining game $(\mathcal{U}, \mathbf{d})$ is such that $u_1 = u_2$ and $d_1 = d_2$, then $f_1(\mathcal{U}, \mathbf{d}) = f_2(\mathcal{U}, \mathbf{d})$.

- *Invariance to equivalent payoff representation:* If a bargaining problem $(\mathcal{U}, \mathbf{d})$ is transformed into another bargaining problem $(\mathcal{U}', \mathbf{d}')$ in such a way that $u_i' = \alpha_i u_i + \beta_i$ and $d_i' = \alpha_i d_i + \beta_i$, then $f_i(\mathcal{U}', \mathbf{d}') = \alpha_i f_i(\mathcal{U}, \mathbf{d}) + \beta_i$.
- *Independent of irrelevant alternatives:* With two bargaining problems denoted by $(\mathcal{U}, \mathbf{d})$ and $(\mathcal{U}', \mathbf{d})$ such that $\mathcal{U}' \subseteq \mathcal{U}$, if $\mathbf{f}(\mathcal{U}, \mathbf{d}) \in \mathcal{U}'$, then $\mathbf{f}(\mathcal{U}', \mathbf{d}) = \mathbf{f}(\mathcal{U}, \mathbf{d})$.

While the above description of the Nash bargaining solution is for two players, the generalized Nash bargaining solution for an N player bargaining game can be obtained as a solution of the optimization problem defined as follows:

$$\max_{(u_1,\dots,u_N)} \prod_{i=1}^{N} (u_i - d_i)^{\rho_i} \tag{7.14}$$

where ρ is the bargaining power of a player i for $\sum_{i=1}^{N} \rho_i = 1$. In other words, the bargaining power is the weight for the negotiation capability of each player. The player with higher bargaining power will have an advantage in getting a higher payoff from the bargaining solution.

Coalitional game: A coalitional game is a branch of a cooperative game to model cooperative behavior of players. A coalitional game defines a set of players $\mathcal{N}$ whose objectives are to participate in the game as groups. The coalition $\mathcal{S} \subseteq \mathcal{N}$ is a group with an agreement among the players to act as a single entity. The coalition $\mathcal{S}$ has a value $v(\mathcal{S})$, which determines the worth of the coalition. Therefore, in general, a coalition game is defined as a pair $(\mathcal{N}, v)$, where v is the value or mapping, which quantifies the payoffs that the players receive in the game. The coalitional game can have transferable utility (TU) or non-transferable utility (NTU). In the TU coalitional game, the values of coalitions can be apportioned and allocated to the players arbitrarily. On the other hand, in an NTU coalitional game, there are restrictions on the payoff distributions among players.

The coalitional game can be either a canonical coalitional game or a coalition formation game.

Canonical coalitional game: In a canonical coalitional game, players benefit from forming big coalitions. In particular, it is assumed that the payoffs when the players are in coalitions are always higher than or equal to that of when the players are in smaller coalitions. This is referred to as the superadditivity property, which is formally defined as follows for two coalitions $\mathcal{S}_1$ and $\mathcal{S}_2$:

$$v(\mathcal{S}_1 \cup \mathcal{S}_2) \geq v(\mathcal{S}_1) + v(\mathcal{S}_2), \quad \forall \mathcal{S}_1, \mathcal{S}_2 \subset \mathcal{N}, \mathcal{S}_1 \cap \mathcal{S}_2 = \emptyset. \tag{7.15}$$

Since the canonical coalitional game has the superadditivity property, the aim is to divide and allocate the value among players in a grand coalition (i.e., a coalition of all the players). The payoff allocation should be done to ensure that none of the players has incentive to deviate from the grand coalition. One of the most commonly used solution concept for payoff allocation is the core. The core of the TU canonical coalitional game is defined as follows:

$$\mathcal{C} = \left\{ \mathbf{x} : \sum_{i \in \mathcal{N}} x_i = v(\mathcal{N}) \text{ and } \sum_{i \in \mathcal{S}} x_i \geq v(\mathcal{S}), \forall \mathcal{S} \subseteq \mathcal{N} \right\}. \tag{7.16}$$

The core ensures that the players cannot gain higher payoff by splitting from the grand coalition, since any payoff allocation $\mathbf{x} = (x_1, \ldots, x_N)$ in the core will be always higher than or equal to the payoff allocation of any subcoalition $\mathcal{S} \subset \mathcal{N}$. However, the core is not guaranteed to exist. Also, if the core exists, it is always an infinite set. Therefore, there are some refined solution concepts for the payoff allocation in canonical coalitional game. One of solution concepts is the Shapley value.

The Shapley value $\phi(v)$ is defined as follows:

$$\phi_i(v) = \sum_{\mathcal{S} \subseteq \mathcal{N} \setminus \{i\}} \frac{|\mathcal{S}|!(N - |\mathcal{S}| - 1)!}{N!} (v(\mathcal{S} \cup \{i\}) - v(\mathcal{S})) \qquad (7.17)$$

where $v(\mathcal{S} \cup \{i\}) - v(\mathcal{S})$ is the marginal contribution of every player in a coalition $\mathcal{S}$. $\frac{|\mathcal{S}|!(N - |\mathcal{S}| - 1)!}{N!}$ is the weight indicating the probability that the player i will join the coalition $\mathcal{S}$ with a random order. Note that the Shapley value has no relation to the core. In particular, the Shapley value may not be in the core. Also, even though the core is empty, the Shapley value may exist.

Coalition formation game: A coalition formation game focuses on the self-interest behavior of a player in making cooperation with other players. In particular, the players will form coalitions to act as groups strategically to maximize their individual payoffs. Therefore, in the coalition formation game, a grand coalition may not be stable, in which the game may not possess the superadditivity property. The objective of the coalition formation game is to find the stable coalitions formed by players. Let the collection of coalitions be defined as $\Omega = \{\mathcal{S}_1, \ldots, \mathcal{S}_{|\Omega|}\}$, where $\mathcal{S}_i \subseteq \mathcal{N}$ is a mutual disjoint coalition and $\bigcup_{i=1}^{|\Omega|} \mathcal{S}_i = \mathcal{N}$. The collection is basically a partition of the grand coalition $\mathcal{N}$.

A distributed coalition formation game is considered, in which the players make decisions to form coalitions based on their individual payoffs. The most commonly used algorithm is the merge-and-split. In the merge-and-split algorithm, the preference relation $\rhd$ is defined to compare two collections $\Omega_1 = \{\mathcal{R}_1, \ldots, \mathcal{R}_{|\Omega_1|}\}$ and $\Omega_2 = \{\mathcal{S}_1, \ldots, \mathcal{S}_{|\Omega_2|}\}$. In this case, $\Omega_1 \rhd \Omega_2$ indicates that Ω_1 is preferred to Ω_2. Two common preference relations are the utilitarian order and individual-value order.

- In the utilitarian order, the players prefer to form coalitions in the collection Ω_1 if the total social welfare achieved from Ω_1 is strictly higher than that of Ω_2, that is, $\sum_{i=1}^{|\Omega_1|} v(\mathcal{R}_i) > \sum_{i=1}^{|\Omega_2|} v(\mathcal{S}_i)$.
- The individual-value order is concerned about an individual payoff of each player in a coalition. One example of the individual-value order is the Pareto order, which is defined as follows. Let $\mathbf{x}$ and $\mathbf{y}$ be the payoff allocations by two collections Ω_1 and Ω_2, respectively. Then, $\Omega_1 \rhd \Omega_2$ in the Pareto order if $\mathbf{x} \geq \mathbf{y}$ with at least one x_i of $\mathbf{x}$ such that $x_i > y_i$, where x_i and y_i are the elements of $\mathbf{x}$ and $\mathbf{y}$, respectively.

Based on the preference relation, an algorithm can be developed, which is composed of two actions, that is, merge and split, defined as follows:

- *Merge:* Coalitions $\{\mathcal{S}_1, \ldots, \mathcal{S}_L\}$ can be merged if the resulting coalitions (i.e., collection $\{\bigcup_{i=1}^{L} \mathcal{S}_i\}$) is preferred by the players, that is, $\{\bigcup_{i=1}^{L} \mathcal{S}_i\} \rhd \{\mathcal{S}_1, \ldots, \mathcal{S}_L\}$.

- *Split:* Coalitions $S = \bigcup_{i=1}^{L} S_i$ will be split if the resulting coalitions (i.e., collection $\{S_1, \dots, S_L\}$) is preferred by the players, that is, $\{S_1, \dots, S_L\} \vartriangleright \{\bigcup_{i=1}^{L} S_i\}$.

The merge-and-split algorithm with an individual-value order (e.g., Pareto order) is the dynamic coalition formation, which can perform in a distributed fashion. It has been shown that if there is a stable collection of coalitions, the algorithm can reach that stable coalition regardless of initial coalitions and the order of the merge or split actions. However, naturally, there could be multiple stable collections in a coalition formation game. In this case, the initial coalitions and order of actions of the algorithm could lead to different stable coalitions.

7.1.5 Game Theory and Radio Resource Management in Multi-Tier Networks

Some considerations of applying game theory to model and analyze the resource management problems in the multi-tier networks such as femtocell networks can be summarized as follows:

- *Players and strategies:* As the basic components of a game, the players and their strategies have to be defined. Depending on the resource management issues in the femtocell networks, the players can be identified as the entities which influence or relate to the input, process, or output of the resource management. The players could be physical entities (e.g., MUEs and FUEs, macro base stations and femto access points). Alternatively, the players could be logical/business entities (e.g., providers and owners). After the players of the resource management game are identified, their strategies will be determined based on the context of the resource management. For example, in the UL power control problem, the players could be the macro and femto users which interfere with each other. The strategy will be the transmission power.

- *Payoff:* The payoff of a player is typically defined as utility, which is a function of the strategies and state of players. The utility represents the preference of the players over the performance (e.g., transmission rate) and parameters (e.g., power consumption). The utility function should be defined based on the requirement of the players and objective of the resource management. Specifically, the utility function should be based on the QoS to be provided to the players. For example, in the power control problem, the utility can be defined as the transmission rate minus energy consumption. In many cases, the utility function is defined to simplify the mathematical analysis but can reasonably represent the preference of the actual players in femtocell networks. For example, the utility can be a convex function of a transmission rate, so that the standard technique in the convex optimization can be applied to obtain the equilibrium solution of the game.

- *Type of game:* The type of a game can be chosen based on the behavior of the players and strategies. The game can be noncooperative if players are concerned about their individual payoffs only. On the other hand, the game can be cooperative if players aim to achieve a group objective. The game can be static or dynamic depending on the assumption and objective of resource management. If the resource management aims to maximize the instantaneous performance, a static game, which is easier to analyze, may be suitable. However, if the players are concerned about their long-term performance, the dynamic game should be used. Sometimes, comparison among different types of games can lead to an interesting result. For example, the price of anarchy, defined as the ratio of efficiency from the noncooperative and cooperative games, can determine the performance degradation if the players have self-interest rather than a cooperative behavior. This phenomenon is known as the *tragedy of the commons*.

- *Solution:* For different types of games, different solution concepts exist. For example, in the power control problem, a Nash equilibrium, correlated equilibrium, or dominant-strategy equilibrium can be applied. The solution to be considered in the game depends largely on the objective and nature of the game. For example, the correlated equilibrium can be considered if the players are allowed to coordinate to achieve a more efficient solution than the Nash equilibrium. In addition, the properties of the selected solution of the game should be analyzed (e.g., existence and uniqueness of the correlated and Nash equilibrium). Again, it could be interesting, if different solution concepts are applied and compared in the same game (e.g., how much gain is achieved due to the use of correlated equilibrium over Nash equilibrium).

- *Distributed implementation:* The self-organizing capability is important for small cell networks. To achieve the self-organizing capability, a distributed implementation of the game-based resource management would be required. The major processes involved in a distributed algorithm are observing, analyzing and learning, reasoning, and adaptation. A set of these processes is commonly known as the cognitive cycle [12]. In the observing process, the distributed algorithm has to determine which information can be collected or exchanged in small cell networks. The analyzing and learning process is applied to the observation to obtain the knowledge and state of the system. Then, the reasoning process is used to determine the best action (e.g., throughput optimization). The adaptation process is to perform the selected action to achieve the objective of the algorithm. Most of the distributed algorithms implemented for game-based resource management are iterative. In particular, the aforementioned processes are repeated until the final equilibrium result is reached. In this case, a convergence analysis will be important. Also, due to lack of complete information of the system, the optimal solution may not be achieved by the distributed algorithm. In this case, the efficiency of an algorithm has to be evaluated. As in the algorithm design, an overhead in terms of communication, computation, and storage space should be also quantified.

7.2 GAME FORMULATIONS FOR POWER CONTROL AND SUB-CHANNEL ALLOCATION

The sub-channels and the transmission power are the radio resources which are shared among the macrocells and small cells in OFDMA-based multi-tier networks (e.g.,

LTE networks). Therefore, power control and sub-channel allocation are the most important issues to achieve the optimal network performance [13].

Game theory has been applied to model and analyze the power control and sub-channel allocation. One of the seminal works is [2], which analyzes the interference among macro and femto users. A noncooperative game is formulated to obtain the Nash equilibrium in terms of transmission power, which maximizes the individual utility. Alternatively, [3] considers the sub-channel allocation, in which the femto access points can choose to transmit on selected sub-channels. Instead of a Nash equilibrium, a correlated equilibrium is considered as the solution of this game. It is shown that the correlated equilibrium strategies are more efficient than the Nash equilibrium strategies. The spectrum resource (i.e., sub-channel) allocation is also considered in [4]. However, the allocation is performed in a hierarchical manner. Specifically, the spectrum resource is allocated to the macro users and femto access points first. Then, the femto access point allocate the received resource to femto users. The macro and femto users can request the resources based on their demands and a noncooperative game model is formulated to analyze the resource request strategies. The solution is the dominant-strategy equilibrium, which ensures that the macro and femto users can obtain the highest payoffs.

A game model is proposed in [14] in which the players are the femto access points with the strategy to allocate transmission power to femto users. Instead of considering an achievable rate to be the payoff (e.g., as in [15]), the payoff of the femto access point is the logarithmic function of the achievable rate minus the weighted transmission power (i.e., cost due to the energy consumption). A decentralized power control algorithm for femto users in a closed access femtocell network is proposed to achieve the Nash equilibrium at the steady state. The game is shown to be a supermodular game, which has the following attractive properties. Firstly, the pure strategy Nash equilibrium exists. Also, if the Nash equilibrium is unique, it is also globally stable when the decentralized power control algorithm is used. [16] extends the game model in [14] by considering also the sub-channel allocation and the best reply dynamics is adopted for the distributed implementation, which is proved to converge to the Nash equilibrium (i.e., by means of a potential game). [17] extends the existing game model to consider the average utility of the macro and femto users. Due to random location and mobility of these users, the macrocell and femtocells (i.e., players) must optimize the transmission power (i.e., strategy), such that the average aggregated utility is maximized. A Nash equilibrium is considered as the solution of this game. Similarly, with the randomness of a wireless channel, [18] models the interferer's activity as a two-state Markov chain. Then, a game model is formulated to maximize the payoff defined as the expected transmission rates of the femto access points (i.e., players). To obtain the Nash equilibrium, the game model is transformed into the variational inequality (VI) problem [19], and an iterative gradient projection algorithm (IGPA) [20] is applied. The convergence of this algorithm is then proved.

The power control problem in a multi-tier network can be modeled as a hierarchical game where the macrocell is the leader and the small cells (e.g., femtocells) are the followers [21, 22]. With this hierarchical game model (i.e., Stackelberg game), the macrocell can allocate the transmission power to the macro users before the femtocells allocate the transmission powers to femto users. Based on the transmission power chosen by the macrocell, the femtocells will adjust the transmission powers accordingly. The Stackelberg equilibrium is considered to be the solution of the hierarchical game model.

Dynamic spectrum access based on cognitive radio can be used in small cell networks. In particular, as a secondary user, an SBS can opportunistically use the same channel allocated to the macro users that can be considered as primary users. In this case, the interference from the SBS to the primary users has to be constrained. Reference 23 considers the power control problem in a cognitive radio environment. The aim is to maintain the DL CTI at the target level. Also, to guarantee the performances of macro and small cell users, their outage probabilities must be maintained below the target thresholds. A noncooperative game model is presented, in which the players are macro and femto users. The strategy is the transmission power and the payoff is the throughput, given the constraints on the interference to the primary users and the outage probability. The game is proved to be a supermodular game.

In the following, the detailed formulations of the selected game models for power control and sub-channel allocation are discussed.

7.2.1 Utility-Based Distributed SINR Adaptation

Due to the advantages of limited coordination and control, game theoretic models have been proposed to optimize the power control as an alternative to the centralized optimal power control [2]. The aim is to optimize the transmission powers of macrocell and femto users noncooperatively such that the individual utility (or payoff) is maximized in a distributed fashion (i.e., without control from a centralized controller). A noncooperative game model is formulated with the following players, strategies, and payoff functions. The *players* are macrocell and femto users, the set of which is denoted by $\mathcal{N} = \{0, 1, \ldots, N\}$ (Figure 7.1), where N is the total number of femto users. User $i = 0$ is the macro user, while users $i = 1, \ldots, N$ are femto users. The *strategy* is the transmission power denoted by p_i for user i. The *payoff* is the utility of a user denoted by $U_i(p_i, \mathbf{p}_{-i})$, where $\mathbf{p}_{-i}$ denotes the strategies (i.e., transmission powers) of all the users except user i. The utility of macrocell and femto

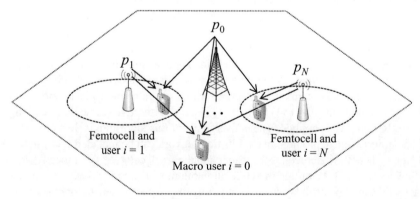

Figure 7.1 Macrocell and femtocell users in noncooperative game presented in [2].

users are defined differently. The macro user wants to achieve the minimum SINR requirement. Therefore, the utility of the macro user considered in [2] is expressed as follows:

$$U_0(p_0, \mathbf{p}_{-0}) = -(\gamma_0 - \Gamma_0)^2 \tag{7.18}$$

where γ_i is the SINR of user i defined as $\gamma_i = p_i h_{i,i}/I_i(\mathbf{p}_{-i})$, $h_{i,i}$ is the channel gain between the transmitter and receiver of user i, and Γ_i is the target SINR. The interference experienced by user i is defined as follows:

$$I_i(\mathbf{p}_{-i}) = \sum_{j \neq i} p_j h_{i,j} + \sigma^2 \tag{7.19}$$

where $h_{i,j}$ is the channel gain between users i and j, and σ^2 is the noise power. The utility of the macro user can be maximized when the macro user uses the lowest power to meet the SINR requirement. The utility in (7.18) is defined to be a concave function.

On the other hand, the utility function of a femto user is based on the reward $R(\gamma_i, \Gamma_i)$ and penalty $\frac{C(p_i, \mathbf{p}_{-i})}{I_i(\mathbf{p}_{-i})}$, that is,

$$U_i(p_i, \mathbf{p}_{-i}) = R(\gamma_i, \Gamma_i) - w_i \frac{C(p_i, \mathbf{p}_{-i})}{I_i(\mathbf{p}_{-i})} \tag{7.20}$$

where w_i is the weighting constant of penalty. The reward function is defined such that the user i associated with a femtocell wants to maximize its individual SINR. On the other hand, a penalty function is defined such that the femto user is discouraged to use excessive power to avoid cross-tier interference (CTI) to the macro user.

For mathematical tractability, several assumptions are made for the utility function in (7.20). Firstly, given the fixed transmission power p_i, the utility is a monotonically increasing concave function of SINR γ_i. Secondly, given the fixed SINR γ_i, the utility is a monotonically decreasing concave function of the transmission power p_i. The rationale behind the first assumption is that the femto user has declining satisfaction if the SINR already exceeds the requirement. For the second assumption, the femto user has an increased penalty for causing more interference. In [2], the reward function is chosen to be

$$R(\gamma_i, \Gamma_i) = 1 - \exp\left(-a_i(\gamma_i - \Gamma_i)\right) \tag{7.21}$$

where a_i is a constant, and

$$C(p_i, \mathbf{p}_{-i}) = -p_i h_{0,i} \tag{7.22}$$

is used for the penalty function.

The solution of the above noncooperative game is the Nash equilibrium for the transmission power denoted by p_i^*. The Nash equilibrium satisfies the condition defined as follows:

$$U_i(p_i^*, \mathbf{p}_{-i}^*) \geq U_i(p_i, \mathbf{p}_{-i}^*) \tag{7.23}$$

for all $p_i \neq p_i^*$ and $i \in \mathcal{N}$. Based on the continuity and strict convexity of the utility function of the transmission power, it is proved that the Nash equilibrium exists for the above game. The Nash equilibrium of the macro user is given by

$$p_0^* = \min \left\{ \frac{I_0(\mathbf{p}_{-0}^*)}{h_{0,0}} \Gamma_0, p_{\max} \right\}. \tag{7.24}$$

The Nash equilibrium of the femto user is given by

$$p_i^* = \min \left\{ \left[\frac{I_i(\mathbf{p}_{-i}^*)}{h_{i,i}} f_i^{-1} \left(-\frac{w_i}{h_{i,i}} \frac{dC(p_i, \mathbf{p}_{-i}^*)}{dp_i} \right) \right]^+, p_{\max} \right\} \tag{7.25}$$

where $[x]^+ = \max(x, 0)$ and

$$f_i(x) = \left[\frac{dR(\gamma_i, \Gamma_i)}{d\gamma_i} \right]_{\gamma_i = x}. \tag{7.26}$$

The proof of existence of the solution in (7.25) is based on the fact that the partial derivative of $U_i(\cdot)$ is monotonically decreasing with increasing p_i. Therefore, the necessary condition for the existence of the local optimal solution can be met, in which the derivative of $U_i(\cdot)$ is equal to zero within $[0, p_{\max}]$. However, if the necessary condition is not met, the transmission power will be chosen to be zero or maximum power if the derivative of the utility function is positive and negative, respectively.

More importantly, given the chosen reward and penalty functions, [2] proves that the SINR equilibrium for the users is unique. The proof is based on using the iterative power control update, that is, $\mathbf{p}^{(t+1)} = g(\mathbf{p}^{(t)})$, for iteration $t + 1$, where $g(\cdot)$ is the power update function. The function is given by

$$p_0^{(t+1)} = \min \left\{ \frac{p_0^{(t)}}{\gamma_0^{(t)}} \Gamma_0, p_{\max} \right\} \tag{7.27}$$

for a macro user, and

$$p_i^{(t+1)} = \min \left\{ \frac{p_i^{(t)}}{\gamma_0^{(t)}} \left[\Gamma_i + \frac{1}{a_i} \ln \left(\frac{a_i h_{i,i}}{w_i h_{0,i}} \right) \right]^+, p_{\max} \right\} \tag{7.28}$$

for femto users. Then, it is proved that the iterative power control update has a fixed point [24]. In addition, the iterative power control update can be used for the distributed implementation, where a femto user only needs to know its own SINR requirement and its channel gain.

However, there could be the case that the macro user may not be able to meet the SINR requirement, which is undesirable. Therefore, an algorithm is proposed for the macro user quality protection. In this case, the algorithm will attempt to achieve the solution close to the required SINR (i.e., the achievable SINR of the macro user becomes $(1 - \epsilon)\Gamma_0$) within a certain number of iterations. This can be achieved by choosing a subset of femto users to reduce their SINR. From the example presented in [2], the macro user can achieve 95% of the initial required SINR.

7.2.2   Multi-Tier Cognitive Cellular Radio Networks

Based on a noncooperative game formulation for the resource management problem in two-tier macrocell–femtocell networks, the Nash equilibrium can be obtained, which maximizes the individual payoff of the players [25]. Alternatively, a game formulation can be developed with an aim to maximizing the global utility function of the network. The benefit of using such a game formulation is that it enables a distributed implementation of channel access by femtocells, which does not require global information and full control of their transmission parameters. The optimal global utility is obtained by defining the payoff of each player (i.e., femto access point) based on different components, that is, its own benefit, fairness, and power consumption.

A different solution concept, namely, the correlated equilibrium is considered in [3]. The advantage of correlated equilibrium over Nash equilibrium is that, at the correlated equilibrium, the players are allowed to coordinate, which could result in a more efficient solution. In the system model considered in [3], the femto access points and macro base station are the secondary and primary users sharing the same pool of spectrum, respectively. The non-overlapping channel sharing scenario is considered, that is, femto access points and macro base station use different channels (i.e., called RBs). B_{FAP} and B_{MBS} RBs are for F femto access points and the macro base station, respectively, where $B_{\mathrm{FAP}} + B_{\mathrm{MBS}} = B$ is the total number of RBs available in the macrocell. Figure 7.2 shows the system model with 3 femto access points and 4 RBs.

A noncooperative game for the femto access point RB allocation is formulated as follows: The *players* are the femto access points. The *strategy* is whether to transmit on the RB f or not. The strategy can be represented by the variable $p_{i,f}$, where $p_{i,f} = 1$ if player i transmits on RB f and $p_{i,f} = 0$ otherwise. Figure 7.2

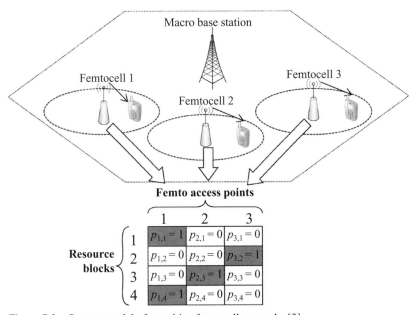

Figure 7.2   System model of cognitive femtocell networks [3].

also shows an example of strategies of players. The *payoff* is defined as the utility function. Firstly, the global utility function is considered which is defined as follows:

$$U_G(\mathbf{p}) = \min_i(\min(r_i/d_i, 1)) \tag{7.29}$$

where r_i is the achievable capacity and d_i is the demand of femto access point i. $\min(r_i/d_i, 1)$ represents the satisfaction level of femto access point i, which should not be greater than one. The achievable capacity can be expressed as follows:

$$r_i = \sum_{f=1}^{B_{FAP}} \log_2\left(1 + \frac{p_{i,f}\hat{h}_{i,f}}{I_{i,f}}\right) \tag{7.30}$$

$$I_{i,f} = \sigma^2 + \sum_{j \neq i} \hat{h}_{i,j,f} p_{j,f} \tag{7.31}$$

where $\hat{h}_{i,f}$ is the channel gain and $p_{i,f}$ is the actual transmission power of femto access point i, $I_{i,f}$ is the total interference, σ^2 is the noise power, and $\hat{h}_{i,j,f}$ is the cross channel gain between femto access points i and j on RB f. $\mathbf{p}$ is composed of the strategy variable $p_{i,f}$. In this case, maximizing the global utility function defined in (7.29) is equivalent to maximizing the performance of the worst femto access point. However, maximizing the global utility function can be done only when the global information about the network is available and the strategies of all the femto access points can be controlled. However, a distributed environment may lack of such capability. To solve this problem, the local utility function of the femto access points (i.e., players of a noncooperative game) is defined to mimic the global utility function, such that maximizing the local utility function will tend to maximize the global utility function. The local utility function of a femto access point i is defined as follows:

$$U_i(\mathbf{p}) = \min(r_i/d_i, 1) - w_{\text{pen}}\frac{1}{d_i}[r_i - d_i]^+ - w_{\text{pow}}\sum_{f=1}^{B_{FAP}} p_{i,f}. \tag{7.32}$$

This local utility function is composed of three components. $\min(r_i/d_i, 1)$ is the self-interest component which is basically the satisfaction level of each femto access point, and $-w_{\text{pen}}\frac{1}{d_i}[r_i - d_i]^+$ is the fairness component which is a negative penalty for the transmission rate exceeding the demand d_i. Here, $-\sum_{f=1}^{B_{FAP}} p_{i,f}$ is the cost component due to the transmission power, w_{pen} and w_{pow} are the weighting factors for the penalty and power cost components, respectively.

The solution considered in [3] is a correlated equilibrium, which is defined based on the policy π (i.e., probability distribution of choosing the strategies). The condition for the correlated equilibrium can be defined as follows:

$$\sum_{\mathbf{p}_{-i}} \pi(\mathbf{p}_i, \mathbf{p}_{-i})U_i(\mathbf{p}_i, \mathbf{p}_{-i}) \geq \sum_{\mathbf{p}_{-i}} \pi(\hat{\mathbf{p}}_i, \mathbf{p}_{-i})U_i(\hat{\mathbf{p}}_i, \mathbf{p}_{-i}) \tag{7.33}$$

where policy $\pi(\mathbf{p}_i, \mathbf{p}_{-i})$ is the correlated equilibrium, for any alternative policy $\pi(\hat{\mathbf{p}}_i, \mathbf{p}_{-i})$. At the correlated equilibrium, no player can deviate to gain a higher expected payoff. In general, the Nash equilibrium can be represented as the correlated equilibrium, when the players make their decisions independently.

To obtain the correlated equilibrium, a decentralized algorithm based on the regret matching procedure [26] is proposed. This algorithm allows the femto access points to adapt their RB allocation strategies dynamically. The convergence of the regret matching procedure algorithm is proved. The proof is based on the fact that the Blackwell's sufficient condition for approachability can be achieved. Specifically, a Markov chain is defined to model the regret update mechanism. The stationary probability of this Markov chain is used to verify the Blackwell's sufficient condition.

In the performance evaluation, a small number of femto access points (i.e., six) is considered to avoid huge computational complexity. However, for a larger network scenario with more number of femto access points, [3] suggests an approach to combine multiple femto access points together into a composite player so that the strategy space can be reduced. A comparison between the proposed regret matching procedure algorithm and the best response dynamics algorithm (i.e., a player chooses the best strategy to maximize its payoff based on the observation in the last iteration) is provided. The result shows that the regret matching algorithm performs better than the best response dynamics algorithm. The reason is that the best response dynamics algorithm does not take the historical outcome of the game into account to update the strategy selection. However, since the equilibrium is not unique, both the algorithms may not converge to a single solution.

7.2.3   On-Demand Resource Sharing in Multi-Tier Networks

An on-demand resource sharing mechanism can be designed taking the users' self-ishness behavior and private traffic characteristic into account [4]. Specifically, the mechanism provides an incentive for the users to reveal the true information (i.e., traffic characteristics) so that this information can be used for efficient and fair resource allocation in two-tier macrocell–femtocell networks. Three main properties of the on-demand resource sharing mechanism proposed in [4] are as follows:

1. The mechanism ensures that all players (i.e., macro and femto users) will reveal their true traffic demand. This is referred to as the *strategy proof* mechanism. Also, the mechanism encourages the players to use the *dominant-strategy equilibrium*, which has the *incentive compatibility* (i.e., gaining the highest payoff).

2. The mechanism is proved to achieve the weighted max–min and proportional fair resource allocation. Also, the dominant-strategy equilibrium is shown to be Pareto efficient.

The system model considered for the on-demand resource sharing mechanism is shown in Figure 7.3. There is one macro base station serving multiple macro users whose set is denoted by $\mathcal{M}$. There is one femto access point, which is denoted by f. This femto access point serves femto users whose set is denoted by $\mathcal{F}$. The total number of resources (e.g., sub-channels in OFDMA) available for both macro and femto users is denoted by R. The macro and femto users have the following attributes.

- *Traffic demand* is denoted by d_i for $i \in \mathcal{M}$ or $i \in \mathcal{F}$.
- *Transmission rates* are denoted by r_i^{m} and r_i^{f} for macro and femto users, respectively.

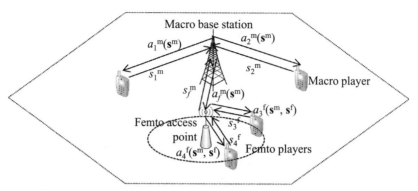

Figure 7.3   System model for on-demand resource sharing in femtocell networks.

- *Resource demands* are denoted by θ_i^m and θ_i^f for macro and femto users, respectively. The resource demand can be calculated from $\theta_i^m = d_i/r_i^m$ and $\theta_i^f = d_i/r_i^f$.
- *Weights* are denoted by w_i^m and w_i^f for macro and femto users, respectively. The weight is assigned by the system (e.g., based on the priority of users) for the resource allocation.

The traffic and resource demands are assumed to be private information of the macro and femto users.

The on-demand resource sharing mechanism works as follows (Figure 7.4). Firstly, the macro users and femto access point send the resource requests denoted by s_i^m where $i \in \mathcal{M} \cup \{f\}$ to the macro base station. The macro base station assigns the resource denoted by $a_i^m(\mathbf{s}^m)$ back to the macro users and the femto access point, where $\mathbf{s}^m$ is composed of s_i^m. Then, the femto users send the resource requests denoted by s_i^f (where $i \in \mathcal{F}$) to the femto access point. Given the resource allocated from the macro base station, the femto access point assigns the resource to femto user i denoted by

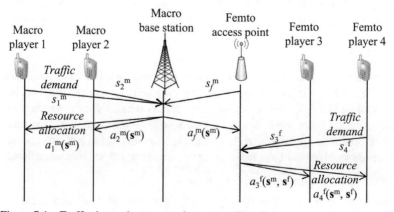

Figure 7.4   Traffic demand request and resource allocation.

$a_i^f(\mathbf{s}^m, \mathbf{s}^f)$, where $\mathbf{s}^f$ is composed of s_i^f. Some assumptions are made for the resource requests. Firstly, the transmission rate of the femto access point is normalized to be one, and the weight of the femto access point is the sum of weights of all the femto users, that is, $w_f^m = \sum_{i \in \mathcal{F}} w_i^f$.

A hierarchical game model is formulated to analyze the behavior of the macro and femto users. This hierarchical game is composed of two sub-games, that is, the macrocell and femtocell games.

- *Macrocell game:* The *macro players* are the macro users and the femto access point, whose set is $\mathcal{M} \cup \{f\}$. Their strategies are the resource requests s_i^m. The *payoff* is the utility defined as $U_i^m(\mathbf{s}^m) = \min(d_i, r_i^m a_i^m(\mathbf{s}^m))$, where $a_i^m(\mathbf{s}^m)$ is the allocated resource to macro player $i \in \mathcal{M} \cup \{f\}$.

- *Femtocell game:* The *femto players* are the femto users, whose set is $\mathcal{F}$. Their strategies are the resource requests s_i^f. The *payoff* is the utility defined as $U_i^f(\mathbf{s}^m, \mathbf{s}^f) = \min(d_i, r_i^f a_i^f(\mathbf{s}^m, \mathbf{s}^f))$, where $a_i^f(\mathbf{s}^m, \mathbf{s}^f)$ is the allocated resource to femto player $i \in \mathcal{F}$. Observe that the allocated resource to the femto users depends on the strategies of the macro players (i.e., $\mathbf{s}^m$).

To obtain the allocated resource, which is an important factor for the payoff of players, the weighted water-filling resource allocation method is proposed [4]. The allocation is divided into two parts, that is, for macrocell and femtocell. For the macrocell, the allocation works as follows:

1. *Sorting:* The macro base station sorts the macro players based on their weighted resource request (i.e., s_i^m / w_i^m for $i \in \mathcal{M} \cup \{f\}$). In this case, $i = 1$ and $i = |\mathcal{M}| + 1$ indicate the macro players with the smallest and the largest weighted resource requests, respectively.

2. *Weighted water filling:* The macro base station allocates the resource R to all macro players. The allocation is performed iteratively for R iterations. In the tth iteration, the macro base station allocates the resource $w_{i'}^m b_t^m$ which is proportional to the weight $w_{i'}^m$ to every macro player i' for $i' = t, t + 1, \dots, |\mathcal{M}| + 1$. The allocation is performed until the macro player has received all the requested resource or until all the resources have been allocated.

Based on this allocation, the allocated resource to macro player i is

$$a_i^m(\mathbf{s}^m) = \begin{cases} s_i^m, & i \leq \hat{i} \\ \dfrac{w_i^m \left(R - \sum_{i'=1}^{\hat{i}} s_{i'}^m \right)}{\sum_{i'=\hat{i}+1}^{|\mathcal{M}|+1} w_{i'}^m}, & \text{otherwise} \end{cases} \tag{7.34}$$

where $\hat{i}$ is the macro player with the largest index that its requested resource is fully allocated. For the femtocell, the allocation works similarly.

Instead of the Nash equilibrium, [4] considers the dominant-strategy equilibrium as the solution of the game. In general, the dominant-strategy equilibrium is stronger but harder to achieve. Nevertheless, it is proved that the equilibrium of the on-demand resource sharing mechanism always exists with the strategy proofness property. Firstly, the dominant strategy s_i^* is defined as the strategy that maximizes the payoff of player i given all possible strategies of other players, that is,

$$U_i(s_i^*, \mathbf{s}_{-i}) \geq U_i(s_i, \mathbf{s}_{-i}), \quad \forall s_i \neq s_i^* \text{ and } \forall \mathbf{s}_{-i}. \tag{7.35}$$

The dominant-strategy equilibrium is then the strategy profile $\mathbf{s}^*$ whose each strategy s_i^* is a dominant strategy for every player i. An attractive property of the dominant-strategy equilibrium, if exists in the game, is that the player can choose the dominant strategy to achieve the dominant-strategy equilibrium without knowing the information of other players. Therefore, it enables a fully distributed implementation and reduces information exchange.

The equilibrium analysis is performed for the on-demand resource sharing mechanism in [4]. The major results can be summarized as follows:

- The proposed game has the truth-revealing dominant strategy equilibrium. This equilibrium is also strategy proof (i.e., $s_i^* = \theta_i$, where θ_i is the true resource demand). This property can be proved by showing that the best strategy of each player is to submit the true resource demand to maximize the payoff. There is no chance that the player can gain a higher payoff by submitting a resource demand lower or higher than the true resource demand.

- The resource allocation at the truth-revealing dominant strategy equilibrium of the game achieves the weighted max–min fairness. Let $\mathbf{a}^*$ denote the resource allocation at such an equilibrium, which is composed of a_i for $i \in \mathcal{M} \cup \{f\}$ or $i \in \mathcal{F}$ for macro or femto players, respectively. The condition of the weighted max–min fairness is $\exists \mathbf{a} \neq \mathbf{a}^*$, $\min(\theta_i, a_i) > \min(\theta_i, a_i^*)$ for some i, and then

$$\frac{\min(\theta_{i'}, a_{i'})}{w_{i'}} < \frac{\min(\theta_{i'}, a_{i'}^*)}{w_{i'}} \leq \frac{\min(\theta_i, a_i^*)}{w_i}. \tag{7.36}$$

- The resource allocation at the truth-revealing dominant strategy equilibrium is weighted proportional fair. Again, let $\mathbf{a}^*$ denote the resource allocation at such an equilibrium. It is the solution of the following optimization problem,

$$\mathbf{a}^* = \max_{\mathbf{a}} \prod_i \min(\theta_i, a_i)^{w_i}, \quad \text{subject to} \sum_i a_i \leq K \tag{7.37}$$

where K is the available resource, which is $K = R$ or $K = a_f^m$ for the macrocell or femtocell, respectively.

- The resource allocation at the truth-revealing dominant strategy equilibrium is Pareto efficient. This Pareto efficiency ensures that none of the players (i.e., either macro or femto players) can increase her payoff without decreasing the payoff of other players, that is, $\exists \mathbf{a} \neq \mathbf{a}^*$, $U_{i'}(a_{i'}) < U_{i'}(a_{i'}^*)$ for some i', then $\exists i$, $U_i(a_i) < U_i(a_i^*)$.

In addition to analyzing the properties of the dominant strategy equilibrium in the game for the on-demand resource sharing mechanism, [4] also discusses about accepting macro users as the open access mode in the femtocell. Specifically, the femtocell grouping scheme is proposed to allow the macro users near to the femto access point (i.e., having the transmission rate to the femto access point higher than that to the macro base station) to join the femtocell. With the grouping scheme, the femto users are divided into the closed access users (i.e., users subscribed to femtocell) and general users. A weighted water-filling resource allocation can be applied to the general femto users and closed access group. Then, the resource is allocated to the femto users in the close access group. In this case, the resource can be reallocated from one femto user group to another group efficiently.

Then, a rate-related weight configuration is proposed. The weight configuration allows the weight of macro or femto users to be adjusted to achieve a better performance. Specifically, less resource is required for the high transmission rate users to meet their traffic demand. In this case, the weight is defined as a function of the transmission rate, that is, $w_i(r_i)$. The key design issue for the weight function is that the weight should be smaller when the transmission rate is higher (i.e., to reduce requested resource) and the weight should be chosen such that the user will receive the higher payoff. In the numerical results, the weight function is chosen to be $w_i(r_i) = (1/r_i)^{1/2}$.

Although the on-demand resource sharing mechanism proposed in [4] is for a single femtocell, its extension for the multiple femto access points is also given. The same weighted water-filling resource allocation can be applied. However, the challenge is at the proofs of different properties (e.g., truth-revealing dominant strategy equilibrium), which require further studies. In addition, the hierarchical game model is not shown to achieve the Stackelberg equilibrium. The Stackelberg equilibrium will be more suitable solution if the macro users have higher priority to obtain the resource.

7.3 GAME FORMULATIONS FOR PRICING

A pricing scheme and incentive mechanism can be adopted in the femtocell networks to control interference [27]. This economic approach can be modeled and analyzed using game theory, especially when there are multiple rational and self-interested entities in the networks. In [27], the interference from femtocells to the macro users is controlled through a pricing mechanism. Specifically, the femto users are charged with a certain price according to the interference that they cause to the macro users. A hierarchical game model is proposed, where the leader player is the macro base station optimizing the price charged to femto users. The femto users as the follower players optimize the transmission power according to the price.

Reference [28] also considers the cognitive radio capability, where the MBS can obtain the spectrum resource from the primary users. The first tier players (i.e., primary users) optimize the price to maximize their revenue. The second tier player (i.e., macro base station) determines the spectrum resource demand and allocation to the macro users and femto access points. The third tier players (i.e., femto access points) optimize the transmission power. A similar game structure is considered in [29], where the game model is divided into three stages. In the first stage, a femto service provider determines the optimal ratio of the resource for an open access mode (i.e., for macro users). In the second stage, a macro service provider determines the price for the spectrum resource leasing to a femto access point. Then, in the third stage, the femto access point optimizes the spectrum demand. A three-stage Stackelberg equilibrium is derived as the solution of this game.

Reference [30] considers a pricing issue in the femtocell networks under the randomness of the interference. In particular, the activity of macro users is modeled as a two-state homogeneous discrete-time Markov chain. The femto access points measure the interference and adjust the transmission power given the price charged per unit of transmission power. The Nash equilibrium is considered to be the solution. Reference [31] considers the pricing issue in the open or close access modes of the femtocell networks. Specifically, the provider can charge a certain price to the users and the users determine their traffic demand and choose to join the macrocell or femtocell (i.e., by connecting to the macro base station or a femto access point, respectively). A two-stage game model is proposed. In the first stage, the provider selects the optimal prices to maximize its revenue. In the second stage, the user determines the traffic demand and chooses to join macrocell or femtocell.

In the following, the detailed formulations of the selected game models for pricing and incentive mechanism in the femtocell networks are discussed.

7.3.1 Price-Based Spectrum Sharing

The utility-based distributed SINR adaptation proposed in [2] considers only the power allocation. Also, the macro and femto users have the same priority, which may not be applicable in some cases. References [32, 27] extend the work in [2] by considering the case that the macro user should be protected from the interference caused by the femto users. Also, a macrocell (i.e., macro base station) can obtain revenue by charging the price from the femto users given the interference caused by the femto users. This is referred to as the price-based spectrum sharing scheme, which is modeled as a Stackelberg game in [27].

Figure 7.5 shows the system model of the femtocell network considered for price-based spectrum sharing [27]. There is one macro base station and there are N femto users associated with the femto access points. The macrocell and femtocells use the same sub-channel. The macro base station can tolerate the interference from

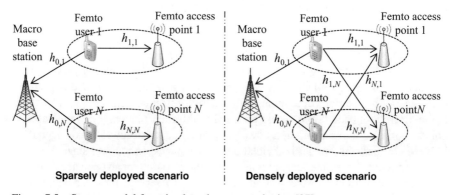

Sparsely deployed scenario **Densely deployed scenario**

Figure 7.5   System model for price-based spectrum sharing [27].

femto users up to $I_{\max}$, that is, $\sum_{i=1}^{N} I_i \le I_{\max}$, where I_i is the interference power from femto user i. The macro base station sets and charges the prices per unit of received interference power from each femto user. Then, based on this price, the femto user adjusts their transmission power to maximize their individual payoffs. The price charged to the different femto users can be different or the same, which are referred to as the non-uniform and uniform pricing schemes, respectively. The price charged to femto user i is denoted by μ_i and μ for the former and latter pricing schemes, respectively.

Two femtocell network scenarios are considered, that is, sparsely and densely deployed scenarios (Figure 7.5). In the sparsely deployed scenario, there is no mutual interference among femtocells, since each femtocell could be deployed far away from each other. On the other hand, in the densely deployed scenario, mutual interference among femtocells exists. In this case, the aggregate interference at the femto access point $i \in \{1, \ldots, N\}$ due to the transmission of femto users associated with other femto access points must be maintained below a specified threshold.

A Stackelberg game model is formulated in [27] and its structure is shown in Figure 7.6. In the Stackelberg game, there are the leader and follower players. The leader can make a decision before the follower. Since the follower can observe the decision of the leader, the leader can optimize its payoff given the belief that the follower will apply the best response strategy (i.e., strategy of the follower that maximizes its own payoff). In the price-based spectrum sharing, the *leader player* is the macro base station, whose *strategy* is the price charged to the femto users. The *follower players* are the femto users (i.e., N femto users), whose *strategy* is the transmission power denoted by p_i. The *payoff* of the leader is defined as the revenue, that is,

$$U_{\text{MBS}}(\mu, \mathbf{p}) = \sum_{i=1}^{N} \mu_i I_i(p_i) \tag{7.38}$$

where μ_i is the price charged to femto user i, and $I_i(p_i)$ is the interference quota that the femto user i is permitted to cause to the macro base station, and hence to be

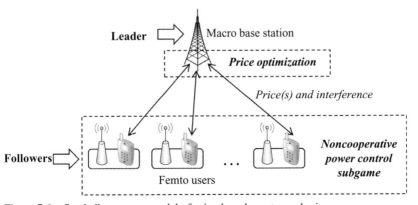

Figure 7.6   Stackelberg game model of price-based spectrum sharing.

charged. This interference quota is given by $I_i(p_i) = h_{0,i} p_i$, where $h_{0,i}$ is the channel gain between femto user i and the macro base station. μ contains all prices charged to all the femto users and $\mathbf{p}$ contains transmission powers of all the femto users. In (7.38), transmission power p_i could depend on the price μ_i (to be discussed later). The *payoff* of the follower is defined as the utility of the transmission rate and price paid to the macro base station, that is,

$$U_i(p_i, \mathbf{p}_{-i}, \mu_i) = w_i \log_2 \left(1 + \frac{p_i h_{i,i}}{\sum_{j \neq i} p_j h_{i,j} + \sigma^2} \right) - \mu_i I_i(p_i). \quad (7.39)$$

Here $h_{i,i}$ and $h_{i,j}$ are the channel gains from femto user i to its corresponding femto access point and from femto user i to the femto access point of femto user j, respectively. w_i is the weighting factor. $\mathbf{p}_{-i}$ contains the transmission powers of all the femto users except femto user i. Notice that the payoff of the femto users in (7.39) does not account for the interference from macro users.

Since the macro base station and femto users are rational players, optimization models can be formulated for them to obtain the optimal strategies given the strategies of other players. The optimization model for the macro base station in the Stackelberg game is expressed as follows:

$$\max_{\mu} \ U_{\text{MBS}}(\mu, \mathbf{p}) \quad (7.40)$$

$$\text{subject to} \ \sum_{i=1}^{N} I_i(p_i) \leq I_{\max}, \quad \mu_i \geq 0.$$

That is, the macro base station optimizes the price charged to the femto users given that the interference must not exceed the threshold $I_{\max}$. Similarly, the optimization model for the femto user in the Stackelberg game is expressed as follows:

$$\max_{p_i \geq 0} U_i(p_i, \mathbf{p}_{-i}, \mu_i). \quad (7.41)$$

The solution of the Stackelberg game is the Stackelberg equilibrium, which ensures that the leader gains the highest payoff, while the followers cannot deviate to use other strategies which are not the Stackelberg equilibrium. Let μ^*, p_i^*, and $\mathbf{p}_{-i}^*$ are the Stackelberg equilibrium for the strategies μ, p_i, and $\mathbf{p}_{-i}$ of the leader and followers, respectively. They satisfy the following conditions:

$$U_{\text{MBS}}(\mu^*, \mathbf{p}^*) \geq U_{\text{MBS}}(\mu, \mathbf{p}^*) \quad (7.42)$$
$$U_i(p_i^*, \mathbf{p}_{-i}^*, \mu_i^*) \geq U_i(p_i, \mathbf{p}_{-i}^*, \mu_i^*) \quad \text{for all } i. \quad (7.43)$$

The Stackelberg equilibrium can be obtained by finding the subgame perfect Nash equilibrium. In this case, there are multiple followers (i.e., femto users). They will play the noncooperative power control subgame (Figure 7.6) given the price set by the leader (i.e., macro base station). In other words, if the price is fixed, p_i^* and $\mathbf{p}_{-i}^*$ are the Nash equilibrium strategies. From the macro base station's perspective, p_i^* and $\mathbf{p}_{-i}^*$ are the best response strategies that the followers will apply given price μ_i. Therefore, as a standard method (i.e., backward induction) to solve for the Stackelberg equilibrium, the optimization model of the follower defined in (7.41) will be solved first. Then, the optimization model of the leader defined in (7.40) will be solved.

In the following, non-uniform and uniform pricing will be presented for a sparsely deployed scenario. Since the mutual interference is avoided in the sparsely deployed scenario, solving for the Stackelberg equilibrium is simpler than that in the densely deployed scenario. Also, a closed-form solution can be obtained.

In the non-uniform pricing scheme, the price could be different for the different femto users, which is denoted by μ_i. With the sparsely deployed scenario, the optimization model of the follower becomes

$$\max_{p_i \geq 0} w_i \log_2 \left(1 + \frac{p_i h_{i,i}}{\sigma^2}\right) - \mu_i h_{0,i} p_i. \tag{7.44}$$

This optimization model is a convex problem, which can be solved using the KKT conditions. The solution (i.e., best response) of the follower is expressed as follows:

$$p_i^* = \left[\frac{w_i}{\mu_i h_{0,i}} - \frac{\sigma^2}{h_{i,i}}\right]^+, \quad i = 1, \ldots, N. \tag{7.45}$$

It can be observed that if the price charged by the macro base station is too high (i.e., $\mu_i > \frac{w_i h_{i,i}}{h_{0,i}\sigma^2}$), the femto user will not transmit.

The optimization model of the leader becomes

$$\max_{\mu} \sum_{i=1}^{N} \left[w_i - \frac{\mu_i h_{0,i}\sigma^2}{h_{i,i}}\right]^+ \tag{7.46}$$

$$\text{subject to } \sum_{i=1}^{N} \left[\frac{w_i}{\mu_i} - \frac{h_{0,i}\sigma^2}{h_{i,i}}\right]^+ \leq I_{\max}, \quad \mu_i \geq 0.$$

However, the optimization model of the leader in (7.46) is non-convex, e.g., due to the constraints. To obtain the globally optimal solution, the optimization model can be transformed into multiple convex sub-problems by using an indicator function defined as follows:

$$\chi_i = \begin{cases} 1, & \mu_i < \frac{w_i h_{i,i}}{h_{0,i}\sigma^2} \\ 0, & \text{otherwise.} \end{cases} \tag{7.47}$$

The indicator function implies whether the femto user should be allowed to transmit or not. Then, the optimization model of the leader becomes

$$\max_{\chi,\mu} \sum_{i=1}^{N} \chi_i \left[w_i - \frac{\mu_i h_{0,i}\sigma^2}{h_{i,i}}\right] \tag{7.48}$$

$$\text{subject to } \sum_{i=1}^{N} \chi_i \left[\frac{w_i}{\mu_i} - \frac{h_{0,i}\sigma^2}{h_{i,i}}\right] \leq I_{\max}$$

$$\mu_i \geq 0, \quad \chi_i \in \{0, 1\} \tag{7.49}$$

where χ contains all χ_i for $i = 1, \ldots, N$. Given χ_i, the above problem is convex, which can be solved to obtain the globally optimal solution (i.e., Stackelberg equilibrium). Now, it is possible to solve for χ_i. This can be done by identifying the interference threshold I_{max} to which χ_i should be one or zero. As shown in [27], the Stackelberg equilibrium for the price can be obtained in the closed form [27, Eq. (21)]. For example, if the transmission of all femto users can meet the interference threshold, the optimal price can be expressed as follows:

$$\mu_i^* = \sqrt{\frac{w_i h_{i,i}}{h_{0,i}\sigma^2}} \frac{\sum_{i=1}^{N} \sqrt{\frac{w_i h_{0,i}\sigma^2}{h_{i,i}}}}{I_{max} + \sum_{i=1}^{N} \frac{h_{0,i}\sigma^2}{h_{i,i}}}, \qquad \text{for } i = 1, \ldots, N. \qquad (7.50)$$

Also, since the Stackelberg equilibrium can be obtained through the indicator function, it is possible to design an admission control scheme. That is, femto user i is admitted and allowed to transmit data if $\mu_i < \frac{w_i h_{i,i}}{h_{0,i}\sigma^2}$. In other words, the macro base station sets the price μ_i such that the femto users want to be admitted or not.

A centralized algorithm to obtain the Stackelberg equilibrium is also proposed. However, it is found that, in this algorithm, the macro base station needs to compute $\frac{w_i h_{i,i}}{h_{0,i}\sigma^2}$ for each femto user. This will incur a substantial amount of overhead. To avoid this overhead, a uniform pricing scheme is proposed in which the macro base station needs to measure the total received interference power $\sum_{i=1}^{N} I_i(p_i)$ only.

In the uniform pricing scheme, the price charged to all the femto users is the same, that is, $\mu = \mu_1 = \cdots = \mu_N$. Given the uniform price μ, the best response strategy of the follower can be expressed similar to that in (7.45). Also, the optimization model of the leader becomes

$$\max_{\mu \geq 0} \sum_{i=1}^{N} \left[w_i - \frac{\mu h_{0,i}\sigma^2}{h_{i,i}} \right]^+ \qquad (7.51)$$

$$\text{subject to} \sum_{i=1}^{N} \left[\frac{w_i}{\mu} - \frac{h_{0,i}\sigma^2}{h_{i,i}} \right]^+ \leq I_{max}. \qquad (7.52)$$

Again, this optimization model of the leader can be solved directly to obtain a unique solution in a closed form. A distributed algorithm is proposed in which only the total received interference power is required to compute the Stackelberg equilibrium for the price of the macro base station. This property is desirable from a practical point of view, since the complexity is much lower than that of the non-uniform pricing scheme. However, only the non-uniform pricing scheme, which takes the individual femto user's parameters into account, can maximize the payoff of the macro base station (i.e, price for each femto user can be customized based on channel quality and weighting factor). On the other hand, a uniform pricing scheme can maximize the sum-rate of the femto users, but the payoff of the leader may not be maximized.

Then, a densely deployed scenario is considered and a similar analysis to obtain the Stackelberg equilibrium is applied. Although there is the bound on the aggregate interference at the femto access point of femto user i, the best response strategy of the femto user i can be obtained in the close form as follows:

$$p_i^* = \left[\frac{w_i}{\mu_i h_{0,i}} - \frac{\sum_{j \neq i} p_j^* h_{i,j} + \sigma^2}{h_{i,i}} \right]^+, \qquad i = 1, \ldots, N. \qquad (7.53)$$

The Stackelberg equilibrium for the case of uniform pricing scheme is obtained in a similar way. Note that the Stackelberg equilibrium in terms of the price of macro base station cannot be obtained in a closed form anymore for the densely deployed scenario.

7.3.2 Energy-Efficient Spectrum Sharing and Power Allocation

While throughput is mostly considered to be a main performance metric for the femto-cell networks, EE is another important performance metric. Reference 28 introduces the energy-efficient spectrum sharing and power allocation scheme for cognitive femtocell networks. A resource allocation scheme similar to the on-demand resource sharing mechanism in [4] is considered. However, the available resource is sold by the primary networks in [28] instead of being always available as in [4].

The system model considered for the energy-efficient spectrum sharing and power allocation is shown in Figure 7.7. There are K primary users (i.e., primary BSs) owning the non-overlapping spectrum resources. The primary BS k sells part of its spectrum resource whose size is denoted by b_k to a cognitive macro base station. This cognitive macro base station pays the price μ_k to the primary BS k. The cognitive macro base station serves M macro users and F femto access points. It is assumed that the spectrum resource is allocated to either macro user j or one of the femto access points i, that is, $\sum_{i=1}^{F} x_{k,i} + \sum_{j=1}^{M} = 1$ for all k. The spectrum

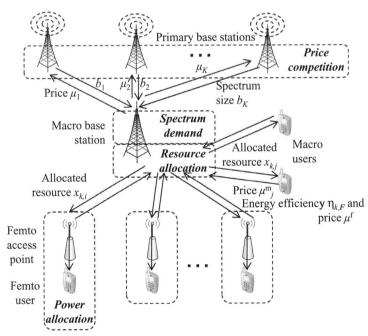

Figure 7.7   Stackelberg game model of energy-efficient spectrum sharing and power allocation.

resource allocation is denoted by $x_{k,j}$ and $x_{k,i}$ for macro user j and femto access point i, respectively. $x_{k,j} = 1$ or $x_{k,i} = 1$ if the spectrum resource from primary BS k is allocated to macro user j or FBS i, respectively. Otherwise, $x_{k,j} = 0$ or $x_{k,i} = 0$.

A hierarchical game theoretic model is developed to analyze the energy-efficient spectrum sharing and power allocation. This game model is composed of three tiers, which can be considered as an extension of the model proposed in [27], considering the pricing issue. The *first-tier players* are the primary networks (i.e., primary BSs). The *second-tier player* is the cognitive macro base station for a macrocell. The *third-tier players* are the femto access points for femtocells. The *strategy* of the first-tier player is the price μ_k charged to the cognitive macro base station. This price is optimized through the "price competition" component as shown in Figure 7.7. The *strategy* of the second-tier player is composed of the size of spectrum b_k to be bought from primary BS k and spectrum allocation $x_{k,j}$ and $x_{k,i}$. The *strategy* of the third-tier player is the transmission power denoted by p_i for femto access point i. The transmission power is optimized through the "resource allocation" component as shown in Figure 7.7. The size of spectrum to be bought and spectrum allocation of the second-tier player is optimized through the "spectrum demand" and "resource allocation," respectively.

The *payoff* of the femto access point i is the utility defined as follows:

$$U_i^f = \sum_{k=1}^{K} x_{k,i}(\lambda_i - \mu^f)b_k\eta_{k,i} \tag{7.54}$$

where λ_i is the revenue for femto access point i (e.g., obtained from a femto user) and μ^f is the price charged by the cognitive macro base station for the allocated resource, for $\lambda_i > \mu^f$. $\eta_{k,i}$ is the EE of femto access point i using spectrum resource from primary BS k. The EE is defined as follows:

$$\eta_{k,i} = \frac{\log_2\left(1 + \frac{h_{k,i}p_i}{\sigma^2}\right)}{p_{\text{cir}} + p_i} \tag{7.55}$$

where $h_{k,i}$ is the channel gain, σ^2 is the noise power, and p_{cir} is the energy consumption of the transceiver circuit. Note that the EE denoted by $\eta_{k,j}$ of macro user j can be defined similarly. It is proved that the payoff function of the femto access point defined in (7.54) is quasiconcave.

The *payoff* of the cognitive macro base station is defined as follows:

$$U^m(\mathbf{b}) = \sum_{k=1}^{K} b_k \left(\sum_{i=1}^{F} \mu^f x_{k,i}\eta_{k,i} + \sum_{j=1}^{M} \mu_j^m x_{k,j}\eta_{k,j} \right)$$
$$- \frac{1}{2} \left(\sum_{k=1}^{K}(b_k)^2 + 2s \sum_{k' \neq k} b_{k'}b_k \right) - \sum_{k=1}^{K} \mu_k b_k \tag{7.56}$$

where $\mathbf{b}$ contains the sizes of spectrum resource b_k bought from the primary BS k. $s \in [-1, 1]$ is the spectrum substitutability factor. If $s = 1$, the femto access point and macro user can switch among the spectrum resources freely. If $s = 0$, they cannot switch among the spectrum resources. If $s < 0$, the spectrum resources used by them are complementary. μ_j^m is the price paid by the macro user j to the cognitive macro base station. The utility of the cognitive macro base station in (7.56) is chosen to be a quadratic function due to its mathematical tractability (e.g., concavity and linear demand function obtained from its derivative).

The *payoff* of the primary BS is the revenue defined as follows:

$$U_k^p(\mu) = w(B_k - b_k)r_k + \mu_k b_k \qquad (7.57)$$

where w is a weighting factor, B_k is the total available spectrum resource of primary BS k, and r_k is the transmission efficiency. μ is composed of the prices of all primary BSs μ_k. The first part of the revenue (i.e., $w(B_k - b_k)r_k$) is from serving primary users while the second part (i.e., $\mu_k b_k$) is from selling the spectrum resource to the cognitive macro base station.

The solution of the hierarchical game model is the Stackelberg equilibrium. The backward induction method is applied to obtain such an equilibrium.

Power allocation of femto access point: Firstly, the optimal transmission power of a FAP to the femto user is obtained. The optimal transmission power denoted by p_i^* is obtained under the condition of function $R_i(p_i)$ defined as follows:

$$R_i(p_i) = \sum_{k=1}^{K} x_{k,i}(\lambda_i - \mu^f)b_k \log_2\left(1 + \frac{h_{k,i}p_i}{\sigma^2}\right). \qquad (7.58)$$

In particular, if $R_i(p_i)$ is strictly concave, there is a unique globally optimal transmission power for the femto access point. The gradient assisted binary search algorithm is used to obtain the optimal solution.

Spectrum demand and resource allocation by cognitive macro base station: Secondly, the spectrum demand of the cognitive macro base station is obtained. By differentiating the payoff of the cognitive macro base station defined in (7.56) and equating it to zero, the linear function for the spectrum demand denoted by b_k^* can be obtained in a closed form [28, Eq. (13)]. Then, the spectrum allocation is performed to allocate each of b_k^* to one of the femto access points or one of the macro users. A greedy-based algorithm is adopted. In particular, the allocation is performed iteratively, and in each iteration, the allocation $x_{k,i}$ or $x_{k,j}$ is chosen such that the revenue of the cognitive macro base station is maximized (i.e., $\max_{i,j}(\mu^f \eta_{k,i}, \mu_j^m \eta_{k,j})$).

Price competition of primary BS: Thirdly, the price of the spectrum resource is determined by the primary BS. The objective is to maximize the revenue defined in (7.57). However, the revenue of one primary BS is affected by the prices offered by the other primary BSs. Therefore, a subgame for the price competition is formulated for the primary BSs. The best response function is defined as $A(\mu_{-k}) = \arg\max_{\mu_k} U_k^p(\mu_k, \mu_{-k})$, where μ_{-k} is composed of the prices of all the primary BSs except the primary BS k. The Nash equilibrium of the primary BSs, given the spectrum demand from the cognitive macro base station, is the solution of $\mu_k^* = A(\mu_{-k}^*) = \arg\max_{\mu_k} U_k^p(\mu_k, \mu_{-k}^*)$. The proof for existence of the Nash equilibrium is based on the nonempty convex and compact subset of Euclidean space for the price μ_k. Also, the revenue function defined in (7.57) is concave. Then, the uniqueness of the Nash equilibrium is proved by show-ing that the best response is a standard function. That is, the standard function has the positivity (i.e., $A(\mu^*_{-k}) > 0$), monotonicity (i.e., for $\mu^*_{-k} > \hat{\mu}^*_{-k}$, then $A(\mu^*_{-k}) \geq A(\hat{\mu}^*_{-k})$), and scalability (i.e., for all $y > 1$, then $yA(\mu^*_{-k}) > A(y\mu^*_{-k})$) properties.

Then, the Stackelberg equilibrium is obtained, which are denoted by $\mu^*_k, b^*_k, x^*_{k,i}$ and $x^*_{k,j}$, and p^*_i. μ^*_k is the optimal price for the primary BS. b^*_k is the optimal spectrum demand, and $x^*_{k,i}$ and $x^*_{k,j}$ are the optimal spectrum allocation of the femto access point i and macro user j, respectively. p^*_i is the optimal transmission power for the femto access point. A gradient iteration algorithm is proposed to obtain the Stackelberg equilibrium. The gradient update in iteration $t + 1$ can be expressed as follows:

$$\mu(t+1) = \mu(t) + \Delta \frac{\partial U^{p}_k(\mu(t))}{\partial \mu_k} \tag{7.59}$$

where Δ is a step size. The simulation results show that the proposed algorithm converges to the Stackelberg equilibrium. However, the mathematical proof of such convergence is open for the future study.

7.4   GAME FORMULATIONS FOR ACCESS CONTROL

In open and closed access modes of a multi-tier network, users can connect to a macro base station or a SBS (e.g., femto access point) if they are in their coverage. The selection can be done by the rational and self-interested users. Alternatively, femtocell and macrocell can control the users' access to maximize their own benefits. These situations can be modeled using game theory. [33] considers the BS (e.g., macro base station and femto access point) selection problem and proposes a game theoretic model to analyze the decision of users. Different preferences of the users in choosing the BS are considered, that is, equal transmission time and equal throughput. If the users are concerned about the equal transmission time, the solution is shown to be the unique Nash equilibrium, which also achieves proportional fairness. If the users are concerned about the equal throughput, there could be multiple Nash equilibria, some of which could be far from the optimal solution. Reference [34] presents an incentive mechanism, namely refund framework, to motivate femto access points to accept and serve macro users. In particular, the mobile BS pays money to allow some of their macro users to connect to the femto access point, off-loading traffic from macrocell. Reference [35] considers the situation where a femto access point can accept macro users who are the non-subscribers of the femtocell based on the utility.

Coalitional game has been applied to model and analyze the access control problem of the femtocell network. In [25], a macro user can choose to transmit data directly to the macro base station or to form a coalition with a femto user and let the femto user relay the macro user's traffic to the femto access point to improve the performance. A coalition formation game model is proposed to analyze the stable coalition among macro and femto users. In [36], to reduce interference, the coalition formation among femto access points is considered as the solution. In particular, the femto access points forming a coalition can coordinate their spectrum resource usage to avoid co-tier interference. In [6], the cell selection of users between macrocell and femtocell is considered as a coalition formation game. Specifically, the user chooses to connect to the macro base station or femto access point based on their optimized transmission power and utility. The stable coalitions are analyzed by using the Markov chain.

In the following, the detailed formulations of the selected game models for access control in the femtocell networks are discussed.

7.4.1   Refunding Framework for Hybrid Access Small Cell Network

Open and closed access modes of small cells have limitations in terms of limited QoS guarantee and low resource utilization, respectively. Therefore, a hybrid access mode is more suitable for a small cell. For example, a femtocell with the hybrid access mode can reserve the radio resource for femto users (i.e., controlled by the femtocell owner) and at the same time can provide service to other users (e.g., unregistered roaming macro users). Therefore, the QoS guarantee can be supported by reducing CTI, and the radio resource utilization can be improved [31]. However, since the femtocell owner is rational, the service provider of the macrocell has to provide an incentive to motivate the femtocell owner to share the femtocell service. In [34], an access control mechanism is considered to address an incentive issue in hybrid access femtocell networks. Specifically, a utility-aware refunding framework is developed for the femtocell owner and service provider to maximize their utilities. A game model is proposed to analyze the equilibrium strategies in terms of the refund (i.e., money) to be paid by the service provider to the femtocell owner and the radio resource (i.e., transmission period) to be allocated for the roaming macro users by the femtocell owner.

Note that the refunding framework proposed in [34] is opposite to the case of the price-based spectrum sharing in [2]. That is, in the price-based spectrum sharing, the macro base station earns revenue from the interference caused by the femto users (i.e., the same sub-channel is used). On the other hand, in the refunding framework, the femtocell owner earns revenue by allowing macro users to access its femtocell (i.e., the different sub-channels are used, and hence there is no CTI). In contrast, the refunding framework is similar to the macro–femto cooperation proposed in [25] in the sense that the femtocell owner can cooperate to allow the macro users to access the femtocell. However, instead of granting a transmission period for the femto users as an incentive, in the refunding framework, the service provider of the macrocell will pay the refund in terms of money to the femtocell owner in return.

The system model considered in the refunding framework is shown in Figure 7.8. There is one macrocell (i.e., one macro base station) and N underlaid femtocells. The macro users subscribed to the service provider can communicate with the macro base station. The registered femto users can communicate with the femto access point. When the macro users move into the coverage of the femto access point, the macro users can communicate with the femto access point (i.e., off-loading), if the femtocell is in a hybrid access mode. It is assumed that the macrocell and femtocells

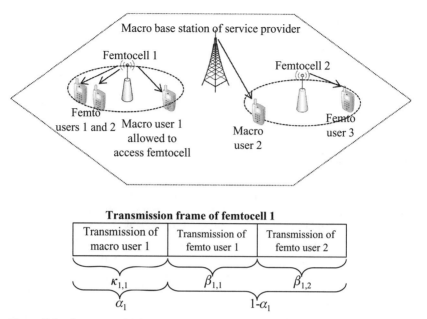

Figure 7.8 System model for refunding framework for hybrid access femtocell network [34].

operate using TDMA for data transmission, and they are allocated with different frequencies. Therefore, the CTI does not occur.

The transmission frame of the femtocell is divided into two parts, that is, one for femto users and another for macro users. α_i is the fraction of the transmission frame (i.e., transmission period) that the owner of femtocell i allocates for the transmission by the macro users. Macro user m is allocated with $\kappa_{i,m}$ fraction of the transmission frame for its own transmission, where $\sum_{m=1}^{M_i} \kappa_{i,m} = \alpha_i$ and M_i is the number of macro users in femtocell i. A simple fair share of the transmission period can be applied, for example, each macro user is allocated with a transmission period equal to $\kappa_{i,m} = \alpha_i/M_i$. The rest of the frame (i.e., $1 - \alpha$) is used for the transmission by the femto users registered to femtocell i. Femto user j is assigned with $\beta_{i,j}$ fraction of the transmission frame. Note that $\alpha_i = 0$ corresponds to the case of a closed access mode (i.e., macro users are not allowed to communicate with the femto access point). Figure 7.8 shows the example of the transmission frame. There are femtocells 1 and 2. Femtocell 1 operates in the hybrid access mode, while femtocell 2 operates in the closed access mode. For femtocell 1, there are one macro user and two femto users. Fractions $\beta_{1,1}$ and $\beta_{1,2}$ of the frame are allocated for the transmission of femto users 1 and 2, respectively, where $\beta_{1,1} + \beta_{1,2} = 1 - \alpha_1$. Since there is only one macro user, the fraction of the frame $\kappa_{1,1} = \alpha_1$ is allocated for the transmission of this macro user.

In the aforementioned system model, the refund mechanism works as follows (Figure 7.9). Since femtocells belong to the femtocell owners, the service provider intending to off-load macro users to the femtocell has to pay the refund (i.e., money) to the femtocell owner. The total refund paid to all the femtocell owners is denoted

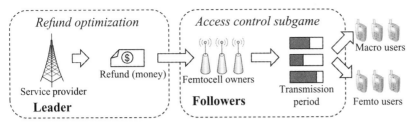

Figure 7.9   Refund mechanism for a hybrid access femtocell network.

by μ. The refund from the service provider is distributed among femtocell owners proportionally to the fractions of a transmission frame allocated to the macro users as follows:

$$\mu_i = \mu \frac{\alpha_i}{\sum_{i=1}^{N} \alpha_i} \qquad (7.60)$$

where μ_i is the refund given to femtocell owner i. The femtocell owners observe the received amount of refund and adjust the size of the transmission period for the macro users.

A Stackelberg game model is formulated to analyze the interaction among the service provider operating the macrocell and the femtocell owners operating different femtocells. The *leader player* is the service provider. The *follower players* are the femtocell owners. The *strategy of the leader* is the refund μ. The *strategy of the follower* is the transmission period α_i allocated to the macro users. The *payoff* of the leader (i.e., service provider) is the benefit from off-loading the macro users to femtocells, expressed as follows:

$$U_{\text{SP}} = w_{\text{M}}(1 - c) - \mu \qquad (7.61)$$

where c is the off-loading rate of macro users and w_{M} is the weighting factor of the macrocell. When the macrocell can off-load the macro users, the macrocell can provide a better QoS performance, yielding higher satisfaction to the users. The off-loading rate is estimated as the sigmoid function, defined as follows:

$$c = \frac{1}{1 + \exp(-a(b - R_m))} \qquad (7.62)$$

where R_m is the transmission rate of macro user m. a and b are the sensitivity parameters due to the QoS increment and reserved traffic demand of the macro users, respectively. The *payoff* of the follower (i.e., femtocell owner) is a function of the transmission rate of femto users and the refund obtained from the service provider is defined as follows:

$$U_i^{\text{FO}} = w_{\text{F}} r_i + \mu_i \qquad (7.63)$$

where r_i is the transmission rate of femto user i and w_{F} is the weighting factor. The transmission rates of the macro user m and femto user i can be obtained from $R_m = \alpha_i C_m$ and $r_i = (1 - \alpha_i)c_i$, respectively. C_m and c_i are the channel rates obtained from $C_m = \log_2(1 + \gamma_m^M)$ and $c_i = \log_2(1 + \gamma_i^F)$, where γ_m^M and γ_i^F are the SINRs of macro and femto users, respectively.

A standard method of backward induction is applied to solve the Stackelberg game. In the first step, an optimization model of the follower is formulated as $\max_{\alpha_i} U_i^{FO}$ and solved for the best response strategy. Then, based on the best response strategies of the followers, the leader formulates and solves another optimization model to obtain the optimal strategy. Since there are multiple femtocell owners as the followers, the optimization model is the noncooperative access control subgame (i.e., to determine the transmission period for the macro users). It is shown that the best response strategy of the femtocell owner i given the strategies of other femtocell owners $\alpha_{i'}$ can be expressed as follows:

$$\alpha_i^* = \sqrt{\frac{\mu \sum_{i' \neq i} \alpha_{i'}}{w_F c_i}} - \sum_{i' \neq i} \alpha_{i'} \tag{7.64}$$

if $\sum_{i' \neq i} \alpha_{i'} \leq \frac{\mu}{w_F c_i}$, and $\alpha_i^* = 0$ otherwise. It is proved that the best response strategy of the femtocell owner is the unique Nash equilibrium of the noncooperative access control subgame. This Nash equilibrium can be expressed as follows:

$$\alpha_i^* = \frac{(N-1)\mu}{w_F \left(\sum_{i'=1}^{N} c_{i'}\right)^2} \left(\sum_{i'=1}^{N} c_{i'} - (N-1)c_i\right) \tag{7.65}$$

if $c_i \leq \frac{\sum_{i'=1}^{N} c_{i'}}{N-1}$, and $\alpha_i^* = 0$ otherwise.

Then, the service provider solves for the refund which is shown to be

$$\mu^* = \frac{1}{\sum_{i=1}^{N} c_i \rho_i} \left\{ b + \frac{1}{a} \ln \left[\left(\frac{w_M a}{2w_F} \sum_{i=1}^{N} c_i \rho_i - 1 \right) \right. \right.$$
$$\left. \left. + \sqrt{\left(\frac{w_M a}{2w_F} \sum_{i=1}^{N} c_i \rho_i - 1 \right)^2 - 1} \right] \right\} \tag{7.66}$$

$$\rho_i = \begin{cases} \dfrac{(N-1)\left(\sum_{i'=1}^{N} c_{i'} - (N-1)c_i\right)}{w_F \left(\sum_{i'=1}^{N} c_{i'}\right)^2}, & \text{if } \sum_{i'=1}^{N} c_{i'} < (N-1)c_i \\ 0, & \text{otherwise.} \end{cases} \tag{7.67}$$

It is observed that the above solution is locally optimal. Clearly, the convexity of the optimization model of the leader cannot be guaranteed, that is, due to the sigmoid function.

Finally, a hybrid access protocol is introduced to implement the above Stackelberg game. The protocol is composed of two steps, that is, optimal refunding for the service provider and access control for the femtocell owners. In the optimal refunding step, the femto access point periodically collects the channel information of the macro and femto users in its coverage. This information is forwarded to the service provider.

The service provider checks whether the current condition will be profitable to off-load the macro users or not (i.e., by checking the payoff given the optimal refund μ^*). If it is profitable, the optimal refund is paid to the femtocell owners. In the access control step, the femtocell owner chooses the transmission period for the macro users. However, if the condition is not favorable (e.g., the channel quality to the macro user is too poor or there are too many connected femto users), the femtocell owner will switch to the closed access mode (i.e., $\alpha_i = 0$).

7.4.2 Selection of Network Tier

With the open access mode, macro users can connect to a femto access point if they are in its coverage. However, the performance of connecting to the femto access point can be degraded due to congestion. To analyze the cell selection behavior of the rational users in a femtocell with the open access mode, a game model is formulated and the Nash equilibrium is analyzed in [35]. The system model considers one macrocell and one femtocell. A user can be a subscriber or non-subscriber of the femtocell. When the subscriber is in the femtocell, this subscriber will connect to the femto access point. However, when the non-subscriber is in the femtocell, this non-subscriber has a choice to connect to the macro base station or the femto access point. The non-subscriber is a rational agent, and hence will decide on its connection to maximize the individual utility.

A noncooperative game model is formulated for the cell selection of the non-subscribers. The *players* are non-subscribers in the coverage of the femtocell. The *strategy* is to connect to either the macro base station or femto access point. The *payoff* is the utility of non-subscriber i, defined as follows:

$$U_i(n) = \begin{cases} u_M(n), & \text{connect to macro base station} \\ u_F(n), & \text{connect to femto access point} \end{cases} \tag{7.68}$$

where $u_M(n)$ and $u_F(n)$ are the utilities of connecting to the macro base station and femto access point, respectively, given the number of non-subscribers n connecting to the femto access point. Firstly, for the case of selecting the macrocell, the utility is expressed as follows:

$$u_M(n) = B \log_2 \left(1 + \frac{p h_m}{\sigma^2} \right) \tag{7.69}$$

where B is the spectrum size, p is the transmission power, h_m is the channel gain to the macro base station, and σ^2 is the noise power. In [35], the transmission power p is chosen such that all the users connected to the macro base station achieve the same SINR. For the case of selecting the femtocell, the utility is expressed as follows:

$$u_F(n) = \left(\frac{1-\alpha}{H+n} \right) \overline{R}(n), \quad \text{where} \tag{7.70}$$

$$\overline{R}(n) = (L - N + n) B \log_2 \left(1 + \frac{p_{max} h_f}{\sigma^2} \right) + \sum_{i \in \mathcal{N}_m} B \log_2 \left(1 + \frac{p_i h_{f,i}}{\sigma^2 + p_m h_{f,m}} \right).$$

Here H is the number of subscribers of the femtocell, L is the total spectrum available for the femto access point, N is the total number of non-subscribers in the coverage of the femtocell, p_{max} is the maximum transmission power, and h_f is the channel gain from the femto user to the femto access point. $\mathcal{N}_m$ is a set of non-subscribers which use the same spectrum as that of macro user (i.e., interference exists). p_i is the transmission power of non-subscriber i, $h_{f,i}$ is the channel gain to the femto access point, p_m is the transmission power of the macro user, and $h_{f,m}$ is the channel gain from the macro user to the femto access point. α is the ratio of the total capacity for the femtocell. If $\alpha = 1$, the femtocell is completely closed and cannot accept the macro users in the femtocell. In contrast, if $\alpha = 0$, the femtocell is completely open and all the macro users in the femtocell will be accepted. The utility of selecting the femtocell defined in (7.70) is composed of two parts. Firstly, $B \log_2 \left(1 + \frac{p_{max} h_f}{\sigma^2} \right)$ is the transmission rate of the non-subscribers accessing the non-interfered spectrum (i.e., only one non-subscriber transmits on one spectrum). Secondly, $\sum_{i \in \mathcal{N}_m} B \log_2 \left(1 + \frac{p_i h_{f,i}}{\sigma^2 + p_m h_{f,m}} \right)$ is the transmission rate of the non-subscribers accessing the interfered spectrum (i.e., non-subscriber and macro user access the same spectrum).

A Nash equilibrium is considered to be the solution of the cell selection game. The existence of the Nash equilibrium is proved by showing that the cell selection game is a potential game. The potential game has the potential function $\Phi(\cdot)$. The potential function satisfies the following condition, $U_i(s_i, \mathbf{s}_{-i}) - U_i(s_i', \mathbf{s}_{-i}) = \Phi(s_i, \mathbf{s}_{-i}) - \Phi(s_i', \mathbf{s}_{-i})$, for all strategies s_i and s_i'. The Nash equilibrium is a pure strategy of this potential game if and only if $\Phi(s_i^*, \mathbf{s}_{-i}^*) \geq \Phi(s_i, \mathbf{s}_{-i}^*)$ for all i and s_i. It is shown that the cell selection game has a potential function, which is defined as follows:

$$\Phi(n) = \sum_{i=1}^{n} u_F(i) + \sum_{i=n}^{N-1} u_M(i). \tag{7.71}$$

Although the cell selection game is proved to have a Nash equilibrium, the algorithm to reach the Nash equilibrium is not discussed. An efficient algorithm would be required, especially in a distributed environment, where the non-subscribers have to select the cell independently.

7.4.3 Coalitional Game for Cooperation among Macrocells and Small Cells

In a two-tier macrocell–femtocell network, a macro user, especially at a cell boundary area, usually suffers from low SINR. On the other hand, femtocells are interference limited. Nevertheless, the cooperation between macro and femto users can alleviate such trade-off. This cooperation under a closed access mode is considered in [25]. In this cooperation scheme, the cooperative femto user (i.e., FUE) can relay the data transmission of the cooperative macro user, consequently improving the SINR. In return, the cooperative macro user grants the fraction of the transmission time to the cooperative femto user. This cross-tier cooperation scheme introduces a mutual benefit for co-channel femtocell networks, by avoiding CTI at the femtocell.

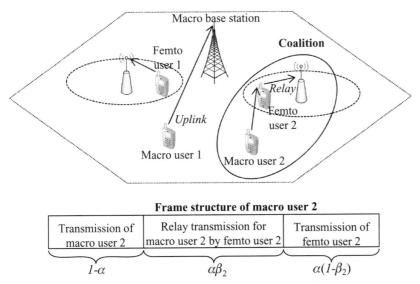

Figure 7.10   Cooperation between macro user 2 and femto user 2 and the frame structure [25].

An example of the cross-tier cooperation is shown in Figure 7.10. There are femto users 1 and 2, and there are two macro users 1 and 2. Femto user 1 does not cooperate with macro user 1. Therefore, in this noncooperative case, macro user 1 has to transmit to the macro base station directly, creating strong interference to a femtocell using the same sub-channel. On the other hand, femto user 2 cooperates with macro user 2. Therefore, the transmission of macro user 2 can be relayed by the femto user 2 to the corresponding femto access point. Also, macro user 2 allows femto user 2 to access part of its transmission frame. In this cooperative case, macro user 2 gains the benefit in terms of better transmission quality and low transmission power due to shorter range transmission. Likewise, femto user 2 gains the benefit in terms of larger transmission bandwidth.

An example of the frame structure due to cooperation is also shown in Figure 7.10. In this case, the frame is divided into three parts. The first part is for the transmission by macro user 2. The size of the first part is $1 - \alpha$ of the frame. The second part is for the femto user to relay the transmission of macro user 2 (e.g., using a decode-and-forward scheme). The size of the second part is $\alpha\beta_i$, where β_i is the proportion that femto user i (i.e., $i = 2$ in this example) uses to relay the transmission from the cooperative macro user. The third part is for the femto user 2 to transmit its own data, and the size of the third part is $\alpha(1 - \beta_i)$. Note that this cooperative scheme is also known as the spectrum leasing, which is originally proposed for a cognitive radio network [37].

To analyze and obtain the solution for the cross-tier cooperation scheme, [25] proposes a coalition game model. The *players* are macro users and femto users. The *strategy* is to cooperate or not. If the macro user cooperates, this user will allow the femto user to transmit in its frame. If the femto user cooperates, this user will relay the transmission from the corresponding cooperative macro user. Let $\mathcal{M}$ denote a set of macro users. Let $\mathcal{N}_f$ denote a set of femto users associated with femto access point f, for $f \in \mathcal{F}$, where $\mathcal{F}$ is the set of femto access points. Each femto access point is allocated with a dedicated sub-channel in OFDMA-based networks. The set of players is $\Psi = \mathcal{M} \cup (\bigcup_{f \in \mathcal{F}} \mathcal{N}_f)$. One femto user can cooperate with multiple macro users. Therefore, if the strategies of the macro and femto users are to cooperate, then the coalition of femto user i denoted by $\mathcal{S}_i \subseteq \Psi$ can be formed. The *payoff* is defined as follows:

$$u_i(\mathcal{S}_i, \Pi) = \frac{r_i^C(\alpha, \beta_i, \Pi)^\delta}{(D_i^C)^{1-\delta}} \tag{7.72}$$

where $r_i^C(\alpha, \beta_i, \Pi)$ is the achievable rate of the macro and femto users from cooperation. This achievable rate depends on the cooperation parameters α and β_i for femto user i, which will be optimized. For example, the achievable rate of the cooperative femto user i can be obtained from $r_i^C(\alpha, \beta_i) = \alpha(1 - \beta)r_i^R$, where r_i^R is the transmission rate of the femto user i. This transmission rate is obtained from

$$r_i^R = \log_2 \left(1 + \frac{h_i p_i}{\sum_{j \in \Phi_{-S_i}} h_{i,j} p_j + \sigma^2} \right) \tag{7.73}$$

where h_i is the channel gain between femto user i and the corresponding femto access point, and $h_{i,j}$ is the channel gain between macro user j and the femto access point of the femto user i. p_i and p_j are the transmission powers of femto and macro users i and j, respectively. Φ_{-S_i} denotes the set of macro users who are not in a coalition S_i using the same sub-channel as that of femto user i, and σ^2 is the noise power.

However, due to the interference, the achievable rate of one user also depends on the cooperation of other users. The cooperation (i.e., coalitions) of all the users in the game is defined as the partition Π, which is a parameter of the payoff defined in (7.72). D_i^C is the transmission delay of user i from cooperation. This transmission delay is composed due to transmission delay from the macro user to the cooperative femto user and from the cooperative femto user to the femto access point. In [25], the delay performance is obtained from an M/D/1 queueing model. Finally, δ is the transmission capacity-delay trade-off parameter in the payoff.

For example, in Figure 7.10, let f_1 and f_2 denote femto users 1 and 2, respectively. Let m_1 and m_2 denote macro users 1 and 2, respectively. The set of players is $\Psi = \{f_1, f_2, m_1, m_2\}$. There is one cooperation, whose coalition is defined as $S_2 = \{f_2, m_2\}$. The partition is defined by $\Pi = \{\{f_1\}, \{m_1\}, \{f_2, m_2\}\}$. Clearly, the achievable rate of macro and femto users m_2 and f_2 will be affected by the transmission of macro and femto users m_1 and f_1, if all of them use the same sub-channel.

Based on the payoff defined in (7.72), it is observed that the coalitional game is in a partition form with NTU. The reason is that the payoff depends on transmission rate and delay which could not be arbitrarily allocated to any player. The solution of this game is to find a stable cooperation scheme, which is defined as the stable coalitions (or stable partition for all the users). Also, it is important to derive the optimal solutions for the cooperation parameters (i.e., α and β_i), and for the transmission parameter (i.e., transmission power p_i) of each user. In [25], the recursive

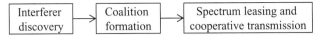

Figure 7.11   Distributed coalition formation algorithm for macro–femto cooperation.

core [38] is considered as a solution of this coalitional game. The recursive core is a set of partitions that the players can form coalitions within such that each player gains the highest payoff. The recursive core is a desirable solution due to the following properties.

- Rationality: A player will never choose to cooperate if its payoff will be degraded.
- Well-definition: If it exists, the recursive core will always be unique.
- Efficiency: All partitions in the recursive core are equivalent in terms of individual payoff of each player.

In short, at the recursive core, none of the players has an incentive to deviate from the current coalition, which is the notion of stability.

For the coalitional game for macro–femto cooperation proposed in [25], the existence of the recursive core is also discussed. It is shown that the recursive core exists. The reason is that there is no case that all possible partitions will result in the same payoff.

Reference 25 also presents a distributed algorithm to reach the recursive core and also to obtain the optimal cooperation and transmission parameters. Figure 7.11 shows the major three steps of the distributed algorithm. In the interference discovery step, the macro user periodically transmits the received signal strength indicator (RSSI) to the macro base station. The potential femtocells to cooperate are identified from the RSSI. A similar procedure is performed by the femto users. Then, for the potential macro and femto users, the payoffs are calculated as in (7.72), given the optimal cooperation and transmission parameters of the femto user. The optimal cooperation and transmission parameters are obtained from solving the following optimization, that is,

$$\alpha^*, \beta_i^*, p_i^{\mathrm{T}*}, p_i^{\mathrm{R}*} = \arg \max_{\alpha, \beta_i, p_i^{\mathrm{T}}, p_i^{\mathrm{R}}} u_i(\mathcal{S}_i, \Pi) \tag{7.74}$$

$$\text{subject to } 0 < \alpha, \beta_i \leq 1, \beta_i p_i^{\mathrm{R}} + (1 - \beta_i) p_i^{\mathrm{T}} < p_{\max} \tag{7.75}$$

where p_i^{T} and p_i^{R} are the transmission powers for the femto user's own transmission and relay transmission, respectively. $p_{\max}$ is the maximum transmission power. After the payoff is computed, in the coalition formation step, the macro and femto users send a request to initiate cooperation to gain the highest payoff. In the spectrum leasing and cooperative transmission step, if the macro and femto users agree to cooperate, they will establish a relay connection and the cooperative macro user will inform and off-load the transmission from the macro base station.

The convergence of the distributed algorithm for the macro–femto cooperation is discussed. It is observed that the payoff will always decrease in each iteration. The algorithm can terminate in the first iteration, if the femto user cannot form any coalition which improves the payoff. Therefore, the stable coalition is reached. The simulation results show that the proposed macro–femto cooperation can achieve significant performance gain. For example, the transmission rate of the macro user increases by two times compared with that for noncooperative case in a network with 200 femtocells.

7.4.4   Cooperative Interference Management

In addition to power control (e.g., [2, 27]), cooperation in sub-channel access is regarded as one of the effective approaches to avoid interference. For a two-tier macrocell–femtocell network, [36] introduces the concept of cooperative interference management by letting femto access points form a coalition to share the spectrum resource (i.e., sub-channels), reducing the co-tier interference. Specifically, if the femto access points form the coalition, they can coordinate their transmissions through using traffic scheduling in a time division duplexing (TDD) mode. The system model considered in [36] is based on an OFDMA-based femtocell network. The set of femto access points is denoted by $\mathcal{F}$. The set of macro users is denoted by $\mathcal{M}$. When the femto access points do not cooperate, they use sub-channels from a set denoted by $\mathcal{K}$ for their transmission. These sub-channels are in the macrocell UL spectrum.

A coalition formation game model is formulated in [36], where the femto access points can be self-organizing to form a coalition and share the sub-channels to avoid the co-tier interference. The *players* are the femto access points. The *strategy* is to form the coalitions. Let $\mathcal{S}$ denote a coalition of the femto access points. To derive the payoff of each player, first a value of a coalition is given. The value of any coalition $\mathcal{S}$ can be expressed as follows:

$$v(\mathcal{S}, \Theta_{\mathcal{F}}) = |\mathcal{S}| \sum_{i \in \mathcal{S}} \sum_{k \in \mathcal{K}} \log_2 \left(1 + \frac{p_{i,k} h_{i,k}}{\sigma^2 + I_{i,k}^{\mathrm{f}} + I_{i,k}^{\mathrm{m}}} \right) \tag{7.76}$$

where $|\mathcal{S}|$ is the number of players in the coalition $\mathcal{S}$. $\Theta_{\mathcal{F}}$ is a set of all coalitions (i.e., partition), defined as $\bigcup_{\mathcal{S} \in \Theta_{\mathcal{F}}} \mathcal{S} = \mathcal{F}$. $p_{i,k}$ and $h_{i,k}$ are the transmission power and channel gain of femto access point i using sub-channel k, respectively. $I_{i,k}^{\mathrm{f}}$ and $I_{i,k}^{\mathrm{m}}$ are the interferences from the other femto access points and from the macro users, respectively. The interference from macro users can be obtained from

$$I_{i,k}^{\mathrm{m}} = \sum_{m \in \mathcal{M}} p_{m,k} h_{m,i,k} \tag{7.77}$$

where $p_{m,k}$ and $h_{m,i,k}$ denote, respectively, the transmission power and channel gain of macro user m to the femto access point i on sub-channel k. The interference from the other femto access points can be obtained from

$$I_{i,k}^{\mathrm{f}} = \sum_{i' \in \mathcal{F} \backslash \mathcal{S}} p_{i',k} h_{i',i,k} \tag{7.78}$$

where $p_{i',k}$ and $h_{i',i,k}$ are the transmission power and channel gain of femto access point i', respectively, which is not in the same coalition $\mathcal{S}$ as that of the femto access point i on sub-channel k. Clearly, the value function defined in (7.76) of the players in one coalition is affected by the interferences from other players in the different coalitions. Therefore, the externality exists in the game.

However, not all the femto access points can form the coalition, since they may not be able to exchange the coalition formation information among each other (e.g., femto access points which are faraway from each other) through a common control channel. The transmission power required for femto access point i to reach the farthest femto access point i' such that the coalition can be formed among them is given as follows:

$$p_i^{\text{req}} = \frac{\gamma_{\min}\sigma^2}{h_{i,i'}} \tag{7.79}$$

where $\gamma_{\min}$ and $h_{i,i'}$ are the minimum required SINR and the channel gain for the communication between femto access points i and i', respectively. σ^2 is the noise power. The transmission power of the femto access point i in a coalition $\mathcal{S}$ must be bounded by

$$\sum_{k \in \mathcal{K}} p_{i,k} \le p_{\mathcal{S},i} = \left[p_{\max} - p_i^{\text{req}}\right]^+. \tag{7.80}$$

Therefore, if $p_{\mathcal{S},i} = 0$, then the value of the coalition $\mathcal{S}$ will be zero (i.e., there is no transmission power left for data transmission).

In the same coalition $\mathcal{S}$, the femto access points must also divide the value of the coalition $v(\mathcal{S}, \Theta_{\mathcal{F}})$. The value of the coalition is assumed to be transferable (i.e., TU). In this case, the Egalitarian fair method is adopted. The *payoff* of the player i in a coalition $\mathcal{S}$ can be expressed as follows:

$$U_i = \frac{1}{|\mathcal{S}|}\left(v(\mathcal{S}, \Theta_{\mathcal{F}}) - \sum_{i' \in \mathcal{S}} v(\{i'\}, \Theta_{\mathcal{F}})\right) + v(\{i\}, \Theta_{\mathcal{F}}). \tag{7.81}$$

Figure 7.12 shows the example of the system model for the coalition formation among femto access points. There are two macro users (i.e., $\mathcal{M} = \{m_1, m_2\}$). There are four femto access points (i.e., $\mathcal{F} = \{f_1, f_2, f_3, f_4\}$). Femto access points f_1 and f_2 form the coalition $\mathcal{S}_1$, while f_3 and f_4 form the coalition $\mathcal{S}_2$. Therefore, the set of coalitions

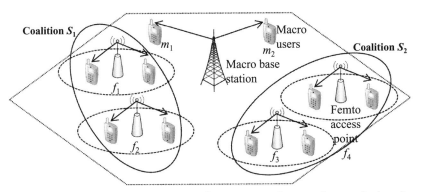

Figure 7.12   Coalition formation among femto access points to reduce co-tier interference.

of all the femto access points is denoted by $\Theta_{\mathcal{F}} = \{\mathcal{S}_1, \mathcal{S}_2\} = \{\{f_1, f_2\}, \{f_3, f_4\}\}$. In this example, the femto access point f_1 may not be able to form a coalition with f_4 since they are far from each other and cannot exchange the coalition formation information.

To obtain stable coalitions, a femto access point cooperation algorithm is proposed. The algorithm has three steps, that is, neighbor discovery, coalition formation, and coalition-level scheduling.

1. *Neighbor discovery:* The femto access points exchange messages to explore all the femto access points in the same network.

2. *Coalition formation:* The following steps are repeated until convergence (i.e., there is no change in the set of coalitions).

   (a) Each femto access point contacts with another femto access point to identify the potential partner to cooperate and form a coalition.

   (b) The payoff U_i from the coalition is calculated as in (7.81) for the potential cooperation.

   (c) Each femto access point forms a coalition which maximizes the payoff U_i

3. *Coalition-level scheduling:*

   (a) All the femto access points in the same coalition exchange the transmission parameters and channel gains.

   (b) The sub-channels are selected for the transmission by the femto access points in the same coalition. The time slots are allocated to the femto access points to share the sub-channels.

   (c) The scheduling information is transferred to the macrocell.

Simulation results show that if the number of femto access points is small, the chance of a coalition to be formed is small due to the limited capability of the femto access points to communicate and exchange the cooperation information with each other. Consequently, the payoff is small. When the number of femto access points increases, the coalitions can be formed to coordinate the sub-channel allocation and access. As a result, the payoff increases. If the number of femto access points is too large, the coalitions with many femto access points will be formed. However, due to the limited available time slots for the scheduling, the sub-channels cannot be efficiently utilized. Therefore, the payoff decreases. In summary, there will be an optimal network size such that the payoff of the femto access point is maximized.

7.4.5   Coalition-Based Access Control

One way to reduce interference in small cell networks is to perform an access control of the macro and femto users (e.g., to connect to a macro base station or a femto access point). The access control method for a small cell should take the individual user's performance into account, which is especially important in a distributed environment. Therefore, the access control problem for a small cell network can be formulated as a coalitional game [6] to find the equilibrium and stable solution for cell association. The system model under consideration is composed of a macro

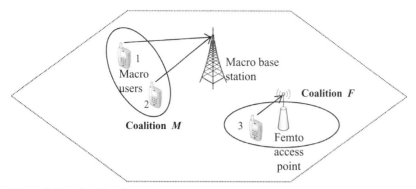

Figure 7.13   Coalition formation among transmitters to connect to the macro base station or femto access point.

base station and femto access point. A set of available sub-channels is $\mathcal{K}$ and a set of transmitters (i.e., in the UL direction) is $\mathcal{N}$, choosing to transmit to the macro base station or femto access point.

A coalition formation game model is presented in [6], in which the *players* are the transmitters. The *strategy* of a player is to choose to transmit to either the macro base station or the femto access point. In particular, two coalitions will be formed, that is, $\mathcal{M}$ and $\mathcal{F}$ for the transmitters transmitting to the macro base station and femto access point, respectively. An example is shown in Figure 7.13. There are three transmitters (i.e., $1, \ldots, 3$). Transmitters 1 and 2 choose to form the coalition to connect to the macro base station (i.e., $\mathcal{M} = \{1, 2\}$), while transmitter 3 forms the singleton coalition to connect to the femto access point (i.e., $\mathcal{F} = \{3\}$).

Similar to the system model in [36], the transmitters in the same coalition will not interfere with each other (e.g., due to the sub-channel allocation and traffic scheduling). To obtain the payoff, the value of a coalition is first derived considering the worst-case performance of the transmitters in each coalition. The worst-case performance is influenced by the jamming effect of the transmitters in the different coalitions. Therefore, the value of a coalition can be obtained as the solution of the following optimization problem:

$$v(\mathcal{F}) = \max_{p_{i,k}} \min_{p_{j,k}} \sum_{i \in \mathcal{F}} u_i(\mathbf{p}) \tag{7.82}$$

$$\text{subject to} \sum_{k \in \mathcal{K}} p_{i,k} \leq p_{\max}, \quad p_{i,k} \geq 0 \tag{7.83}$$

where $u_i(\mathbf{p})$ is the utility of the transmitter i, which is a function of $\mathbf{p}$ for the transmission powers of all the transmitters. $p_{i,k}$ is the transmission power of the transmitter i on a sub-channel k and $p_{\max}$ is the maximum transmission power on all the sub-channels. The value function defined in (7.82) considers the worst case jamming from the transmitters in the different coalition (i.e., $\min\limits_{p_{j,k}}$). The utility function of the transmitter i in coalition $\mathcal{F}$ is defined as follows:

$$u_i(\mathbf{p}) = \sum_{k \in \mathcal{K}} \log_2 \left(1 + \frac{p_{i,k} h_{i,k}}{\sigma^2 \sum_{j \in \mathcal{M}} p_{j,k} h_{j,i,k}} \right) \tag{7.84}$$

where $h_{i,k}$ is the channel gain of the transmitter i to the receiver (i.e., femto access point). $p_{j,k}$ and $h_{j,i,k}$ are the transmission power and channel gain of the transmitter j in the coalition $\mathcal{M}$ to the receiver of the transmitter i, respectively. The value function of the transmitters in the coalition $\mathcal{M}$ can be obtained similarly.

It is proved that the value function of the coalitional game is not necessarily super-additive. The super-additivity of the value function is defined as follows: $v(\mathcal{S}_1 \cup \mathcal{S}_2) \geq v(\mathcal{S}_1) + v(\mathcal{S}_2)$ for two disjoint coalitions $\mathcal{S}_1$ and $\mathcal{S}_2$. Therefore, the grand coalition (i.e., all transmitters will be in the same coalition) may not be stable. Two issues arise here, that is, how the value is distributed among the transmitters in the same coalition and how to form the stable coalitions. For the distribution of the value of the coalition, the Shapley value is applied [8]. The Shapley value, which determines the *payoff* of the player, is defined as follows:

$$U_i = \sum_{\mathcal{S} \subset \mathcal{F}, i \in \mathcal{F}} \frac{(|\mathcal{S}| - 1)!(|\mathcal{F}| - |\mathcal{S}|)}{|\mathcal{F}|} (v(\mathcal{S}) - v(\mathcal{S} \setminus \{i\})). \tag{7.85}$$

A Markov chain model is formulated to analyze the stable coalitions. For the example shown in Figure 7.13, the state transition diagram is shown in Figure 7.14. The state is denoted as the partition $\Theta = \{\mathcal{M}, \mathcal{F}\}$. There are 8 possible states (i.e., 8 possible combinations) for the partition. The transition is assumed to happen when one transmitter changes its strategy. For example, the transition $\{\{1\}, \{2, 3\}\} \to \{\{1, 2\}, \{3\}\}$ happens when the transmitter 2 chooses to transmit to the macro base station, and hence joins the coalition $\mathcal{M} = \{1\}$. The stationary probability is obtained for the Markov chain. The stable coalitions (i.e., partitions) are determined based on the states that the stationary probabilities are larger than zero.

Consider a set of three transmitters $\{1, 2, 3\}$. There are 12 sub-channels and the total transmission power is 10 dBm for all transmitters. The noise power is -90 dBm.

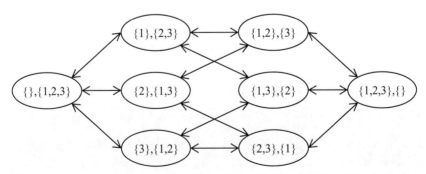

Figure 7.14   State transition diagram of the Markov chain to model coalition formation of access control.

Receiver *s* for coalition *M*

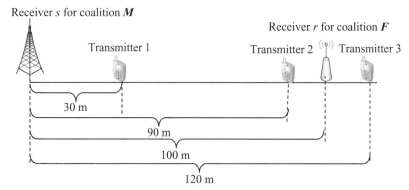

Figure 7.15   Locations of transmitters and receivers.

We assume that all the transmitters and receivers are located along the x-axis. Let the receiver *s* be at the origin, and the receiver *r* be located at 100 m away from the origin. Let the three transmitters be at 30, 90, and 120 m away from the origin, respectively (Figure 7.15). The antenna gain is 17.95 dB. The random fading in each sub-channel is given by an independent and identical Rayleigh random variable with average gain of -10 dB. Finally, the sub-channels are divided evenly among the transmitters within a coalition.

The power allocation is performed and the coalitional value is calculated. The Shapley value of each player under all possible coalition structures ω is given in Table 7.1. We first consider the coalitional values, $v(\mathcal{M})$ and $v(\mathcal{F})$, and apparently they are not super-additive. For instance, $v(\mathcal{M} = \{1\}) = 69.76$ and $v(\mathcal{M} = \{3\}) = 0.45$, while $v(\mathcal{M} = \{1, 3\}) = 53.52$, which is clearly less than the sum $v(\mathcal{M} = \{1\}) + v(\mathcal{M} = \{3\})$. Secondly, we observe that the coalition structures ω_2, ω_3, and ω_5 are not feasible due to the negative payoff obtained by one of the players. That is, at least one of the players will always have an incentive to join a different coalition.

Based on the Markov chain, the transition among coalition structure can be determined. For example, in $\omega_6 = (\{2\}\{1, 3\})$, we have (0.24, 1.55, 14.34), while in $\omega_4 = (\{1, 2\}\{3\})$ we have (70.37, 2.16, 14.24). Since transmitter 1 achieves a higher

TABLE 7.1   Payoff matrix for three transmitter coalition game

Coalition Structure	$v(\mathcal{M})$	$v(\mathcal{F})$	$U_1(\omega)$	$U_2(\omega)$	$U_3(\omega)$
$\omega_1 = (\{1, 2, 3\}, \varnothing)$	177.94	0	102.92	42.53	32.47
$\omega_2 = (\{2, 3\}, \{1\})$	0.96	0.149	0.149	1.03	−0.07
$\omega_3 = (\{1, 3\}, \{2\})$	53.52	40.51	61.41	40.51	−7.89
$\omega_4 = (\{1, 2\}, \{3\})$	72.54	14.24	70.37	2.16	14.24
$\omega_5 = (\{3\}, \{1, 2\})$	0.45	22.21	−9.07	31.28	0.45
$\omega_6 = (\{2\}, \{1, 3\})$	1.55	14.58	0.24	1.55	14.34
$\omega_7 = (\{1\}, \{2, 3\})^*$	69.76	120.96	69.76	73.61	47.34
$\omega_8 = (\varnothing, \{1, 2, 3\})$	0	256.43	42.21	115.58	98.63

payoff in ω_4 than that in ω_6 without downgrading the payoff of transmitter 2, the transmitter 1 in ω_6 has the incentive to split from $\mathcal{F} = \{1, 3\}$ and join $\mathcal{M} = \{2\}$ to form ω_4. Thus $(\{2\}\{1, 3\}) \rightarrow (\{1, 2\}\{3\})$. In this case, we observe that ω_7 is the absorbing state of the corresponding Markov chain. Thus this coalition structure is stable, and therefore, is the required solution for our access control problem.

7.5   FUTURE RESEARCH DIRECTIONS

In this chapter, game theoretic approaches to address the RRM in small cell networks have been presented. First, an overview of game theory including the motivation and different types of games has been introduced. A review of the game models developed for solving three major RRM issues (i.e., power control and sub-channel allocation, pricing, and access control) have been provided. There are still a number of significant issues related to the game theoretic RRM in small cell networks which require further investigation. Some possible future research directions are outlined below.

- *Incomplete information game:* In the existing game models for resource management of small cell networks, the information about the other players is usually assumed to be completely available. However, in reality, some information is not available, for example, the channel gains of other players to the macro base stations. To address this issue, an incomplete information game model should be developed. By using the Bayesian theorem, the belief on the parameters of other players could be built and used to optimize the strategy. A solution in terms of the Bayesian Nash equilibrium can be obtained for such an incomplete information game.

- *Delay and loss of backhaul communication:* Most of the game models proposed in the literature assume that the information exchange among players is immediate and perfect. In particular, there is no delay in making a decision. However, in small cell networks, the connections between macro base stations and SBSs are based on a broadband backhaul, whose transmission delay could be substantial and transmission loss can occur. Such a delay and loss must be taken into account when the players choose their strategies. Also, the impact of delay and loss on the payoffs of the players and system efficiency needs to be analyzed.

- *Integration with multi radio access technology (multi-RAT):* In typical small cell networks, macro base stations and SBSs are assumed to operate using the same technology (e.g., LTE). However, the multi-RAT technology will be used to off-load the users' traffic from the cellular networks to unlicensed networks (e.g., Wi-Fi networks). When the multi-RAT technology is deployed, competition exists and game theory can be applied to model and analyze the operation of multi-RAT small cell networks. Also, different RATs can be operated by different service providers. Competition and cooperation strategies for the service providers can be analyzed using game theoretic models.

REFERENCES

1. Z. Han, D. Niyato, W. Saad, T. Basar, and A. Hjørungnes, *Game Theory in Wireless and Communication Networks: Theory, Models, and Applications*. Cambridge University Press, 2011.
2. V. Chandrasekhar, J. G. Andrews, T. Muharemovic, Z. Shen, and A. Gatherer, "Power control in two-tier femtocell networks," *IEEE Transactions on Wireless Communications*, vol. 8, no. 8, pp. 4316–4328, August 2009.
3. J. W. Huang and V. Krishnamurthy, "Cognitive base stations in LTE/3GPP femtocells: A correlated equilibrium game-theoretic approach," *IEEE Transactions on Communications*, vol. 59, no. 12, pp. 3485–3493, December 2011.
4. C.-H. Ko and H.-Y. Wei, "On-demand resource-sharing mechanism design in two-tier OFDMA femtocell networks," *IEEE Transactions on Vehicular Technology*, vol. 60, no. 3, pp. 1059–1071, March 2011.
5. W. Saad, Z. Han, M. Debbah, A. Hjørungnes, and T. Basar, "Coalitional game theory for communication networks: A tutorial," *IEEE Signal Processing Magazine – Special Issue on Game Theory*, vol. 26, no. 5, pp. 77–97, September 2009.
6. S. Guruacharya, D. Niyato, and D. I. Kim, "Access control via coalitional power game," in *Proceedings of IEEE Wireless Communications and Networking Conference (WCNC)*, pp. 2824–2828, April 2012.
7. D. Fudenberg and J. Tirole, *Game Theory*. The MIT Press, August 1991.
8. R. B. Myerson, *Game Theory: Analysis of Conflict*. Harvard University Press, September 1997.
9. J. Friedman, *Game Theory with Applications to Economics*. Oxford University Press, 1990.
10. J. Rosen, "Existence and uniqueness of equilibrium points for concave n-person games," *Econometrica: Journal of the Econometric Society*, vol. 33, no. 3, pp. 520–534, 1965.
11. D. Monderer and L. Shapley, "Potential Games," *Games and Economic Behavior*, vol. 14, no. 1, pp. 124–143, 1996.
12. E. Hossain, D. Niyato, and Z. Han, *Dynamic Spectrum Access and Management in Cognitive Radio Networks*. Cambridge University Press, 2009.
13. A. Attar, V. Krishnamurthy, and O. N. Gharehshiran, "Interference management using cognitive base-stations for UMTS LTE," *IEEE Communications Magazine*, vol. 49, no. 8, pp. 152–159, August 2011.
14. E. J. Hong, S. Y. Yun, and D.-H. Cho, "Decentralized power control scheme in femtocell networks: A game theoretic approach," in *Proceedings of International Symposium on Personal, Indoor and Mobile Radio Communications (PIMRC)*, pp. 415–419, September 2009.
15. S. Barbarossa, S. Sardellitti, A. Carfagna, and P. Vecchiarelli, "Decentralized interference management in femtocells: A game-theoretic approach," in *Proceedings of International Conference on Cognitive Radio Oriented Wireless Networks & Communications (CROWN-COM)*, June 2010.
16. L. Giupponi and C. Ibars, "Distributed interference control in OFDMA-based femtocells," in *Proceedings of IEEE International Symposium on Personal Indoor and Mobile Radio Communications (PIMRC)*, pp. 1201–1206, September 2010.
17. W.-Chih Hong and Z. Tsai, "On the femtocell-based MVNO model: A game theoretic approach for optimal power setting," in *Proceedings of IEEE Vehicular Technology Conference (VTC-Spring)*, May 2010.

18. S. Barbarossa, A. Carfagna, S. Sardellitti, M. Omilipo, and L. Pescosolido, "Optimal radio access in femtocell networks based on Markov modeling of interferers' activity," in *Proceedings of IEEE International Conference on Acoustics, Speech and Signal Processing (ICASSP)*, pp. 3212–3215, May 2011.
19. F. Facchinei and J. S. Pang, *Finite-Dimensional Variational Inequalities and Complementarity Problems-Volume I*, Springer-Verlag, New York Inc., December 2003.
20. D. P. Bertsekas and J. N. Tsitsiklis, *Parallel and Distributed Computation: Numerical Methods*. Athena Scientific, 1989.
21. S. Guruacharya, D. Niyato, E. Hossain, and D. I. Kim, "Hierarchical competition in femtocell-based cellular networks," in *Proceedings of Global Telecommunications Conference (GLOBECOM)*, December 2010.
22. Z. Huang, Z. Zeng, and H. Xia, "Hierarchical power game with dual-utility in two-tier OFDMA femtocell networks," in *Proceedings of International Conference on Wireless Communications, Networking and Mobile Computing (WiCOM)*, September 2011.
23. Q. Li, Z. Feng, W. Li, Y. Liu, and P. Zhang, "Joint access and power control in cognitive femtocell networks," in *Proceedings of International Conference on Wireless Communications and Signal Processing (WCSP)*, November 2011.
24. R. D. Yates, "A framework for uplink power control in cellular radio systems," *IEEE Journal on Selected Areas in Communications*, vol. 13, no. 7, pp. 1341–1347, September 1995.
25. F. Pantisano, M. Bennis, W. Saad, and M. Debbah, "Spectrum leasing as an incentive towards uplink macrocell and femtocell cooperation," *IEEE Journal on Selected Areas in Communications*, vol. 30, no. 3, pp. 617–630, April 2012.
26. S. Hart and A. Mas-Colell, "A simple adaptive procedure leading to correlated equilibrium," *Econometrica*, vol. 68, no. 5, pp. 1127–1150, September 2000.
27. X. Kang, R. Zhang, and M. Motani, "Price-based resource allocation for spectrum-sharing femtocell networks: A Stackelberg game approach," *IEEE Journal on Selected Areas in Communications*, vol. 30, no. 3, pp. 538–549, April 2012.
28. R. Xie, F. R. Yu, and H. Ji, "Energy-efficient spectrum sharing and power allocation in cognitive radio femtocell networks," in *Proceedings of IEEE INFOCOM*, pp. 1665–1673, March 2012.
29. Y. Yi, J. Zhang, Q. Zhang, and T. Jiang, "Spectrum leasing to femto service provider with hybrid access," in *Proceedings of IEEE INFOCOM*, pp. 1215–1223, March 2012.
30. S. Barbarossa, S. Sardellitti, and A. Carfagna, "Pricing mechanisms for interference management games in femtocell networks based on Markov modeling," in *Proceedings of Future Network & Mobile Summit (FutureNetw)*, June 2011.
31. S. Yun, Y. Yi, D.-H. Cho, and J. Mo, "Open or close: On the sharing of femtocells," in *Proceedings of IEEE INFOCOM*, pp. 116–120, April 2011.
32. X. Kang, Y.-C. Liang, and H. K. Garg, "Distributed power control for spectrum-sharing femtocell networks using Stackelberg game," in *Proceedings of IEEE International Conference on Communications (ICC)*, June 2011.
33. L. Jiang, S. Parekh, and J. Walrand, "Base station association game in multi-cell wireless networks," in *Proceedings of Wireless Communications and Networking Conference (WCNC)*, pp. 1616–1621, March–April 2008.
34. Y. Chen, J. Zhang, and Q. Zhang, "Utility-aware refunding framework for hybrid access femtocell network," *IEEE Transactions on Wireless Communications*, vol. 11, no. 5, pp. 1688–1697, May 2012.
35. J.-S. Lin and K.-T. Feng, "Game theoretical model and existence of win-win situation for femtocell networks," in *Proceedings of IEEE International Conference on Communications (ICC)*, June 2011.

36. F. Pantisano, M. Bennis, R. Verdone, and M. Latva-aho, "Interference management in femtocell networks using distributed opportunistic cooperation," in *Proceedings of IEEE Vehicular Technology Conference (VTC Spring)*, May 2011.
37. O. Simeone, I. Stanojev, S. Savazzi, Y. Bar-Ness, U. Spagnolini, and R. Pickholtz, "Spectrum leasing to cooperating secondary ad hoc networks," *IEEE Journal on Selected Areas in Communications*, vol. 26, no. 1, pp. 203–213, January 2008.
38. L. Kóczy, "A recursive core for partition function form games," *Theory and Decision*, vol. 63, no. 1, pp. 41–51, August 2007.
39. C.-Y. Wang and H.-Y. Wei, "Revenue extraction in overlay macrocell-femtocell system under shared spectrum model," in *Proceedings of International Symposium on Applied Sciences in Biomedical and Communication Technologies (ISABEL)*, November 2010.

CHAPTER 8

RESOURCE ALLOCATION IN CDMA-BASED MULTI-TIER HETNETS

RRM for CDMA wireless cellular networks has been an active research topic over the past several years. To motivate research issues in multi-tier wireless HetNets, we first review some important resource allocation algorithms, which have been proposed for homogeneous wireless networks. Development of distributed power control for CDMA cellular networks is an interesting research topic that has achieved great success so far. In 1993, Foschini and Miljanic proposed one of the most popular power control algorithms, which is usually referred to as Foschini–Miljanic power control algorithm today [1]. This power control scheme can be implemented distributively, which converges to a Pareto-optimal fixed point supporting predetermined target SINRs using minimum transmission powers whenever possible [2]. Following this great discovery, Yates developed an elegant analytical framework in his seminal paper [3], namely the standard function approach, which can be used to establish convergence and to design power control algorithms. It was later shown in [4, 5] that the Foschini–Miljanic power control algorithm can be indeed designed by using noncooperative game theory with an appropriate payoff function. The Foschini–Miljanic power control algorithm aims to achieve predetermined target SINRs, which, therefore, does not attain high system throughput in lightly loaded networks. In highly loaded networks, different user removal and admission control strategies were proposed [6–8] to resolve congestion that aims to support the largest number of users with their required QoS.

By using game theory, researchers proposed various distributed power control algorithms considering different design objectives and QoS constraints such as power saving, outage constraints, and multiuser diversity exploitation [4, 5, 9–15]. In particular, Sung and Leung have proposed an opportunistic power control strategy that guarantees convergence and enables exploitation of multiuser diversity to enhance the system throughput [16, 17]. Moreover, they developed a more general framework than that in [3] to prove the convergence of their power control scheme, based on the so-called two-sided scalable functions. The opportunistic power control strategy, however, cannot provide minimum QoS support for users as the Foschini–Miljanic

Radio Resource Management in Multi-Tier Cellular Wireless Networks, First Edition.
Ekram Hossain, Long Bao Le, and Dusit Niyato
© 2014 John Wiley & Sons, Inc. Published 2014 by John Wiley & Sons, Inc.

power control scheme. In [18, 19], a dynamic joint BS association and power control algorithm was proposed based on the Foschini–Miljanic power control scheme, and its convergence was also established. In this algorithm, each user will iteratively associate with a BS requiring the minimum transmission power while performing the standard Foschini–Miljanic power updates.

Interference and RRM becomes much more complex in the multi-tier heterogeneous cellular network, which must address the following issues.

- Different tiers of the HetNets may have different priorities in accessing the radio spectrum. Therefore, RRM must be performed in such a way that the QoS of users in the prioritized tiers is protected against spectrum access from other tiers of lower priority. For example, in a two-tier macro–femto HetNet, femto users can only be allowed to utilize the leftover capacity beyond what is needed to support the required QoS of macro users.

- Fair spectrum sharing among users and load balancing among different tiers are desired objectives in the resource management. Here, both co-tier interference and CTI must be appropriately resolved to attain high system throughput while providing fairness guarantees for the multi-tier HetNets. Also, optimizing the trade-off between co-channel interference and spectrum reuse gain must be considered in the multicell and multi-tier resource allocation problem to maximize resource utilization and throughput performance.

- Mass deployment of small cells such as femtocells in the future HetNet is needed to provide sufficient wireless capacity for many emerging broadband wireless applications. Hence, development of decentralized resource allocation algorithms that require low coordination overhead and signaling is a crucial research issue. In addition, such decentralized resource allocation algorithms are desired to achieve optimal performance or at least some guaranteed fraction of the optimal performance (e.g., network throughput or network utility).

Most existing works on HetNets focus on approximated performance analysis for large random networks using stochastic geometry [20–23], design of heuristic transmission power setting, or control schemes [20, 24, 25]. While the stochastic geometry approach can be useful in quantifying approximate performance of very large HetNets, it would not be sufficient for designing practical and implementable resource allocation algorithms for HetNets. In addition, most existing algorithms do not provide efficient QoS protection mechanisms for prioritized network tiers (e.g., macro tier in a macro–femto two-tier network) and allow autonomous and distributed implementation with strong convergence and performance guarantees. Therefore, a lot of further efforts are needed to resolve open research issues in the coming years.

In this chapter, we present some exemplary resource allocation algorithms for CDMA-based multi-tier HetNets. We will demonstrate how game theory and optimization theory can be employed to devise distributed algorithms with different trade-offs between signaling overhead and efficiency. All presented algorithms provide robust QoS protection for macro users and converge to desirable operating points. Firstly, we review existing literature on resource allocation and QoS support in single-tier homogeneous wireless cellular networks. Secondly, a price-based noncooperative game theory approach is taken to design a dynamic spectrum sharing algorithm for two-tier macro–femto networks under the closed access mode. We describe how different utility and cost functions are chosen for macro and femto users to achieve some desirable design objectives. Thirdly, a general framework for joint BS association and power control for wireless HetNets is presented and its convergence is established for a broad class of power control algorithms satisfying some well-defined properties. We then discuss an efficient hybrid power control (HPC) algorithm for the proposed framework exploiting the advantages of well-known Target-SINR-tracking Power Control (TPC) and Opportunistic Power Control (OPC) power control schemes. An adaptation mechanism for the proposed framework using the HPC algorithm and its application to design a hybrid spectrum access scheme for the two-tier HetNet are also presented. Fourthly, we demonstrate how optimization theory can be employed to devise a distributed Pareto-optimal power control algorithm for two-tier HetNets with QoS protection for macro users in terms of minimum SINR requirements. Fairness among femto users is achieved by formulating the resource allocation problem as a utility maximization problem using a general α-fair utility function. Finally, the chapter ends with summary of the presented algorithms and some future research directions.

8.1 POWER CONTROL AND RESOURCE ALLOCATION TECHNIQUES FOR HOMOGENEOUS CDMA NETWORKS

In CDMA-based wireless cellular networks, simultaneous transmissions of all users occur on the same frequency spectrum. Therefore, inter-user interference can significantly degrade the network performance if not controlled appropriately. Power control is the key technique for interference management and QoS support in CDMA cellular networks. Since interference happens for different users in the same cell and for users in different cells (i.e., intracell interference and ICI), distributed power control algorithms are strongly desired so that users can autonomously adapt their transmission powers to cope with the corresponding received interference. In this section, we review some popular power control algorithms and their design principles, which have been proposed for homogeneous single-tier CDMA cellular networks.

Let p_i be the transmission power of user i and σ_i be the power of additive white Gaussian noise measured in the spectrum bandwidth at the receiving end of user i. Also, denote the channel gain from the transmitter of user i to his or her receiver by $h'_{i,i}$, and that from the transmitter of user j to the receiver of user $i \neq j$ by $h_{i,j}$. Then, the received SINR of user i can be written as

$$\gamma_i = \frac{P_G h'_{i,i} p_i}{\sum_{j \neq i} h_{i,j} p_j + \sigma_i} \tag{8.1}$$

where P_G is the processing gain of the system.

Note that the first term in the denominator of (8.1) includes both intracell interference and ICI and this SINR expression applies to both UL and DL scenarios. For notational convenience, let $h_{i,i} = P_G h'_{i,i}$ where the processing gain P_G is absorbed into the channel gain $h'_{i,i}$. The received SINR of user i can then be expressed as

$$\gamma_i = \frac{h_{i,i} p_i}{\sum_{j \neq i} h_{i,j} p_j + \sigma_i}. \tag{8.2}$$

Design of power control algorithms would typically consider some SINR constraints for QoS support and fairness among users. We will describe some well-known power control algorithms and their design principles in the following.

8.1.1 Target-SINR-Tracking Power Control

In CDMA-based cellular wireless standards such as IS-95 and WCDMA, a minimum SINR is required at the receiver so that a minimum data rate can be supported. Maintenance of such minimum SINR targets is well-justified for the voice service to achieve a certain desired BER for a given fixed rate. Given a desired threshold Γ_i for user i, the corresponding SINR constraints can be expressed as

$$\gamma_i \geq \Gamma_i, \ i \in \mathcal{N} \tag{8.3}$$

where $\mathcal{N}$ denotes the set of users. Due to interference coupling among users, the capacity of the CDMA cellular network is limited by the interference. As a result, it may not be possible to support all SINR constraints in (8.3) even if each user has infinite power budget [6–8]. When these SINR constraints are feasible, Foschini and Miljanic have proposed an elegant distributed power control algorithm that converges to a fixed point where each user achieves his or her target SINR exactly [1, 2]. The iterative power update in this Foschini–Miljanic algorithm is given as

$$p_i(t + 1) = \frac{\Gamma_i}{\gamma_i(t)} p_i(t) \tag{8.4}$$

where $p_i(t)$ and $\gamma_i(t)$ denote the transmission power and achieved SINR of user i in iteration t, respectively. In order to update transmission power in each iteration, each user only needs to obtain his or her own current SINR. Therefore, this power control algorithm is fully distributed, which converges very fast to a Pareto-optimal fixed point requiring the minimum power for each user. This algorithm is also referred to as TPC in the literature. Unfortunately, if the SINR constraints (8.3) are not feasible, then the power update (8.4) diverges to the infinite power without convergence. In addition, there is a maximum power constraint for each user in the UL transmission. Let P_i be the maximum UL power budget of user i. Then we have the following power-constrained TPC update rule:

$$p_i(t + 1) = \min \left\{ P_i, \frac{\Gamma_i}{\gamma_i(t)} p_i(t) \right\}. \tag{8.5}$$

It is known that this power-constrained TPC scheme always converges to an equilibrium for both feasible and infeasible systems. In addition, the transmission powers of some users will converge to their maximum powers in an infeasible system where their target SINRs may not be achieved. One important and interesting related problem here is to determine a minimum subset of users to be removed for an infeasible system so that the target SINRs of the remaining users can be supported. This problem is indeed NP-hard [6]; therefore, determining the set of users of minimum size to remove typically requires exponential computational complexity. To resolve this challenge, several heuristic user removal algorithms were proposed to solve this problem in the literature [6, 8]. An elegant distributed Pareto-optimal user removal algorithm was proposed in [26], which was shown to be able to remove a minimum number of users distributively for an infeasible system.

In 1995, Roy Yates proposed an interesting framework to prove the convergence of any iterative power control algorithm of the form $p_i(t + 1) = J_i(p(t))$ where t represents the iteration index and $J_i(p(t))$ denotes the power update function for user i, which is a function of transmission powers of all users in the network. Let $J(p(t))$ be a vector whose ith element $J_i(p(t))$ be the power update function corresponding to user i. Yates showed that any such power control algorithm converges to a fixed point under both synchronous and asynchronous power updates if $J_i(p(t))$ is a standard function that satisfies the following properties.

- Positivity: $J(p(t)) \succ 0$.
- Monotonicity: If $p \succeq p'$, then $J(p) \succeq J(p')$.
- Scalability: For all $\alpha > 1$, $\alpha J(p) \succ J(\alpha p')$.

The TPC strategy indeed satisfies these properties. Therefore, its convergence can be established by using the above result. In [16], Sung and Leung extended the Yates's convergence framework where they showed that any power control algorithm whose power update function satisfies the more general two-sided scalability property converges.

In the works described above, a snapshot channel model is assumed to establish the convergence of the power control algorithms where channel gains are assumed to remain static during the power updates. Extensions of these convergence results to the varying channels have been performed under the stochastic power control framework in [27].

8.1.2   Power Control Design from Game Theoretic View

Game theory has been shown to be a useful tool in designing distributed power control algorithms. In particular, each user can determine his or her transmission power by acting as a player in a noncooperative game. One popular approach within this design framework is to allow each user to iteratively update his or her transmission power as the best response given the transmission powers of other users. There are some design aspects to consider for such a design approach. Firstly, the choice of payoff functions for users plays a central role since it governs the convergence behavior of the underlying power control strategy. Secondly, as the payoff function is given, then investigation of the convergence properties and characterization of possible fixed points are the key research tasks. Specifically, it is often of interest to study the existence and uniqueness of the Nash equilibrium for the game, the efficiency of the Nash equilibrium, and the convergence of the proposed power control strategy. Thirdly, the payoff function needs to be reverse engineered so that the underlying power control strategy can guarantee convergence to an efficient Nash equilibrium and can be easily implemented distributively.

It turns out that several existing power control algorithms can be indeed designed under this noncooperative game framework. Such a game theoretic design view can be illustrated by finding the corresponding payoff function of the underlying game. We describe some important power control algorithms in some detail in the following. Let us start by considering a noncooperative game where the payoff function chosen by each user i is

$$U_{\text{tot},i}^{\text{TPC}} = (\gamma_i - \Gamma_i)^2 \tag{8.6}$$

where γ_i represents the SINR of user i given in (8.2) and Γ_i denotes the target SINR of user i. To find the best response for user i given the transmission powers of other users, we can set the derivative of $U_{\text{tot},i}^{\text{TPC}}$ with respect to p_i equal to zero, which gives the necessary optimal condition. After some simple manipulations, we arrive at the following relationship

$$p_i = \Gamma_i \frac{\sum_{j \neq i} h_{i,j} p_j + \sigma_i}{h_{i,i}}. \tag{8.7}$$

By using this relationship where the transmission powers in the left-hand side and the right high side correspond to iterations $t + 1$ and t, respectively, we obtain the following iterative power control strategy

$$p_i(t+1) = \Gamma_i \frac{\sum_{j \neq i} h_{i,j} p_j(t) + \sigma_i}{h_{i,i}} = \frac{\Gamma_i}{\gamma_i(t)} p_i(t), \tag{8.8}$$

which is exactly the well-known TPC scheme presented before.

Consider another noncooperative game with the following payoff function

$$U_{\text{tot},i}^{\text{OPC}} = \sqrt{\gamma_i} - \lambda_i p_i. \tag{8.9}$$

Again, by setting the derivative of $U_{\text{tot},i}^{\text{OPC}}$ with respect to p_i equal to zero, we have following

$$p_i = \frac{1/(2\lambda_i)^2}{R_i} \tag{8.10}$$

where $R_i = (\sum_{j \neq i} h_{i,j} p_j + \sigma_i)/h_{i,i}$ denotes the effective interference. From this relationship, we can obtain the following iterative power update strategy

$$p_i(t+1) = \frac{\zeta_i}{R_i(t)} \tag{8.11}$$

where $\zeta_i = 1/(2\lambda_i)^2$. This is indeed the OPC algorithm proposed by Leung and Sung in [17]. The more general payoff function of the form $U_{\text{tot},i} = U_i(p) - C_i(p)$ can be chosen to design a power control game where $U_i(p)$ and $C_i(p)$ denote the utility and cost functions, respectively. For example, $C_i(p)$ can represent pricing function that can balance between achievable utility and utilized radio resources. Such price-based power control design has been adopted in many existing works (e.g., see [11–13]).

A distributed power control algorithm that realizes the best response typically converges to a Nash equilibrium of the underlying game. However, such a Nash equilibrium may not be the most efficient operating point for the network since there may exist other operating points that strictly dominate it. Achieving a non-dominated operating point may require a centralized power control strategy, which would not be desirable for practical large-scale networks [11]. Fortunately, it is possible to design Pareto-optimal power control algorithms using, for example, optimization theory, even though such algorithms usually require more signaling overhead (e.g., see [28–30]). In particular, distributed resource allocation algorithms can be devised to solve certain network utility maximization problems that can provide fairness among active users.

8.1.3 Joint Base Station Association and Power Control

In practice, users have to choose BSs to connect with in addition to performing power control to achieve the desired QoS in CDMA cellular networks. The BS association and power control tasks can be performed at different or the same time scales. Users can typically make BS association decisions based on some metrics such as channel gains or SINRs corresponding to potential BSs. When the average values of such metrics are employed, BS association decisions are less frequently made, which, therefore, require less signaling overhead, compared with the case where BS association and power control are updated at the same fast time scale.

One elegant dynamic algorithm in which BS association and power control tasks operate at the same time scale aiming to achieve minimum transmission powers was proposed in [18, 19] for homogeneous cellular networks. This algorithm can be viewed as an extension of the Foschini–Miljanic power control algorithm where BS association decisions are integrated with the power updates. We describe this algorithm for the UL in the following. To proceed, let b_i denote the BS that user i associates with and $h_{b_i,j}$ denote the channel gain from user j to the associated BS b_i of user i. Then, the received SINR of user i can then be written as

$$\gamma_i(b_i) = \frac{h_{b_i,i} p_i}{\sum_{j \neq i} h_{b_i,j} p_j + \sigma_{b_i}}. \tag{8.12}$$

Suppose each user i wishes to achieve a target SINR of Γ_i. Then, the transmission power required by user i to achieve this target SINR when it is associated with BS k can be expressed as

$$p_i(k) = \Gamma_i \frac{\sum_{j \neq i} h_{k,j} p_j + \sigma_k}{h_{k,i}} \tag{8.13}$$

where we explicitly indicate the dependence of the transmission power to the associated BS k. It was proposed in [18, 19] that the BS association and transmission power are updated in each iteration as

$$b_i = \underset{k}{\operatorname{argmin}} \, p_i(k) \tag{8.14}$$

$$p_i = p_i(b_i) \tag{8.15}$$

where we neglect the iteration index for simplicity. In practice, each user only needs to choose one BS in a small set of nearby BSs to associate with. Therefore, this algorithm can be implemented in a distributed manner since each user is required to estimate the transmission powers to achieve his or her target SINR corresponding to potential BS associations in each iteration. This BS association and power control algorithm was shown to converge to a fixed point that achieves the target SINRs for all users with minimum transmission power whenever feasible [18, 19]. Moreover, such convergence holds under both synchronous and asynchronous updates.

For data-oriented wireless networks, efficient utilization of radio resources to enhance the system throughput would be more relevant than achieving fixed target SINRs for active users. Moreover, for wireless HetNets, it is desirable to design flexible, decentralized BS association and resource allocation mechanisms that can effectively manage interference, provide QoS support for users of different tiers, and enable controllable data off-loading from macrocells to small cells. Therefore, lots of further efforts are needed to fulfill these design objectives.

8.2   GAME THEORETIC BASED POWER CONTROL FOR TWO-TIER CDMA HETNETS

Power control is the main mechanism for interference management and QoS provisioning in CDMA two-tier HetNets, where users of different tiers transmit on the same frequency band. For this kind of network, macro users typically have higher priority in accessing the spectrum; therefore, their minimum QoS performance must be protected against the CTI from the femto tier. Also, distributed power control algorithms with little overhead are desirable since the backhaul links based on wired broadband links such as DSL may have limited capacity. In this section, we demonstrate the design of such a power control algorithm by using a noncooperative game theory approach. Specifically, users in the two tiers will have different payoff functions and each user will iteratively play a so-called best-response strategy, which allows it to maximize its payoff given the current strategies (i.e., transmission power levels) chosen by other users.

We consider a two-tier wireless network where a macrocell serving N_m macro users is underlaid with J femtocells using CDMA. Assume that femtocell i has N_i users and define $N_f = \sum_{i=1}^{J} N_i$. The association of the femto users with their closest femto access points is fixed during the runtime of the underlying power and admission control processes. Denote the set of all users by $\mathcal{N}$ and the set of macro and femto users by $\mathcal{N}_m$ and $\mathcal{N}_f$, respectively. For simplicity, we describe the model in the UL scenario even though it is applicable for the DL as well. We consider a snapshot model where the channel gains are assumed to remain unchanged during the runtime of the power and admission control algorithms. Then, the received SINR of user $i \in \mathcal{N}$ is also given in (8.2) where the first term in the denominator of (8.2) includes both co-tier interference and CTI, that is, *aggregated* interference from all macro and femto users except the considered user i (which can be either a macro or femto user). We assume that the prioritized macro user $i \in \mathcal{N}_m$ requires that $\gamma_i \geq \Gamma_i$ for a given desired threshold Γ_i in order to have his or her ongoing operation robustly protected.

We can employ a utility function $U_i(\gamma_i)$ and a cost function $C_i(p_i)$ to represent the degree of satisfaction of user $i \in \mathcal{N}$ to the service quality and the cost incurred, respectively. It is the interest of user $i \in \mathcal{N}$ to maximize his or her own net utility, defined as

$$U_{\text{tot},i} = U_i(\gamma_i) - C_i(p_i). \tag{8.16}$$

In fact, (8.16) is a standard way to define the payoff function for network entities (i.e., wireless users and BSs). Given the transmission powers of other users, such an optimization can be accomplished by dynamic power adaptation performed at individual links.

Assume that $U_i(\gamma_i)$ is a strictly concave function with respect to p_i, whereas $C(p_i)$ is convex in that same variable. The necessary condition for maximization of (8.16) can be obtained by taking the derivative of $U_{\text{tot},i}$, which is also strictly concave in p_i, and equating to zero as

$$\frac{dU_{\text{tot},i}}{dp_i} = \frac{dU_i}{d\gamma_i}\frac{d\gamma_i}{dp_i} - \frac{dC_i}{dp_i} = 0. \tag{8.17}$$

Upon noting that $d\gamma_i/dp_i = h_{i,i}/I_i = \gamma_i/p_i$, we have

$$U_i'(\gamma_i) = \frac{p_i}{\gamma_i}C_i'(p_i) = \frac{I_i}{h_{i,i}}C_i'(p_i) \tag{8.18}$$

where $I_i = \sum_{j\neq i} h_{ij} p_j + \sigma_i$ is the total noise and interference power at the receiving side of user $i \in \mathcal{N}$. From (8.18), the optimal target SINR can be derived as

$$\widehat{\gamma}_i = f_i^{-1}\left(\frac{I_i}{h_{i,i}}C_i'(p_i)\right) \tag{8.19}$$

where $f_i(\gamma_i) = U_i'(\gamma_i)$. By choosing the transmission power satisfying (8.19), user i plays the so-called best response strategy according to the noncooperative game theory. Under such a design, the following iterative power update rule is employed [4]:

$$p_i(t+1) = \widehat{\gamma}_i(t)\frac{I_i(t)}{h_{i,i}(t)} = \frac{\widehat{\gamma}_i(t)}{\gamma_i(t)}p_i(t) \tag{8.20}$$

where $\gamma_i(t)$ is the actual SINR of user i at iteration t. In fact, (8.20) represents a more general power-control rule compared with the well-known Foschini–Miljanic power update strategy given in (8.4). Specifically, the minimum required SINR Γ_i on the right-hand side of (8.4) is replaced by an adaptive SINR threshold $\widehat{\gamma}_i(t)$ in (8.19).

In what follows, we will show how to choose appropriate functions $U_i(\gamma_i)$ and $C_i(p_i)$, together with their operating parameters, to design efficient distributed power and admission control algorithms for both macro users and femto users. The key aspect that makes the existing algorithms (such as those in [6,7]) unsuitable for our current purpose is that the minimum SINRs of the prioritized macro users should be maintained all the times. Accordingly, the transmission powers of newly deployed femto users must be properly controlled or, if needed, may even be removed for the sake of protecting the macro users.

8.2.1 Guaranteeing QoS for Macro Users

The work in [4] recommends the use of a sigmoid utility function and a linear cost function. For the current design, by employing similar utility and cost functions for the macro users and via properly tuning their control parameters, we can develop an efficient and robust power control algorithm that is capable of maintaining the minimum SINR requirements for these users. Specifically, we select the following utility and cost functions for a macro user $i \in \mathcal{N}_m$:

$$U_i(\gamma_i) = \frac{1}{1 + \exp[-b_i(\gamma_i - c_i)]} \tag{8.21}$$

$$C_i(p_i) = a_i^{(m)} p_i. \tag{8.22}$$

Here, b_i and c_i, respectively, control the steepness and the center of the sigmoid function, whereas $a_i^{(m)}$ is the pricing coefficient.

The function $U_i(\gamma_i)$ in (8.21) naturally captures the value of the service offered to user i. Upon noting that $U_i(0) = 0$, $U_i(\infty) = 1$ and that $U_i(\gamma_i)$ is increasing with respect to γ_i, it is clear that user i is more and more satisfied with the offered service as the quality, expressed in terms of the achieved SINR γ_i, improves. On the other hand, power is itself a valuable system resource. The linear cost in (8.22) is chosen to reflect the expenses of power consumption to the user, while still retaining the simplicity of subsequent analysis. As shown later, the use of dynamic values of $a_i^{(m)}$ may significantly affect the resulting equilibrium of the developed algorithms.

Importantly enough, the choice of sigmoid function allows for the design of efficient schemes that guarantee the minimum SINRs imposed by the macro users. Using (8.21) and (8.22), the expression in (8.18) can be rewritten as

$$U_i'(\gamma_i) = f_i(\gamma_i) = \frac{a_i^{(m)} I_i}{h_{i,i}}. \tag{8.23}$$

From this relationship, it is straightforward to see that the optimal SINR target is

$$\widehat{\gamma}_i = f_i^{-1}\left(\frac{a_i^{(m)} I_i}{h_{i,i}}\right). \tag{8.24}$$

With the utility function defined in (8.21), an analytical form of (8.24) can be obtained as follows [4]:

$$\widehat{\gamma}_i = c_i - \frac{1}{b_i} \ln\left[\frac{b_i h_{i,i}}{2a_i^{(m)} I_i} - 1 - \sqrt{\left(1 - \frac{b_i h_{i,i}}{2a_i^{(m)} I_i}\right)^2 - 1}\right]. \tag{8.25}$$

Now, the line that goes through the origin and is tangent to the utility curve $U_i(\gamma_i)$ can be expressed as $U_i(\gamma_i) = U_i'(\gamma_i)\gamma_i$. At the tangent point $\gamma_{i,u}$, it is clear that

$$U_i(\gamma_{i,u}) = U_i'(\gamma_{i,u})\gamma_{i,u}. \tag{8.26}$$

Since the cost function in (8.22) can also be rewritten as $C_i(p_i) = \left(a_i^{(m)} I_i / h_{i,i}\right)\gamma_i$, it is required that $a_i^{(m)} I_i / h_{i,i} \leq U_i'(\gamma_{i,u})$ for a nonnegative total utility as shown in

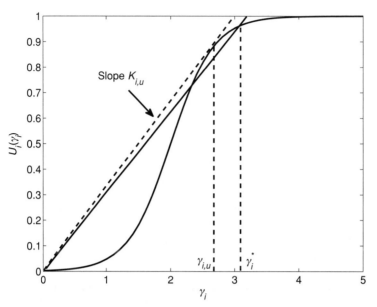

Figure 8.1   Illustration of parameter setting for payoff function.

Figure 8.1. On the other hand, the necessary and sufficient condition for $\widehat{\gamma}_i$ in (8.25) to achieve $U_{\text{tot},i} \geq 0$ is $\widehat{\gamma}_i \geq \gamma_{i,u}$; otherwise, macro user i simply suppresses his or her transmission and still gains zero total payoff. Therefore, by setting

$$\gamma_{i,u} = \Gamma_i, \tag{8.27}$$

we can ensure that any active macro user (i.e., whose transmission power is strictly positive) will attain his or her minimum SINR target. In other words, an active macro user i will eventually achieve SINR $\widehat{\gamma}_i \geq \gamma_{i,u} = \Gamma_i$ under this design. Figure 8.2 illustrates this parameter setting and the achievable SINR $\widehat{\gamma}_i = \gamma_i^*$ at the equilibrium. Some manipulations of (8.26) and (8.27) give [4]

$$c_i = \Gamma_i - \frac{\ln(b_i\Gamma_i - 1)}{b_i}. \tag{8.28}$$

Upon substituting this value of c_i to (8.25), we finally arrive at

$$\widehat{\gamma}_i = \Gamma_i - \frac{\ln(b_i\Gamma_i - 1)}{b_i} - \frac{1}{b_i}\ln\left[\frac{b_ih_{i,i}}{2a_i^{(m)}I_i} - 1 - \sqrt{\left(1 - \frac{b_ih_{i,i}}{2a_i^{(m)}I_i}\right)^2 - 1}\right]. \tag{8.29}$$

In Figure 8.2, we show the operating range of an active macro user i. With a sufficiently large b_i, function $U_i'(\cdot)$ becomes very steep; therefore, the resulting γ_i of user i will be very close to its SINR threshold Γ_i. Also clear from Figure 8.2 is that if the minimum required SINRs of all the macro users are feasible, we can make them all active by setting $a_i^{(m)}$ sufficiently small. Specifically, given its total received

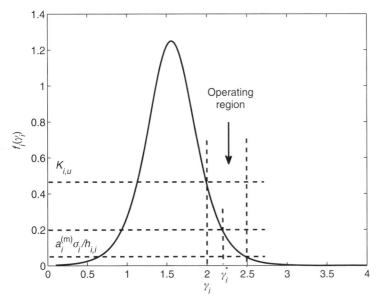

Figure 8.2   Illustration of the equilibrium solution for $\Gamma_i = 2$, $b_i = 5$, $K_{i,u} = U_i'(\Gamma_i)$, $a_i = a_i^{(m)}$. © [2012] IEEE.

interference and noise power I_i, macro user $i \in \mathcal{N}_m$ is active if

$$a_i^{(m)} < h_{i,i} U_i'(\Gamma_i)/I_i. \tag{8.30}$$

We will propose a design example of the utility and cost functions for femto users and the corresponding power and admission control algorithm in the following.

8.2.2   Power Adaptation and Admission Control Algorithm

Given the macro users' QoS requirements already supported, the specific choice of utility and cost functions for femto users allows us to achieve our design objectives, through which certain user satisfaction metrics can be attained. Notice that if the femto users also wish to maintain their respective QoS requirements, the operation of these users may cause network congestion, hence badly affecting the performance of the macro users. In such cases, the femto users should be penalized by appropriately regulating their operating parameters. We will describe one design example for the femto users. In particular, we will propose a joint power adaptation and admission control algorithm and establish its convergence condition.

Suppose that femto user $i \in \mathcal{N}_f$ also requires a minimum SINR Γ_i to maintain the quality of its applications. While a higher SINR at the receiving end of any femto links implies more reliability and better services, this usually requires more transmission power, which in turn leads to a higher cross-interference induced to the macrocell. Such an observation motivates us to consider the following net utility for femto user i (similar to that in [5]):

$$U_{\text{tot},i} = -(\gamma_i - \Gamma_i)^2 - a_i^{(f)} p_i, \quad \forall i \in \mathcal{N}_f. \tag{8.31}$$

Although maximizing the first term on the right-hand side of the above equation, that is, the utility $U_i(\gamma_i) = -(\gamma_i - \Gamma_i)^2$, enforces the SINR γ_i of femto user i to be as close as possible to the SINR target Γ_i, the resulting γ_i^* at the equilibrium may actually be less than Γ_i. Nevertheless, it has been shown in [5] that by allowing a reasonable deviation from the target SINR, a significant reduction in the transmission power (and hence the resulting interference) can be achieved. Given its lower access priority, this type of soft QoS provisioning is totally acceptable for femto user i. On the other hand, the cost function, $C_i(p_i) = a_i^{(f)} p_i$, penalizes the expenditure of transmission power, which potentially creates undue interference to the macrocell as well as other femto users. Here, $a_i^{(f)}$ is the pricing coefficient of such penalization.

Algorithm 8.1   POWER AND ADMISSION CONTROL ALGORITHM

1: Set $p_i = 0$, $\forall i \in \mathcal{N}$; initialize the set of active femto users $\mathcal{A}_f = \mathcal{N}_f$; and set $t = 1$.
2: Each macro user $i \in \mathcal{N}_m$ measures $h_{i,i}$ and $I_i(t)$, and calculates $\widehat{\gamma}_i(t)$ by (8.29).
3: **if** $\widehat{\gamma}_i(t) \geq \Gamma_i$ **then**
4:    Macro user $i \in \mathcal{N}_m$ updates its power: $p_i(t+1) = \frac{I_i(t)\widehat{\gamma}_i(t)}{h_{i,i}}$.
5: **else if** $\widehat{\gamma}_i(t) < \Gamma_i$ and $\left|\mathcal{A}_f\right| > 0$ **then**
6:    Macro user $i \in \mathcal{N}_m$ updates its power: $p_i(t+1) = \frac{I_i(t)\Gamma_i}{h_{i,i}}$.
7:    Each femto user $j \in \mathcal{A}_f$ updates its pricing coefficient: $a_j^{(f)} = k_j^{(f)} a_j^{(f)}$, where $k_j^{(f)} > 1$ are predetermined scaling factors.
8: **end if**
9: Each femto user $j \in \mathcal{A}_f$ measures $h_{j,j}(t)$ and $I_j(t)$ and calculates $\widehat{p}_j$ as
$$\widehat{p}_j = \frac{I_j(t)\Gamma_i}{h_{j,j}} - \frac{a_j^{(f)} I_j(t)^2}{2h_{j,j}^2}.$$
10: Femto user $j \in \mathcal{A}_f$ updates its power: $p_j(t+1) = \widehat{p}_j$.
11: **if** $\widehat{p}_j h_{j,j}/I_j(t) < \underline{\gamma}_j^{(f)}$ **then**
12:    If $t = nT^{(f)}$, then with a small probability α, user j sets $p_j(t+1) = 0$ and removes himself from the set of active femto users: $\mathcal{A}_f = \mathcal{A}_f - \{j\}$.
13: **end if**
14: Any femto access point with no active femto user informs the macrocell through a dedicated signaling channel.
15: Set $t = t + 1$, go to step 2 and repeat until convergence.

Now, applying the result in (8.18) to these particular utility and cost functions yields:

$$\gamma_i = \Gamma_i - \frac{a_i^{(f)} I_i}{2h_{i,i}}. \tag{8.32}$$

Because $U_i(\gamma_i)$ is a concave function in p_i, so is $U_{\text{tot},i}$, $\forall i \in \mathcal{N}_f$. The power value that globally maximizes $U_{\text{tot},i}$ can thus be computed as

$$p_i = \max\left(\frac{I_i \Gamma_i}{h_{i,i}} - \frac{a_i^{(f)} I_i^2}{2h_{i,i}^2}, 0\right). \tag{8.33}$$

Based upon the power update rule in (8.33), a joint power adaptation and admission control algorithm is now developed that is capable of providing soft QoS for the femto users. In this algorithm, we gradually increase pricing coefficients $a_i^{(f)}$ of all active femto users when there exists an active macro user with "soft" SINR target $\widehat{\gamma}_i(t)$ below the required SINR target (i.e., $\widehat{\gamma}_i(t) < \Gamma_i$). This algorithm lends itself to a distributed implementation since each user i only needs to estimate (i) its received interference power $I_i(t)$ and (ii) its own channel gain $h_{i,i}(t)$ in order to update its transmission power in each iteration. In step 12 of the proposed algorithm, if a particular femto user j has his/her SINR falling below the minimum required threshold $\underline{\gamma}_j^{(f)}$, it is removed with a small probability α at most once in every $T^{(f)}$ iterations. This *conservative* user removal scheme is employed to avoid eliminating too many femto users unnecessarily.

Theorem 8.1.   Assume that $x f_i^{-1}(x)$ is an increasing function, $\forall i \in \mathcal{N}_m$, the proposed algorithm converges to an equilibrium, at which point

$$p_i^* = \frac{I_i^*}{h_{i,i}} f_i^{-1}\left(\frac{a_i^{(m)} I_i^*}{h_{i,i}}\right), \quad i \in \mathcal{A}_m \tag{8.34}$$

$$p_i^* = \frac{I_i^* \Gamma_i}{h_{i,i}} - \frac{a_i^{(f)} (I_i^*)^2}{2h_{i,i}^2}, \quad i \in \mathcal{A}_f. \tag{8.35}$$

Moreover, all active macro users $i \in \mathcal{N}_m$ have their SINR γ_i^* satisfying $\gamma_i^* \geq \Gamma_i$.

Proof: The convergence of the proposed power updates in this case can be proved using the standard function technique [3]. For the macro users, [4] maintains that $A_i(\mathbf{p}(t))$, $i \in \mathcal{N}_m$ is a standard function if $x f_i^{-1}(x)$ is an increasing function where $A_i(\mathbf{p}(t))$ denotes the corresponding power update function for macro user i. Indeed, this is true if b_i is chosen to be sufficiently large.

For the femto users, it has been shown in [5] that the power updates for such users [see (8.33)] satisfy the requirements of a standard function if the following conditions hold for all $i \in \mathcal{N}_f$:

$$I_i < \frac{h_{i,i} \Gamma_i}{a_i^{(f)}}, \tag{8.36}$$

$$p_i \leq \frac{\Gamma_i^2}{2a_i^{(f)}}. \tag{8.37}$$

Indeed, conditions in (8.36) and (8.37) can be enforced by the admission control mechanism in the proposed algorithm. If the network is congested enough, the transmission powers of certain users will diverge to some large values, creating a large amount of interference $I_i(t)$ to other users. Note that the power update for $i \in \mathcal{N}_f$ satisfies $\gamma_i(t+1) = \Gamma_i - a_i^{(f)} I_i(t)/(2h_{i,i})$. Therefore, if $I_i(t)$ is sufficiently large so that $\gamma_i(t+1) < \underline{\gamma}_i^{(f)}$, femto user i will be removed, which in turn relieves the network congestion. Together with the proper tuning of pricing coefficient $a_i^{(f)}$, (8.36) and (8.37) are eventually satisfied.

Since the power updates of both the macro users and the femto users are standard functions, Algorithm 8.1 converges to an equilibrium. In addition, all active macro users $i \in \mathcal{A}_m$ at that equilibrium must also have their SINR $\gamma_i^* \geq \Gamma_i$. ∎

Readers can refer to [31] for another design with a different femto payoff function that achieves an efficient femto throughput-power trade-off. In general, the Nash equilibrium achieved by the underlying power control game may not be efficient [11]. However, power control algorithms designed by using the game theoretic approach usually require very low signaling overhead and they can be implemented efficiently in a distributed manner. We will present an alternative Pareto-optimal power control algorithm by using optimization theory in Section 8.4 of this chapter.

8.2.3 Numerical Examples

We present numerical results to illustrate the convergence and performance of the aforementioned power and admission control algorithm. The network setting and user placement in our examples are shown in Figure 8.3, where macro and femto users

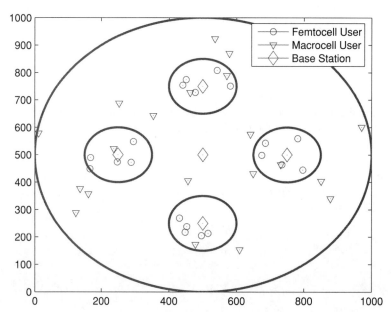

Figure 8.3 Network topology and user placement in two-tier femtocell networks (network dimension in the horizontal and vertical axes is in meter (m)). © [2012] IEEE.

are randomly deployed inside circles of radii of 500 m and 100 m, respectively. Also, it is assumed that the number of users serviced by individual femto access points is identical. For each network realization, channel gains from the transmitter of user j to the receiver of user i are calculated as d_{ij}^{-3} with d_{ij} being their corresponding geographical distance. In addition, other system parameters are chosen as follows: processing gain, $P_G = 100$; power of Gaussian noise, $\sigma_i = 10^{-10}$; and $b_i = 10$, $\forall i \in \mathcal{N}_m$ in the sigmoid utility function of macro users.

The pricing coefficients of all macro users are set equal to $a_i^{(m)} = 1$, $\forall i \in \mathcal{N}_m$ in all simulation results. Moreover, the same initial pricing coefficients $a_j^{(f)}$ and scaling parameters $k_j^{(f)}$ (denoted as k_f in the corresponding figures) are selected for all femto users. The values of SINR targets for macro $\Gamma_i^{(m)}$ and for femto users $\Gamma_i^{(f)}$ can be found in the caption of every plot. The results presented in each figure correspond to one particular network realization, chosen with the intention to demonstrate certain features of the proposed algorithms.

In Figures 8.4 and 8.5, we display the evolutions of SINRs for all users in the low- and high-load scenarios, respectively. In particular, these figures confirm that the presented algorithm actually converges wherein the SINR requirements of all macro users are met at the equilibrium. Also, the convergence speed appears to be slower when the network becomes more congested. In both scenarios, the algorithm enables macro users to just utilize the right fraction of network capacity to meet their own QoS requirements, leaving the leftover to be shared by femto users. Additionally, Figure 8.4 shows that in the lowly loaded regime, the achieved SINRs of femto users are slightly below their corresponding requirements while performance of all macro users is well protected. When network congestion starts building up, the algorithm smoothly reduces the SINRs of femto users so that macro users can eventually reach

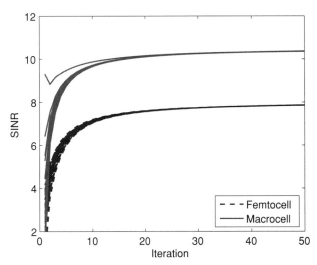

Figure 8.4   SINR evolution in low load for $N_m = 10$, $N_f = 40$, $k_f = 1.5$, $\Gamma_i^{(m)} = 10$, $\Gamma_i^{(f)} = 8$, and $a_i^{(f)} = 10^4$. © [2012] IEEE.

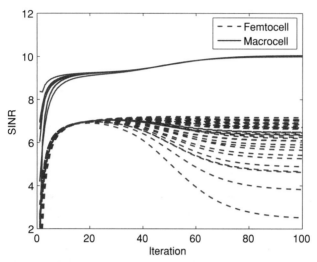

Figure 8.5 SINR evolution in high load for $N_m = 10$, $N_f = 40$, $k_f = 1.5$, $\Gamma_i^{(m)} = 10$, $\Gamma_i^{(f)} = 8$, and $a_i^{(f)} = 10^4$. © [2012] IEEE.

their desired SINR targets as can be observed in Figure 8.5. These results confirm that the presented algorithm is able to offer QoS support for femto users to the extent that network capacity allows.

8.3 JOINT BASE STATION ASSOCIATION AND POWER CONTROL FOR CDMA HETNETS

In the multi-tier HetNets, efficient design of BS association and power control algorithms plays a crucial role in achieving desirable QoS and load balancing for users of different tiers. In addition, such design can be deployed to control the desirable level of data off-loading from macro tier to the high-capacity femto tier in two-tier HetNets. It is also the key in developing an open or hybrid access strategy for two-tier macro–femto HetNets based on which macro users can choose one macro base station or femto access point to associate with.

In this section, we consider the joint BS association and power control problem for the UL of a CDMA-based two-tier HetNet. We assume that there are N users, labeled as $1, 2, \ldots, N$, transmitting information to M BSs, labeled as $1, 2, \ldots, M$, on the same spectrum. The sets of all users and BSs are denoted by $\mathcal{N}$ and $\mathcal{M}$, respectively. Each of these BSs can belong to one of the available cell types (e.g., macrocells, microcells, picocells, and femtocells) in a heterogeneous cellular network. One particular case of interest is the two-tier macro–femto cellular network where $\mathcal{N} = \mathcal{N}_m \cup \mathcal{N}_f$ and $\mathcal{M} = \mathcal{M}_m \cup \mathcal{M}_f$ where $\mathcal{N}_m, \mathcal{N}_f$ denote the sets of macro users and femto users, respectively and $\mathcal{M}_m, \mathcal{M}_f$ denote the sets of macro base stations and femto access points, respectively. We will consider the general scenario in this section. Assume that each user i communicates with only one BS at any time, which is denoted as $b_i \in \mathcal{M}$. However, users can change their associated BSs over time. Note that if $b_i \equiv b_j$, then users i and j are associated with the same BS. Let D_i be the set of BSs that user i can be associated with, which are nearby BSs of user i in practice. Note that we have $D_i \subseteq \mathcal{M}$ and $b_i \in D_i$. We are interested in developing a base station association (BSA) strategy, which determines how each user i will choose one BS $b_i \in D_i$ to communicate with based on the channel state information and interference.

Let the transmission power of user i be p_i, whose maximum value is P_i, that is, $0 \leq p_i \leq P_i$. We arrange transmission powers of all users in a vector, which is denoted as $\mathrm{p} = (p_1, p_2, \ldots, p_N)$. Let $h_{i,j}$ be the channel power gain from user j to BS i, and σ_i be the noise power at BS i. Then, the SINR of user i at BS b_i can be written as

$$\gamma_i(\mathrm{p}) = \frac{h_{b_i,i} p_i}{\sum_{j \neq i} h_{b_i,j} p_j + \sigma_{b_i}} = \frac{p_i}{R_i(\mathrm{p}, b_i)} \tag{8.38}$$

where, for simplicity, we absorb the processing gain P_G into $h_{b_i i}$ as before. $R_i(\mathrm{p}, b_i)$ is the effective interference of user i, which is defined as

$$R_i(\mathrm{p}, b_i) = \frac{\sum_{j \neq i} h_{b_i,j} p_j + \sigma_{b_i}}{h_{b_i,i}} \tag{8.39}$$

where $R_i(\mathrm{p}, k)$ is the effective interference experienced by user i at BS k. We will also write $R_i(\mathrm{p})$ instead of $R_i(\mathrm{p}, b_i)$ when there is no confusion. Suppose that each user i requires the minimum QoS in terms of a target SINR Γ_i, which is expressed as

$$\gamma_i(\mathrm{p}) \geq \Gamma_i, \quad i \in \mathcal{N}. \tag{8.40}$$

Our design objective is to develop distributed joint BSA and power control algorithms that can maintain the SINR requirements in (8.40) (whenever possible) while exploiting the multiuser diversity gain to increase the system throughput. Moreover, we aim to achieve these design objectives for a heterogeneous wireless environment where there are different kinds of users with differentiated QoS targets (e.g., voice and data users) and potentially different cell types (e.g., macrocells, microcells, picocells, and femtocells). In particular, voice users would typically require some fixed target SINR Γ_i while data users would seek to achieve a higher target SINR Γ_i for broadband multimedia applications.

8.3.1 Base Station Association and Power Control Algorithm

We develop a general minimum effective interference BSA and power control algorithm and establish its convergence. Specifically, we will focus on a general iterative power control algorithm where each user i in the network performs the following power update $p_i(t+1) = J_i(\mathrm{p}(t))$ where t denotes the iteration index and $J_i(\cdot)$ is the power update function (p.u.f.). In fact, according to Sung and Leung [16, 17], this kind of iterative power control algorithm converges if its corresponding p.u.f. is a *two-sided scalable* (2.s.s.) *function*, which is indeed a more general technique than the standard function approach proposed by Yates in [3]. The challenges involved in designing such a power control algorithm include the following. We have to ensure its p.u.f. is *two-sided scalable*; it fulfills our design objectives for the heterogeneous cellular network; and it can be implemented in a distributed manner. In addition, we seek to design a power control algorithm that can be integrated with an efficient BSA mechanism. Toward this end, we give the definition of a two-sided scalable function in the following.

Definition 8.1. A p.u.f. $J(p) = \left[J_1(p), \ldots, J_N(p)\right]^T$ is 2.s.s. with respect to p if, for all $a > 1$ and any power vector p' satisfying $\frac{1}{a}p \leq p' \leq ap$, we have

$$(1/a)J_i(p) < J_i(p') < aJ_i(p), \ \forall i \in \mathcal{N}. \tag{8.41}$$

We will consider a joint BSA and power control algorithm where the p.u.f. of the power control scheme satisfies the 2.s.s. property stated in Definition 8.1. Under this design, each user chooses his or her "best" BS and updates its transmission power in the distributed manner. Specifically, the proposed algorithm is described in the following where each user chooses the BS that results in minimum effective interference and updates its power by using a 2.s.s. p.u.f. $J(p) = J'(R(p)) = [J_1'(R_1(p)), \ldots, J_N'(R_N(p))]$, where $R(p) = [R_1(p), \ldots, R_N(p)]$. The proposed algorithm can be implemented distributively with any distributed power control algorithm. To realize the BSA, each user needs to estimate the effective interference levels for different nearby BSs of interest. Then, each user i can choose one BS k in D_i with minimum value of $R_i(p, k)$. This algorithm ensures that each user experiences low effective interference and therefore high throughput at convergence.

Algorithm 8.2 BASE STATION ASSOCIATION AND POWER CONTROL ALGORITHM

1: Initialization: $p_i(0) = 0$ for all user $i \in \mathcal{N}$, $b_i(0)$ is set as the nearest BS of user i.
2: Iteration t: Each user i ($i \in \mathcal{N}$) performs the following.
 - Calculate the effective interference at BS $k \in D_i$ as follows:

$$R_i^{(t)}(p(t-1), k) = \frac{\sum_{j \neq i} h_{k,j} p_j(t-1) + \sigma_k}{h_{k,i}}. \tag{8.42}$$

 - Choose the BS $b_i^{(n)}$ with the minimum effective interference, that is,

$$b_i(t) = \mathrm{argmin}_{k \in D_i} R_i^{(t)}(p(t-1), k) \tag{8.43}$$

$$R_i^{o(t)}(p(t-1)) = \min_{k \in D_i} R_i^{(t)}(p(t-1), k) \tag{8.44}$$

$$= R_i^{(t)}(p(t-1), b_i(t)). \tag{8.45}$$

 - Update the transmission power for the chosen BS as follows:

$$p_i(t) = J_i'(R_i^{o(t)}(p(t-1))) \tag{8.46}$$

where $J_i'(R_i(p))$ is the p.u.f. with respect to $R_i(p)$.

3: Increase t and go back to step 2 until convergence.

Note that we have expressed the p.u.f.s with respect to both p and $R(p)$. For the expression with respect to $R(p)$, the ith element of the p.u.f. is denoted as $J_i'(R_i(p))$ where $R_i(p)$ is the effective interference experienced by user i given in (8.39). Therefore, we have $J(p) = J'(R(p))$, which depends on both the transmission powers of all users and the BS that each user is associated with. We establish the convergence of this algorithm in the following by utilizing the 2.s.s. function approach. Toward this end, we recall the convergence result for any power control algorithm that employs a bounded 2.s.s. p.u.f. in the following lemma, which was proved in [16].

Lemma 8.1.   Assume that $J(p)$ is a 2.s.s. function, whose elements $J_i(p)$ are bounded by zero and P_i, that is, $0 \leq J_i(p) \leq P_i$. Consider the iterative power update $p_i(t+1) = J_i(p(t))$, $\forall i$ where t denotes the iteration index. Then, we have the following results.

1. The p.u.f. $J(p)$ has a unique fixed point corresponding to a transmission power vector p^* that satisfies $p^* = J(p^*)$.

2. Given an arbitrary initial power vector $p(0)$, the power control algorithm based on p.u.f. $J(p)$ converges to that unique fixed point p^*.

We are now ready to state one important result for the joint BSA and power control operations described in the proposed algorithm in the following theorem.

Theorem 8.2.   Assume $J(p)$ and $J'(R(p))$ are arbitrary 2.s.s. p.u.f.s with respect to p and $R(p)$, respectively. Then, the proposed BSA and power control algorithm converges to an equilibrium.

Proof: According to Lemma 8.1, a power control algorithm converges if the corresponding p.u.f. is 2.s.s. Hence, we can prove Theorem 8.2 by showing that the p.u.f. $J^o(p) = [J_1^o(p), \ldots, J_N^o(p)]$ is a 2.s.s. function with respect to p, where $J_i^o(p) = J_i'(R_i^o(p))$ and

$$R_i^o(p) = \min_{k \in D_i} R_i(p, k), \forall i \in \mathcal{N}. \tag{8.47}$$

Recall that we have assumed $J'(R(p))$ is 2.s.s. with respect to $R(p)$ and $J(p) = J'(R(p))$. Therefore, $J'(R^o(p))$ is a 2.s.s. function with respect to $R^o(p)$. Consequently, the theorem is proved if we can show that the following statement holds. If $\frac{1}{a}p \leq p' \leq ap$, then we have

$$\frac{1}{a}R_i^o(p) \leq R_i^o(p') \leq aR_i^o(p), \forall i \in \mathcal{N}. \tag{8.48}$$

We will prove (8.48) by contradiction. Let b_i and b_i' be the BSs chosen by user i corresponding to $R_i^o(p)$ and $R_i^o(p')$, respectively. From our assumption $\frac{1}{a}p \leq p' \leq ap$ for $a > 1$, we have $\frac{1}{a}p_j \leq p_j' \leq ap_j$, $\forall j \in \mathcal{N}$. Performing some simple manipulations, we can obtain

$$(1/a)R_i\left(p, b_i'\right) \le R_i^o(p') \le aR_i(p, b_i'),$$
$$(1/a)R_i^o(p) \le R_i(p', b_i) \le aR_i^o(p). \tag{8.49}$$

Now suppose that (8.48) is not satisfied for $\frac{1}{a}p \le p' \le ap$. Then, we must have $\frac{1}{a}R_i^o(p) > R_i^o(p')$ or $R_i^o(p') > aR_i^o(p)$. Let us consider these possible cases in the following.

- If $\frac{1}{a}R_i^o(p) > R_i^o(p')$, then using the results in (8.49), we have

$$R_i^o(p) > R_i\left(p, b_i'\right). \tag{8.50}$$

- Similarly, if $R_i^o(p') > aR_i^o(p)$, then using the results in (8.49), we have

$$R_i^o(p') > R_i(p', b_i). \tag{8.51}$$

Both (8.50) and (8.51) indeed result in contradiction to the definition of $R_i^o(p)$ and $R_i^o(p')$, which is given in (8.44). Hence, we have $\frac{1}{a}R_i^o(p) \le R_i^o(p') \le aR_i^o(p)$ if $\frac{1}{a}p \le p' \le ap$, which completes the proof of the theorem. ∎

The result in this theorem implies that if the p.u.f.s J(p) and J'(R(p)) are 2.s.s., then the BSA strategy described in (8.43) and (8.44) results in a composite 2.s.s. p.u.f., which corresponds to the proposed joint BSA and power control operation. In other words, the proposed BSA scheme preserves the 2.s.s. property of the employed p.u.f.s J(p).

8.3.2   Hybrid Power Control Algorithm

To complete the design for the proposed joint BSA and power control framework, we need to develop a distributed power control strategy, which is employed in (8.46). In general, the performance of a power control algorithm depends on how we design the corresponding 2.s.s. p.u.f. J(p). We will conduct such design by using game theory in the following.

8.3.2.1   *Game-Theoretic Formulation*   We define a following power control game.

- Players: The set of mobile users $\mathcal{N}$.
- Strategies: Each user i chooses transmission power in the set $[0, P_i]$.
- Payoffs: User i is interested in maximizing the following payoff function

$$U_i(p) = -\alpha_i(p_i - \xi_i R_i(p)^{\frac{x}{x-1}})^2 - (p_i - \Gamma_i R_i(p))^2 \tag{8.52}$$

where Γ_i denotes the target SINR for user i. x is a special parameter whose desirable value will be revealed later. α_i and ξ_i are nonnegative control parameters, that is, $\alpha_i, \xi_i \ge 0$, which will be adaptively adjusted to achieve our design objectives.

This game-theoretic formulation arises naturally in the considered heterogeneous wireless network, where mobile users tend to be selfish and only interested in maximizing their own benefits. Using this formulation, we will develop an iterative power control algorithm in which each user maximizes his or her own payoff in each iteration given the chosen power levels from other users in the previous iteration (i.e., each user plays the *best response* strategy). To devise such an algorithm, each user i chooses the power level, which is obtained by setting the first derivative of the underlying user's payoff function to zero.

In fact, by maximizing the payoff function given in (8.52), each user i strikes a balance between achieving the SINR target Γ_i and exploiting its potential favorable channel condition to increase its SINR. In particular, by maximizing $-(p_i - \Gamma_i R_i(\mathrm{p}))^2$, each user i attempts to reach his or her target SINR Γ_i. Moreover, it can be shown that the best response strategies achieved by maximizing the first term $-\alpha_i (p_i - \xi_i R_i(\mathrm{p})^{\frac{x}{x-1}})^2$ and maximizing $\gamma_i^x - \lambda_i p_i$ are the same, for $\xi_i = (\lambda_i / x)^{\frac{1}{x-1}}$ and $0 < x < 1$, where λ_i represents the pricing coefficient of user i. Also, the best-response-based power control scheme with payoff function $\gamma_i^x - \lambda_i p_i$ for $x = 1/2$ is the well-known OPC algorithm [16, 17]. In addition, if each user i plays the best-response strategies using $U_i(\mathrm{p})$, where $\alpha_i = 0$, then we obtain the well-known Foschini–Miljanic TPC algorithm [1]. Therefore, our chosen payoff function in (8.52) aims to design a HPC strategy that exploits the advantages of both OPC and TPC algorithms.

Proposed HPC Algorithm. We are now ready to develop a HPC algorithm corresponding to the payoff function in (8.52). Specifically, we can obtain the power update rule for the HPC algorithm according to the best-response strategy of the underlying payoff function. After some manipulations, we can obtain the following best response under the chosen payoff function (8.52):

$$p_i = I_i^H(\mathrm{p}) \triangleq \frac{\alpha_i \xi_i R_i(\mathrm{p})^{\frac{x}{x-1}} + \Gamma_i R_i(\mathrm{p})}{\alpha_i + 1}. \tag{8.53}$$

Considering the maximum power constraints, the HPC algorithm employs the following iterative power update:

$$p_i(t+1) = J_i^H(\mathrm{p}(t)) = \min\left\{P_i, I_i^H(\mathrm{p}(t))\right\} \tag{8.54}$$

where t denotes the iteration index and $I_i^H(\mathrm{p})$ is given in (8.53). Here, parameters α_i control the desirable performance of the proposed HPC algorithm. Specifically, by setting $\alpha_i = 0$, user i actually employs the standard Foschini–Miljanic TPC algorithm to achieve its target SINR Γ_i while if $\alpha_i \to \infty$, user i attempts to achieve a higher SINR.

It can be observed that each user i only needs to calculate or estimate the effective interference $R_i(\mathrm{p})$ to update its power in the proposed HPC algorithm. User i can fulfill this if he or she has information about the total interference and noise power (i.e., the denominator of (8.38)) and the channel power gain $h_{b_i,i}$ according to SINR expression in (8.38), since the current transmission power level p_i is readily available. In addition, the channel power gains $h_{b_i,i}$ can be estimated by the BS and sent back to each user by using the pilot signal and any standard channel estimation technique [32]. Moreover, the total interference and noise power for each user can be estimated as follows. Each BS estimates the total receiving power and then broadcasts this value to its connected users. Each user i can calculate the total interference and noise power by subtracting his or her received signal power (i.e., $h_{b_i,i} p_i$) from the total receiving power broadcast by the BS. Therefore, calculation of the effective interference only requires the standard channel estimation of $h_{b_i,i}$ and estimation of the total receiving power at the BS. The estimation of the effective interference R_i (p) and, therefore, the proposed power control algorithm given in (8.54) can be implemented in a distributed fashion.

Convergence of HPC Algorithm. We establish the convergence condition for the proposed HPC algorithm by using the *"two-sided scalable function"* approach given in Definition 8.1. We state a sufficient condition under which the p.u.f. J^H (p) in (8.54) is 2.s.s. in the following theorem.

Theorem 8.3. If the parameter x of functions I_i^H (p) given in (8.53) satisfies $0 < x \le 1/2$, then the p.u.f. J_i^H (p) given in (8.54) is 2.s.s. Also, the proposed HPC algorithm in (8.54) converges to the Nash equilibrium of the underlying power control game.

Proof: The convergence of the proposed HPC algorithm follows immediately from the results of Lemma 8.1 if we can prove that its p.u.f. J^H (p) $= \left[J_1^H(\text{p}), \ldots, J_N^H(\text{p}) \right]$ in (8.54) is 2.s.s. with respect to p. In addition, the resulting equilibrium (i.e., the power vector at convergence) is the Nash equilibrium of the power control game defined in Section 8.3.2.1 since users play the best-response strategy. The detailed proof that J^H (p) is 2.s.s. is available in [33]. ∎

The proposed HPC scheme will be employed in the proposed algorithm presented in Section 8.3.1. Note, however, that this proposed HPC scheme can be used as a stand-alone power control algorithm.

8.3.3   Hybrid Power Control Adaptation Algorithm

The equilibrium achieved at convergence by the proposed BSA and power control algorithm depends on several design parameters, namely α_i and ξ_i for $i \in \mathcal{N}$. In what follows, we develop decentralized mechanisms to adjust these parameters so that target SINRs of all users can be achieved whenever possible while attempting to achieve higher SINRs, and therefore, higher system throughput. The proposed adaptive mechanism comprises two updating operations in two different time scales, i.e., running the joint BSA and HPC algorithm to achieve the Nash equilibrium point in the small time scale and updating α_i and ξ_i for all users to achieve a more desirable Nash equilibrium in the large time scale. Let $\Delta = \{\alpha_i | i \in \mathcal{N}\}$ and $\Xi = \{\xi_i | i \in \mathcal{N}\}$ be the sets of α and ξ parameters of all users in p.u.f. of the HPC scheme, respectively.

As discussed before, the TPC scheme is a special case of the proposed HPC scheme with $\alpha_i = 0$ for $i \in \mathcal{N}$. It is known that the TPC scheme is able to support all the users' target SINRs expressed in (8.40), as long as the system is feasible (i.e., the SINR requirements in (8.40) can be supported) [8]. However, the TPC scheme fails

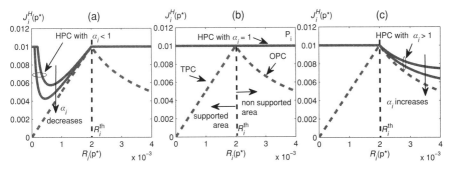

Figure 8.6 Relationship between hybrid power control p.u.f. $J_i^H(R_i(p))$ and $R_i(p)$ for different values of α_i ($P_i = 0.01$ W) where $R_i(p^*) \leq R_i^{th}$ for supported users and $R_i(p^*) > R_i^{th}$ for nonsupported users. © [2013] IEEE.

to achieve high system throughput when the system is feasible and lightly loaded. In addition, the TPC scheme may not be able to support the largest possible number of users when the system is infeasible. Our objective is to develop an adaptive strategy for the proposed HPC algorithm to overcome these limitations of the TPC scheme.

Toward this end, if user i is a voice user who is only interested in maintaining his or her target SINR Γ_i, then we can simply set $\alpha_i = 0$ in (8.52). For each data user i, we will fix ξ_i while adaptively updating power and α_i in two different time scales to achieve the design objectives. Specifically, each data user i will run the HPC and BSA algorithms for a given α_i until convergence (i.e., in the small time scale). Then it updates α_i accordingly (i.e., in the large time scale). Now, it can be verified that the best response corresponding to payoff function $\gamma_i^x - \lambda_i p_i$ can be written as

$$p_i = R_i(p)^{\frac{x}{x-1}} (\lambda_i/x)^{\frac{1}{x-1}} = \xi_i R_i(p)^{\frac{x}{x-1}} \qquad (8.55)$$

where $\xi_i = (\lambda_i/x)^{\frac{1}{x-1}}$. To set the value for ξ_i, suppose that data user i would need to use his or her maximum power P_i to reach the target SINR Γ_i. Then, the value of ξ_i can be found by using the result in (8.55) as follows.

$$\xi_i = \left(P_i/\Gamma_i^x\right)^{\frac{1}{1-x}} \qquad (8.56)$$

where we have substituted $\gamma_i = \Gamma_i$ to (8.55). We now describe how to update α_i in Δ. In the HPC scheme, the p.u.f. $J_i'^H(R_i(p)) = J_i^H(p)$ depends on the effective interference $R_i(p)$ and the value of α_i. Let p^* denote the power vector at convergence, which is obtained by running the proposed HPC algorithm for a particular vector α. We illustrate the relationship between $J_i'^H(R_i(p^*))$ and $R_i(p^*)$ in Figure 8.6 where $R_i^{th} = P_i/\Gamma_i$ is a threshold value for the effective interference of user i. We state the relationship between $\gamma_i(p^*)$, $R_i(p^*)$, and α_i in the following lemma whose proof is available in [33].

Lemma 8.2. Assume that $\xi_i = \left(P_i/\Gamma_i^x\right)^{\frac{1}{1-x}}$ and $0 < x \leq 1/2$. Then, we have

1. If $R_i(p^*) > R_i^{th}$, then $\gamma_i(p^*) < \Gamma_i, \forall \alpha_i \geq 0$.
2. If $R_i(p^*) \leq R_i^{th}$, then $\gamma_i(p^*) \geq \Gamma_i, \forall \alpha_i \geq 0$.

3. If $R_i(\mathrm{p}^*) < R_i^{\mathrm{th}}$, then we have

- $\gamma_i(\mathrm{p}^*) = \Gamma_i$ iff $\alpha_i = 0$.
- $I_i^H(\mathrm{p}^*)$ decreases if α_i decreases.

It can be seen from Figure 8.6 that if $\alpha_i < 1$, the HPC curves become closer to the TPC curve as α_i decreases whereas HPC curves become closer to the TPC curve as α_i increases if $\alpha_i > 1$. This is because $I_i^H(\mathrm{p})$ in the p.u.f. of the proposed HPC scheme is a weighted sum of those in the TPC and OPC schemes. Recall that our design objectives are to satisfy the SINR requirements for all users expressed in (8.40) (whenever possible) while enhancing the system throughput. Let user i be a supported user (nonsupported user) if his or her SINR is greater than (less than) his or her target SINR, which occurs as his or her effective interference $R_i(\mathrm{p}^*)$ is less (greater) than the threshold R_i^{th}, respectively. Note that a supported user i will have his or her SINR greater than the target SINR if $\alpha_i > 0$. We refer to such a supported user as a potential user in the following. In fact, any potential user i can reduce his or her transmission power to enhance the SINRs of other users (since this reduces the effective interference experienced by other users) as being implied by the results in Lemma 8.2. Therefore, by adjusting α_i, each user i can vary his or her SINR and assist other users in improving their SINRs.

Algorithm 8.3   HYBRID POWER CONTROL ADAPTATION ALGORITHM

1: Initialization.
   - Set $\mathrm{p}^{(0)} = 0$, i.e., $p_i^{(0)} = 0, \ \forall i \in \mathcal{N}$.
   - Set $\Delta^{(0)}$ as $\alpha_i^{(0)} = 0$ for voice user and $\alpha_i^{(0)} = \alpha_0 \ (\alpha_0 \gg 1)$ for data user.
   - Set $\overline{N}^* = |\overline{\mathcal{N}}^{(0)}|$ and $\Delta^* = \Delta^{(0)}$.
2: Iteration l.
   - Run HPC algorithm until convergence with $\Delta^{(l)}$.
   - If $|\overline{\mathcal{N}}^{(l)}| > \overline{N}^*$, set $\overline{N}^* = |\overline{\mathcal{N}}^{(l)}|$ and $\Delta^* = \Delta^{(l)}$.
   - If $\underline{\mathcal{N}}^{(l)} = \varnothing$ or $\overline{\mathcal{N}}^{(l)} = \varnothing$, then go to step 4.
   - If $\underline{\mathcal{N}}^{(l)} \neq \varnothing$ and $\overline{\mathcal{N}}^{(l)} \neq \varnothing$, then run the *"updating process"* as follows.
   % *Start of updating process* %
   - For user $i \in \underline{\mathcal{N}}^{(l)}$, set $\alpha_i^{(l+1)} = \alpha_i^{(l)}$.
   - For user $i \in \overline{\mathcal{N}}^{(l)}$, set $\alpha_i^{(l+1)} = k_\alpha \alpha_i^{(l)}$ where $0 < k_\alpha < 1$ is a scaling factor.
   % *End of updating process* %
3: Increase l and go back to step 2 until there is no update request for $\Delta^{(l)}$.
4: Set $\Delta := \Delta^*$ and run the HPC algorithm until convergence.

We exploit this fact to develop the HPC adaptation algorithm, which is described in the algorithm presented in Section 8.3.3. Let $\overline{\mathcal{N}}^{(l)}$ and $\underline{\mathcal{N}}^{(l)}$ be the sets of supported and nonsupported users in iteration l, respectively. We use $\overline{N}^*$ to keep the number of supported users during the course of the algorithm. In each iteration, each user i will run the proposed HPC algorithm and slowly update α_i based on the achieved equilibrium. All data users i initially set α_i to be a sufficiently large value so that we reach the first equilibrium that favors strong users. For each nonsupported user

i, we maintain its parameter α_i to be the same to keep its power updating process stable. In addition, user i can save his or her parameter α_i at the instant when the number of nonsupported users decreases and reload this value later as the global update process terminates (in step 4). In addition, design of the *"updating process"* can be done so that the parameters α_i of supported users are slowly decreased to zero if all nonsupported users cannot be assisted to achieve their target SINRs.

The proposed HPC adaptation algorithm can achieve desirable performance, which is described in the following. Let $\overline{N}_{\mathsf{HPC}}$ and $\overline{N}_{\mathsf{TPC}}$ be the number of supported users due to the proposed HPC adaptation algorithm and the TPC algorithm, respectively. Then, we have $\overline{N}_{\mathsf{HPC}} \geq \overline{N}_{\mathsf{TPC}}$. Also, if the network is feasible (i.e., all SINR requirements can be fulfilled by the TPC algorithm), then all users achieve their target SINRs by using HPC algorithm. Also, there exists feasible users who achieve SINRs higher than the target values under the HPC algorithm. The proofs of these results are given in [33]. This theorem implies that the proposed HPC algorithm achieves better performance than the traditional TPC algorithm for both infeasible and feasible systems. Specifically, the HPC algorithm can support at least the same number of users while it can achieve higher SINRs and therefore higher total throughput than that from the TPC algorithm.

8.3.4 Application to Two-Tier Macrocell–Femtocell Networks

The proposed distributed HPC algorithm can be applied to the two-tier macro–femto network. For the two-tier network using open or hybrid access, macro users are allowed to connect to nearby femto access points to enhance their performance and to reduce the interference to other users in the network. However, macro users' connections to a particular femto access point without an appropriate control may significantly degrade the throughput of femto users in the underlying femtocell, which would not be desirable. To resolve this issue, we assume that each femto user i has two target SINRs, for example, $\Gamma_i^{[1]}$ and $\Gamma_i^{[2]}$ where $\Gamma_i^{[1]} < \Gamma_i^{[2]}$. Femto users can attempt to reach the higher target SINRs as long as the network condition allows. Otherwise, they seek to maintain at least the lower target SINRs when it is possible.

Each femto user i can adaptively track the higher target SINR by taking the following procedure. It employs the proposed HPC algorithm to achieve the lower target SINR $\Gamma_i^{[1]}$ in the first phase. If this is successfully fulfilled, it attempts to achieve the higher target SINR $\Gamma_i^{[2]}$ in the second phase. To achieve this objective, femto user i simply sets parameters ξ_i and R_i^{th} by using the higher target SINR $\Gamma_i^{[2]}$. Then, it runs the proposed HPC algorithm again. As femto users at a particular femtocell attempt to achieve higher target SINR, the macro users associated with the underlying femtocell have to decrease their transmission powers and therefore achieve lower SINRs. Hence, the proposed HPC algorithm enables us to attain flexible spectrum sharing between femto users and macro users at each femtocell.

8.3.5 Numerical Examples

We present illustrative numerical results to demonstrate the performance of the proposed algorithms. The network setting and user placement for our simulations are illustrated in Figure 8.7, where macro users and femto users are randomly located

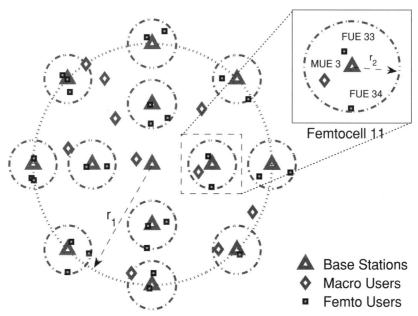

Figure 8.7   Simulated two-tier macrocell–femtocell network. © [2013] IEEE.

inside circles of radii of $r_1 = 1000\ m$ and $r_2 = 50\ m$, respectively. Assume that there are heavy walls at the boundaries of the femtocells. We fix $N_m = 10$ and randomly choose the number of femto users in each femtocell from one to three. Then, eight users of either tier are set as voice users randomly. The channel power gain h_{ij} is chosen according to the path loss $L_{ij} = A_i \log_{10}(d_{ij}) + B_i + C \log_{10}(\frac{f_c}{5}) + W_l n_{ij}$, where d_{ij} is the distance from user j to BS i. (A_i, B_i) are set as $(36, 40)$ and $(35, 35)$ for macro base station and femto access points, respectively. $C = 20$, $f_c = 2.5$ GHz. W_l is the wall-loss value. n_{ij} is the number of walls between BS i and user j. Other parameters are set as follows (unless stated otherwise): $x = 0.5$; processing gain, $P_G = 128$; maximum power, $P_j = 0.01\ W$, $\forall j \in \mathcal{N}$; noise power $\sigma_i = 10^{-13}\ W$, $\forall i \in \mathcal{M}$. We set n_{ij} equal to the number of cell boundaries that the corresponding signal traverses and the wall-loss value $W_l = 12$dB. The SINRs presented in all figures are in linear scale except that those in Figure 8.10 are in dB scale.

Figure 8.8 shows the convergence of the HPC algorithm when we set $\Gamma_i = 6$ for all users. This figures confirms that HPC algorithm converges to an equilibrium for which some users exactly achieve their target SINRs while others attain SINRs higher than their target values. This set of results corresponds to the feasible system where all SINR requirements can be supported. This figure also shows that the proposed HPC algorithm not only attains all SINR requirements but also enables strong users to achieve high throughput performance.

Figure 8.9 illustrates the average spectral efficiency achieved by different schemes versus the target SINR. Here, the spectral efficiency of user i is calculated as $\log_2(1 + \Gamma_i)$ (b/s/Hz). To obtain the results, we randomly generate user locations and obtain results for this fixed topology. We sequentially add more femto

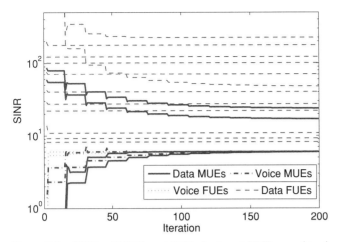

Figure 8.8   SINRs of MUEs and FUEs for target SINRs equal to six. © [2013] IEEE.

users to obtain different points on each curve. For reference, we also present the average spectral efficiency of a TPC max–min scheme in which we slowly increase the target SINR for all users as long as the system is still feasible under the TPC scheme (this scheme is denoted as TPC max–min). As this figure demonstrates, our HPC algorithm attains higher average spectral efficiency than those of both traditional TPC and TPC max–min schemes but lower than that of the OPC algorithm. In particular, when the network is lowly loaded ($\Gamma_i < 8$), our proposed algorithm attains much higher spectral efficiency than those of the traditional TPC and TPC max–min schemes. Moreover, when the network load is higher ($\Gamma_i \geq 8$), the gaps between our proposed algorithm and the two TPC schemes become smaller. This is because the HPC algorithm attempts to maintain the target SINRs for all users as the TPC schemes.

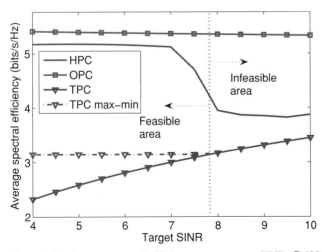

Figure 8.9   Average spectral efficiency versus target SINR. © [2013] IEEE.

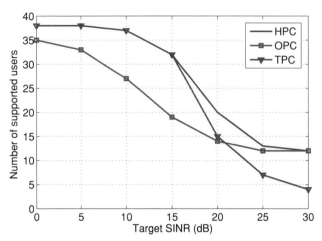

Figure 8.10   Number of supported users in hybrid power control, target-SINR-tracking power control (TPC) and opportunistic power control schemes versus target SINR. © [2013] IEEE.

Figure 8.10 illustrates the number of supported users for different schemes. It can be seen that our proposed HPC algorithm can maintain the SINR requirements for the larger number of users compared with TPC and OPC algorithms. In particular, our proposed algorithm can support roughly the same number of users as the TPC algorithm for low target SINRs. However, our proposed algorithm performs better than the TPC scheme for higher target SINRs. In fact, almost all users utilize the maximum power under the TPC scheme at high target SINRs, which prevent all of them from achieving their target SINRs. On the other hand, in our proposed scheme, nonsupported users tend to use as small transmission power as possible (i.e., high values of α_i). This enables more users to attain their target SINRs. The figure also shows that the proposed HPC algorithm performs better than the OPC scheme for all values of the target SINR.

8.4   DISTRIBUTED PARETO-OPTIMAL POWER CONTROL FOR TWO-TIER CDMA HETNETS

Resource allocation algorithms developed by using the game theory approach such as those presented in the previous sections may not be efficient in general since the achieved Nash equilibrium may not optimally utilize the radio resources. In fact, distributed optimal resource allocation algorithms can be designed by using optimization theory. Moreover, such design can enable us to achieve fair resource sharing under different fairness criteria by maximizing some particular utility function, for example, the α-fair utility function proposed in [34]. We demonstrate the optimal power control design for two-tier CDMA HetNets in this section where the proposed distributed power control algorithm allows macro users to achieve their target SINRs while enabling femto users to share the leftover capacity in a fair and Pareto-optimal manner. This design is motivated and based on the results in [28], which was originally proposed for homogeneous cellular networks.

We again consider the UL transmissions in a two-tier wireless network, in which N_m macro users establish communication links to its servicing macro base station while N_f femto users also transmit to their respective femto access points. We assume that all the macro users and femto users share the same radio frequency band by CDMA. The closed access mode is considered where the macro users are not permitted to connect to any femto access point. Also the association of any user with his or her own BS is fixed during the runtime of the underlying power control. Without loss of generality, denote the sets of macro users and femto users by $\mathcal{N}_m = \{1, \ldots, N_m\}$ and $\mathcal{N}_f = \{N_m + 1, \ldots, N_m + N_f\}$, respectively. The set of all users is then simply $\mathcal{N} = \mathcal{N}_m \cup \mathcal{N}_f$ whose cardinality is $N = N_f + N_m$. It is assumed here that the time scale of network topology changes is negligible compared with that of power adaptation.

Denote by b_i the serving BS of user $i \in \mathcal{N}$, which is either a macro user or a femto user. For brevity, the path between user i and its servicing BS b_i shall be referred to as link i. Also, let $h_{k,j}$ be the absolute channel gain from user j to BS k, and define its corresponding normalization as $\bar{h}_{k,j} = h_{k,j}/h_{b_j,j}$. To represent the normalized channel gain from user j to the serving BS b_i of user i, we define an $(N_f + N_m) \times (N_f + N_m)$ channel matrix $\mathbf{H}$ whose (i, j)th entry is

$$H_{i,j} = \begin{cases} 0, & \text{if } i = j, \\ 1, & \text{if } b_i = b_j, \ i \neq j, \\ \bar{h}_{b_i,j}, & \text{if } b_i \neq b_j. \end{cases} \tag{8.57}$$

Suppose that user j transmits to his or her serving BS b_j, and let $\bar{p}_j$ be the received power at b_j by that transmission. Since $h_{b_j,j}$ is the channel gain from user j to BS b_j, it is clear that user j must have transmitted at a power level $\bar{p}_j/h_{b_j,j}$. At any BS k, this signal appears with a power $h_{k,j}(\bar{p}_j/h_{b_j,j}) = \bar{h}_{k,j}\bar{p}_j$. The transmission of other user j results in an interference power of $\bar{h}_{b_i,j}\bar{p}_j = H_{i,j}\bar{p}_j$ to link i. The total interference plus noise at BS b_i that serves user $i \in \mathcal{N}$ on link i can be expressed as

$$q_i = \sum_{j=1}^{N_f+N_m} H_{i,j}\bar{p}_j + \sigma_i \tag{8.58}$$

where σ_i is the noise power at the receiving end of link i. We assume that $\boldsymbol{\sigma} = [\sigma_1, \ldots, \sigma_N]^T \neq \mathbf{0}$. In a matrix form, (8.58) can also be written as

$$\mathbf{q} = \mathbf{H}\mathbf{p} + \boldsymbol{\sigma} \tag{8.59}$$

where $\mathbf{p}$ and $\mathbf{q}$ are vectors whose ith elements are $\bar{p}_i$ and q_i, respectively. Let $\bar{\gamma}_i = P_G \bar{p}^{(i)}/q^{(i)}$ denote the SINR at link $i \in \mathcal{N}$, where P_G is the system processing gain. For notational convenience, we define the normalized SINR at link i as $\gamma_i = \bar{\gamma}_i/P_G$. It is then easy to see that

$$\mathbf{p} = \mathbf{D}(\boldsymbol{\gamma})\mathbf{q} \tag{8.60}$$

where $\mathbf{D}(\boldsymbol{\gamma}) = \mathrm{diag}(\gamma_1, \ldots, \gamma_N)$. By substituting (8.60) to (8.59) and after some simple algebra, we obtain the following [28]:

$$\mathbf{q} = \mathbf{H}\mathbf{D}(\boldsymbol{\gamma})\mathbf{q} + \boldsymbol{\sigma}, \tag{8.61}$$

$$\mathbf{p} = \mathbf{D}(\boldsymbol{\gamma})\mathbf{H}\mathbf{p} + \mathbf{D}(\boldsymbol{\gamma})\boldsymbol{\sigma}. \tag{8.62}$$

Since we do not consider totally isolated groups of links that are not interacting with each other, it is practical to assume that both matrices $\mathbf{H}\mathbf{D}(\boldsymbol{\gamma})$ and $\mathbf{D}(\boldsymbol{\gamma})\mathbf{H}$ are primitive.[1]

Our goal is to devise a jointly optimal power allocation $\mathbf{p}$ and SINR assignment $\boldsymbol{\gamma}$ for the two types of users (i.e., macro users and femto users) with different service priorities and design objectives. The prioritized macro users with higher access rights demand that their ongoing services be, at least, unaffected regardless of any femtocell deployment. A set of minimum SINRs $\boldsymbol{\Gamma}^{(m)} = [\Gamma_1, \ldots, \Gamma_{N_m}]^T$ prescribed by the macro users must therefore be supported in the first place, i.e.,

$$\gamma_i \geq \Gamma_i, \quad \forall i \in \mathcal{N}_m \tag{8.63}$$

where Γ_i is the normalized target SINR corresponding to the actual SINR $P_G \Gamma_i$ required by macro user i. Note that a general QoS $\boldsymbol{\Gamma}^{(m)}$ can be translated to different specific requirements. For instance, a higher value of $\boldsymbol{\Gamma}^{(m)}$ means that a higher throughput, a lower BER, and a shorter time delay are guaranteed for macro user i.

Our design objective is to maximize the sum utility of all femto users. The utility function $U_i(\gamma_i)$ (which is typically an increasing function) represents the value that femto user $i \in \mathcal{N}_f$, who is assigned with SINR γ_i, contributes to the overall network. The higher the SINR, the greater the contribution. Depending on the type of utility functions, fairness, an important system-wide objective, can also be achieved. Proportional fairness and max–min fairness are among the most common metrics used in practice to characterize how competing users share system resources. The α-fair function proposed by [34] provides a useful means to enforce these two types of fairness, in that it generalizes proportional fairness and includes arbitrarily close approximations of max–min fairness. Specifically, we consider the following utility for femto user $i \in \mathcal{N}_f$:

$$U_i(\gamma_i) = \begin{cases} \log(\gamma_i), & \text{if } \alpha = 1 \\ (1-\alpha)^{-1}(\gamma_i)^{1-\alpha}, & \text{if } \alpha \geq 0 \text{ and } \alpha \neq 1. \end{cases} \tag{8.64}$$

Here, $\alpha = 1$ corresponds to proportional fairness whereas $\alpha \to \infty$ gives max–min fairness.

Let $\rho(\mathbf{X})$ denote the spectral radius of matrix $\mathbf{X}$, that is, the maximum modulus eigenvalue of $\mathbf{X}$. Given channel matrix $\mathbf{H}$, the specific value of $\rho(\mathbf{H}\mathbf{D}(\boldsymbol{\gamma}))$ indicates whether a certain SINR $\boldsymbol{\gamma}$ is supportable or not. In particular, it is required by [2]

---

[1]A nonnegative matrix is called *primitive* if it is irreducible and has only one eigenvalue of maximum modulus [35, Definition 8.5.0].

that $\rho\big(\mathbf{HD}(\boldsymbol{\gamma})\big) < 1$ for the existence of a feasible power vector $\mathbf{p} \succ \mathbf{0}$. In the limit $\rho\big(\mathbf{HD}(\boldsymbol{\gamma})\big) = 1$, an infinite amount of transmission power would be needed to attain $\boldsymbol{\gamma}$. For $\rho\big(\mathbf{HD}(\boldsymbol{\gamma})\big) > 1$, the network can be regarded so congested that only removing certain users and/or lowering the SINR targets can help relieve such congestion. Considering a practical non-congested network with attainable target SINRs, we insist that $\rho\big(\mathbf{HD}(\boldsymbol{\gamma})\big) \leq \bar{\rho}$ where $0 \leq \bar{\rho} < 1$, for the existence of a feasible solution with $0 < p_i < \infty$, $\forall i \in \mathcal{N}$.

Given $\bar{\rho} \in [0, 1)$, we are interested in the following problem:

$$\max_{\boldsymbol{\gamma}, \, \mathbf{p}} \quad \sum_{i \in \mathcal{N}_f} U_i(\gamma_i)$$

$$\text{subject to} \quad \rho\big(\mathbf{HD}(\boldsymbol{\gamma})\big) \leq \bar{\rho}, \tag{8.65}$$

$$\gamma_i \geq \Gamma_i, \; \forall i \in \mathcal{N}_m,$$

$$\boldsymbol{\gamma} \in \mathbb{R}_+^{N_f + N_m}, \; \mathbf{p} \in \mathbb{R}_+^{N_f + N_m}.$$

It can be shown that a larger value of $\bar{\rho}$ corresponds to a larger feasible set, and in turn a potentially higher utility. Therefore, it is desirable to choose $\bar{\rho}$ to be as close to 1 as possible while ensuring that $\boldsymbol{\gamma}$ be supportable.

The problem in (8.65) is not convex because the set $\big\{ \boldsymbol{\gamma} \in \mathbb{R}_+^{N_f + N_m} \mid \rho\big(\mathbf{HD}(\boldsymbol{\gamma})\big) \leq \bar{\rho} \big\}$ is not convex. However, if we let $\widetilde{\boldsymbol{\gamma}} = \log \boldsymbol{\gamma}$, then its equivalence $\big\{ \widetilde{\boldsymbol{\gamma}} \in \mathbb{R}^{N_f + N_m} \mid \rho\big(\mathbf{HD}(e^{\widetilde{\boldsymbol{\gamma}}})\big) \leq \bar{\rho} \big\}$ is actually a convex set [36]. Through such a change of variable and upon denoting $\widetilde{\Gamma}_i = \log(\Gamma_i)$, the following equivalent problem of (8.65) is considered instead:

$$\max_{\widetilde{\boldsymbol{\gamma}}, \, \mathbf{p}} \quad \sum_{i \in \mathcal{N}_f} U_i(\widetilde{\gamma}_i)$$

$$\text{subject to} \quad \rho\big(\mathbf{HD}(e^{\widetilde{\boldsymbol{\Gamma}}})\big) \leq \bar{\rho}, \tag{8.66}$$

$$\widetilde{\gamma}_i \geq \widetilde{\Gamma}_i, \; \forall i \in \mathcal{N}_m,$$

$$\widetilde{\boldsymbol{\gamma}} \in \mathbb{R}^{N_f + N_m}, \; \mathbf{p} \in \mathbb{R}_+^{N_f + N_m}.$$

In this case, the utility function becomes

$$U_i(\widetilde{\gamma}_i) = \begin{cases} \widetilde{\gamma}_i, & \text{if } \alpha = 1 \\ (1 - \alpha)^{-1} e^{(1 - \alpha)\widetilde{\gamma}_i}, & \text{if } \alpha \geq 0 \text{ and } \alpha \neq 1 \end{cases} \tag{8.67}$$

which is increasing, twice-differentiable, and concave with respect to $\widetilde{\gamma}_i$. The problem in (8.66) is a convex optimization program. However, due to the complicated coupling in the feasible region, centralized algorithms are typically needed to resolve this kind of problem. Given the nature of two-tier networks where central coordination and processing are usually inaccessible, we aim at developing optimal solutions that can be distributively implemented by individual users.

8.4.1 Distributed Pareto-Optimal SINR Assignment

We perform the following matrix and vector partitions: $\mathbf{p} = [\mathbf{p}_m^T, \mathbf{p}_f^T]^T; \mathbf{q} = [\mathbf{q}_m^T, \mathbf{q}_f^T]^T; \widetilde{\boldsymbol{\gamma}} = [\widetilde{\boldsymbol{\gamma}}_m^T, \widetilde{\boldsymbol{\gamma}}_f^T]^T; \boldsymbol{\sigma} = [\boldsymbol{\sigma}_m^T, \boldsymbol{\sigma}_f^T]^T;$ and

$$\mathbf{H} = \begin{bmatrix} \mathbf{H}_{11} & \mathbf{H}_{12} \\ \mathbf{H}_{21} & \mathbf{H}_{22} \end{bmatrix} \tag{8.68}$$

where $\mathbf{q}_m, \mathbf{p}_m, \boldsymbol{\sigma}_m \in \mathbb{R}_+^{N_m}, \widetilde{\boldsymbol{\gamma}}_m \in \mathbb{R}^{N_m}; \mathbf{q}_f, \mathbf{p}_f, \boldsymbol{\sigma}_f \in \mathbb{R}_+^{N_f}, \widetilde{\boldsymbol{\gamma}}_f \in \mathbb{R}^{N_f}; \mathbf{H}_{11} \in \mathbb{R}_+^{N_m \times N_m}; \mathbf{H}_{12} \in \mathbb{R}_+^{N_m \times N_f}; \mathbf{H}_{21} \in \mathbb{R}_+^{N_f \times N_m};$ and $\mathbf{H}_{22} \in \mathbb{R}_+^{N_f \times N_f}$.

Proposition 8.1. The optimal solution of (8.66) lies on the following boundary:

$$\partial \mathcal{F}_{\bar{\rho}} = \left\{ \widetilde{\boldsymbol{\gamma}} = [\widetilde{\boldsymbol{\gamma}}_m, \widetilde{\boldsymbol{\gamma}}_f]; \text{ subject to } \rho\big(\mathbf{HD}(e^{\widetilde{\boldsymbol{\gamma}}})\big) = \bar{\rho} \text{ and } \widetilde{\boldsymbol{\gamma}}_m = \widetilde{\boldsymbol{\Gamma}}_m \right\}. \tag{8.69}$$

Proof: Suppose that at optimality, there is some $\widetilde{\gamma}_m^{(i)}$ such that $\widetilde{\gamma}_m^{(i)} > \widetilde{\Gamma}_m^{(i)}$. Since $\widetilde{\gamma}_m^{(i)} = \log\big(\bar{p}_m^{(i)}/q_m^{(i)}\big)$, one can reduce $\bar{p}_m^{(i)}$ to have $\widetilde{\gamma}_m^{(i)} = \widetilde{\Gamma}_m^{(i)}$ without violating the constraint. On the other hand, such a reduction in power leads to a lesser amount of interference perceived by all the femto users. Due to the monotonic property, this implies an increase in $\sum_{i \in N_f} U_i(\widetilde{\gamma}_i)$, and hence contradicts the assumption of optimality. Therefore, $\widetilde{\boldsymbol{\gamma}}_m = \widetilde{\boldsymbol{\Gamma}}_m$ holds. Moreover, it can be proved that $\rho\big(\mathbf{HD}(e^{\widetilde{\boldsymbol{\gamma}}})\big) = \bar{\rho}$ at the optimum. This completes the proof. ∎

Proposition 8.1 implies that the search space for Pareto-optimal SINRs in this case is reduced to simply $\mathbb{R}^{N_f}$. To unveil the complicated coupling between $\widetilde{\boldsymbol{\gamma}}_m$ and $\widetilde{\boldsymbol{\gamma}}_f$ in the relation $\rho(\mathbf{HD}(e^{\widetilde{\boldsymbol{\gamma}}})) = \bar{\rho}$, the following result is now in order.

Proposition 8.2. Suppose that we are operating on $\partial \mathcal{F}_{\bar{\rho}}$ and that $\rho(\mathbf{H}_{11}\mathbf{D}(e^{\widetilde{\boldsymbol{\Gamma}}_m})) < \bar{\rho}$. Then, $\rho(\mathbf{HD}(e^{\widetilde{\boldsymbol{\gamma}}})) = \rho(\mathbf{FD}(e^{\widetilde{\boldsymbol{\gamma}}_f}))$ holds, where $\mathbf{F}$ is a positive matrix defined as

$$\mathbf{F} = \mathbf{H}_{21}\mathbf{D}(e^{\widetilde{\boldsymbol{\Gamma}}_m})\big[\bar{\rho}\mathbf{I}_{N_m} - \mathbf{H}_{11}\mathbf{D}(e^{\widetilde{\boldsymbol{\Gamma}}_m})\big]^{-1}\mathbf{H}_{12} + \mathbf{H}_{22}. \tag{8.70}$$

Proof: Let $\mathbf{s}$ be the left eigenvector of $\mathbf{HD}(e^{\widetilde{\boldsymbol{\gamma}}})$ with associated eigenvalue $\bar{\rho} = \rho\big(\mathbf{HD}(e^{\widetilde{\boldsymbol{\gamma}}})\big) \in [0, 1)$. Therefore, $\mathbf{s}^T \mathbf{HD}(e^{\widetilde{\boldsymbol{\gamma}}}) = \bar{\rho}\mathbf{s}^T$, which can also be explicitly expressed as

$$[\mathbf{s}_m^T, \mathbf{s}_f^T] \begin{bmatrix} \mathbf{H}_{11} & \mathbf{H}_{12} \\ \mathbf{H}_{21} & \mathbf{H}_{22} \end{bmatrix} \mathbf{D}(e^{\widetilde{\boldsymbol{\Gamma}}_m}, e^{\widetilde{\boldsymbol{\gamma}}_f}) = \bar{\rho} \, [\mathbf{s}_m^T, \mathbf{s}_f^T] \tag{8.71}$$

where $\mathbf{s}_m^T = [s_1, \ldots, s_{N_m}]$ and $\mathbf{s}_f^T = [s_{N_m+1}, \ldots, s_{N_m+N_f}]$.

By [35, Th. 5.6.9 and Cor. 5.6.16], the assumption $\rho\big(\mathbf{H}_{11}\mathbf{D}(e^{\widetilde{\boldsymbol{\Gamma}}_m})\big) < \bar{\rho}$ ensures that $\big[\bar{\rho}\mathbf{I}_M - \mathbf{H}_{11}\mathbf{D}(e^{\widetilde{\boldsymbol{\Gamma}}_m})\big]^{-1} = (1/\bar{\rho})\sum_{k=0}^{\infty} \big[(1/\bar{\rho})\mathbf{H}_{11}\mathbf{D}(e^{\widetilde{\boldsymbol{\Gamma}}_m})\big]^k$ exists and is positive componentwise. After some algebraic manipulations, one arrives at

$$\mathbf{s}_f^T \Big\{ \mathbf{H}_{21}\mathbf{D}(e^{\widetilde{\boldsymbol{\Gamma}}_m})\big[\bar{\rho}\mathbf{I}_M - \mathbf{H}_{11}\mathbf{D}(e^{\widetilde{\boldsymbol{\Gamma}}_m})\big]^{-1}\mathbf{H}_{12} + \mathbf{H}_{22} \Big\}\mathbf{D}(e^{\widetilde{\boldsymbol{\gamma}}_f}) = \bar{\rho}\mathbf{s}_f^T. \tag{8.72}$$

With $\mathbf{F}$ defined in (8.70), the left-hand side of (8.72) is actually $\mathbf{s}_f^T \mathbf{FD}(e^{\widetilde{\gamma}_f})$. This implies that $\bar{\rho}$ is also an eigenvalue of $\mathbf{FD}(e^{\widetilde{\gamma}_f})$. Since $\mathbf{HD}(e^{\widetilde{\gamma}})$ is primitive, one must have $\mathbf{s} \succ \mathbf{0}$, from which $\mathbf{s}_f \succ \mathbf{0}$. Apparently, $\mathbf{F}$ is a positive matrix; hence, so is $\mathbf{FD}(e^{\widetilde{\gamma}_f})$. By Perron's theorem [35, Th. 8.2.11], $\bar{\rho}$ is the unique eigenvalue of maximum modulus of $\mathbf{FD}(e^{\widetilde{\gamma}_f})$.   ∎

The assumption $\rho(\mathbf{H}_{11}\mathbf{D}(e^{\widetilde{\Gamma}_m})) < \bar{\rho}$ in Proposition 8.2 can be justified by first noting that the channel matrix $\mathbf{H}$ is reduced to simply $\mathbf{H}_{11}$ if there is no femtocell deployed, and then applying the condition for the existence of a feasible power vector $\mathbf{p}_m = [\bar{p}_1, \ldots, \bar{p}_{N_m}]^T \succ \mathbf{0}$ in that case [2].

Essentially, Propositions 8.1 and 8.2 characterize the following Pareto-optimal SINR boundary:

$$\partial\mathcal{F}_{\bar{\rho}} = \{\widetilde{\gamma} = [\widetilde{\gamma}_m, \widetilde{\gamma}_f]; \text{ subject to } \rho(\mathbf{HD}(e^{\widetilde{\gamma}_f})) = \bar{\rho} \text{ and } \widetilde{\gamma}_m = \widetilde{\Gamma}_m\} \quad (8.73)$$

of the problem in (8.66). For every point on $\partial\mathcal{F}_{\bar{\rho}}$, it is impossible to increase the SINR of any one femto link without simultaneously reducing the SINR of some other femto links.

The finding of $\mathbf{F}$ in Proposition 8.2 also reveals that the performance of the femto users depends not only on the structure of the femtocell network (as reflected in $\mathbf{H}_{22}$), but also on the interaction between themselves with the macro users (as represented by $\mathbf{H}_{21}$ and $\mathbf{H}_{12}$). Moreover, the existence of $\mathbf{F}$ is conditional upon the particular values of $\mathbf{H}_{11}$ and $\widetilde{\Gamma}_m$, that is, $\rho(\mathbf{H}_{11}\mathbf{D}(e^{\widetilde{\Gamma}_m})) < \bar{\rho}$. It is somewhat an expected result because macro users have an absolutely higher priority in accessing the radio resource. Such a condition also confirms that femto users can attain their Pareto-optimal SINRs only if the performance of macro users is, at least, unaffected.

The fact that $\mathbf{F}$ is a positive matrix is critical, since it paves the way to adapt the load-spillage parametrization [28] to find all points on $\partial\mathcal{F}_{\bar{\rho}}$. Nevertheless, it is important to point out here that, thanks to Propositions 8.1 and 8.2, one has to only deal with matrix $\mathbf{F}$ in a N_f-dimensional space instead of the original $(N_f + N_m) \times (N_f + N_m)$ channel matrix $\mathbf{H}$. Also, note that, $\mathbf{F}$ does not need to be primitive in the following result, unlike the strict condition on completely connected (i.e., primitive) matrices specifically required by [28].

Proposition 8.3.   For a fixed $\widetilde{\Gamma}_m$, an SINR vector $\widetilde{\gamma} = [\widetilde{\gamma}_m, \widetilde{\gamma}_f]$ lies on the boundary $\partial\mathcal{F}_{\bar{\rho}}$ defined in (8.73) if and only if there exist $\mathbf{s}_f \succ \mathbf{0}$ in $\mathbb{R}^{N_f}$ and $\bar{\rho} \in [0, 1)$ such that

$$\widetilde{\gamma}_m = \widetilde{\Gamma}_m, \tag{8.74}$$

$$\mathbf{s}_f^T \mathbf{FD}(e^{\widetilde{\gamma}_f}) = \bar{\rho}\mathbf{s}_f^T. \tag{8.75}$$

Proof: Similar to that of [28, Lem. 3], the idea of this proof is based on Perron's theorem [35, Th. 8.2.11]. If there is some $\mathbf{s}_f \succ \mathbf{0}$ and $\bar{\rho} \in [0, 1)$ satisfying (8.75), then $\mathbf{s}_f$ is a positive left eigenvector, associated with eigenvalue $\bar{\rho}$, of the positive matrix $\mathbf{FD}(e^{\widetilde{\gamma}_f})$. By Perron's theorem, $\bar{\rho}$ is a unique positive eigenvalue with maximum modulus, that is, $\bar{\rho} = \rho(\mathbf{FD}(e^{\widetilde{\gamma}_f}))$. Along with $\widetilde{\gamma}_m = \widetilde{\Gamma}_m$ in (8.74), the corresponding $\mathbf{\Gamma}$ is on $\partial\mathcal{F}_{\bar{\rho}}$.

Conversely, if $\tilde{\boldsymbol{\gamma}} \in \partial \mathcal{F}_{\bar{\rho}}$, then $\tilde{\boldsymbol{\gamma}}_m = \tilde{\boldsymbol{\Gamma}}_m$ and $\rho\left(\mathbf{FD}(e^{\tilde{\boldsymbol{\gamma}}_f})\right) = \bar{\rho} < 1$. Let $\mathbf{s}_f$ be the left eigenvector associated with eigenvalue $\bar{\rho}$. Again, by Perron's theorem, $\mathbf{s}_f \succ \mathbf{0}$ and $\bar{\rho} > 0$ since $\mathbf{FD}(e^{\tilde{\boldsymbol{\gamma}}_f})$ is a positive matrix. $\blacksquare$

Using Proposition 8.3, we can now parameterize all $\tilde{\boldsymbol{\gamma}}_f$ on the boundary $\partial \mathcal{F}_{\bar{\rho}}$ as follows. If we let

$$\mathbf{v}_f = \mathbf{F}^T \mathbf{s}_f, \tag{8.76}$$

then (8.75) becomes $\mathbf{v}_f^T \mathbf{D}(e^{\tilde{\boldsymbol{\gamma}}_f}) = \bar{\rho} \mathbf{s}_f^T$. From which,

$$\tilde{\gamma}_f^{(i)} = \log\left(\bar{\rho}\, s_f^{(i)}/v_f^{(i)}\right), \qquad i = 1, \ldots, N_f. \tag{8.77}$$

After multiplying the right-hand side of (8.75) by $\mathbf{F}$ and using (8.76), it is clear that $\mathbf{v}_f^T \mathbf{D}(e^{\tilde{\boldsymbol{\gamma}}_f})\mathbf{F} = \bar{\rho}\mathbf{v}_f^T$, that is, $\mathbf{v}_f$ is a left eigenvector associated with eigenvalue $\bar{\rho}$ of $\mathbf{D}(e^{\tilde{\boldsymbol{\gamma}}_f})\mathbf{F}$. Furthermore, it can be shown that $\bar{\rho} = \rho\left(\mathbf{D}(e^{\tilde{\boldsymbol{\gamma}}_f})\mathbf{F}\right) = \rho\left(\mathbf{FD}(e^{\tilde{\boldsymbol{\gamma}}_f})\right)$ [35, Th. 1.3.20].

Once $s_f^{(i)}$ is known, the computation of $\tilde{\gamma}_f^{(i)}$ in (8.77) requires $v_f^{(i)}$ to be found by (8.76). However, as $\mathbf{F}$ involves a matrix inverse operation [see (8.70)], it is not yet straightforward to find $v_f^{(i)}$ distributively. Using (8.70), we rewrite (8.76) as

$$\mathbf{v}_f^T = \mathbf{s}_f^T \mathbf{H}_{21} \mathbf{D}(e^{\tilde{\boldsymbol{\Gamma}}_m})\left[\bar{\rho}\mathbf{I}_M - \mathbf{H}_{11}\mathbf{D}(e^{\tilde{\boldsymbol{\Gamma}}_m})\right]^{-1}\mathbf{H}_{12} + \mathbf{s}_f^T \mathbf{H}_{22} \tag{8.78}$$

and define $\mathbf{s}_m \in \mathbb{R}_+^M$ such that

$$\mathbf{s}_m^T = \mathbf{s}_f^T \mathbf{H}_{21} \mathbf{D}(e^{\tilde{\boldsymbol{\Gamma}}_m})\left[\bar{\rho}\mathbf{I}_M - \mathbf{H}_{11}\mathbf{D}(e^{\tilde{\boldsymbol{\Gamma}}_m})\right]^{-1}. \tag{8.79}$$

Proposition 8.4.   Given an initialization $\mathbf{s}_m^T[0] \succ \mathbf{0}$, $\mathbf{s}_m^T$ can be realized by the following update:

$$\mathbf{s}_m^T[t+1] = \frac{1}{\bar{\rho}}\mathbf{s}_m^T[t]\mathbf{H}_{11}\mathbf{D}(e^{\tilde{\boldsymbol{\Gamma}}_m}) + \frac{1}{\bar{\rho}}\mathbf{s}_f^T \mathbf{H}_{21}\mathbf{D}(e^{\tilde{\boldsymbol{\Gamma}}_m}). \tag{8.80}$$

Proof: By recursively substituting into (8.80) and taking the limit $t \to \infty$, we have that

$$\lim_{t \to \infty} \mathbf{s}_m^T[t] = \mathbf{s}_m^T[0] \lim_{t \to \infty}\left[\frac{1}{\bar{\rho}}\mathbf{H}_{11}\mathbf{D}(e^{\tilde{\boldsymbol{\Gamma}}_m})\right]^t$$
$$+ \frac{1}{\bar{\rho}}\mathbf{s}_f^T \mathbf{H}_{21}\mathbf{D}(e^{\tilde{\boldsymbol{\Gamma}}_m}) \lim_{t \to \infty}\left\{\sum_{i=0}^{t-1}\left[\frac{1}{\bar{\rho}}\mathbf{H}_{11}\mathbf{D}(e^{\tilde{\boldsymbol{\Gamma}}_m})\right]^i\right\}. \tag{8.81}$$

Recall that if $\rho(\mathbf{H}_{11}\mathbf{D}(e^{\tilde{\boldsymbol{\Gamma}}_m})) < \bar{\rho} < 1$, then it is clear that $\lim_{t \to \infty}[(1/\bar{\rho})$ $\mathbf{H}_{11}\mathbf{D}(e^{\tilde{\boldsymbol{\Gamma}}_m})]^t = 0$. By [35, Th. 5.6.9 and Cor. 5.6.16], $[\mathbf{I}_M - (1/\bar{\rho})\mathbf{H}_{11}$ $\mathbf{D}(e^{\tilde{\boldsymbol{\Gamma}}_m})]^{-1} = \sum_{t=0}^{\infty}[(1/\bar{\rho})\mathbf{H}_{11}\mathbf{D}(e^{\tilde{\boldsymbol{\Gamma}}_m})]^t$ exists and is positive componentwise.

As such, (8.81) becomes

$$
\lim_{t \to \infty} s_m^T[t] = \frac{1}{\rho} s_f^T H_{21} D(e^{\tilde{\Gamma}_m}) \left\{ \sum_{i=0}^{\infty} \left[\frac{1}{\rho} H_{11} D(e^{\tilde{\Gamma}_m}) \right]^i \right\}
$$

$$
= \frac{1}{\rho} s_f^T H_{21} D(e^{\tilde{\Gamma}_m}) \left[I_M - \frac{1}{\rho} H_{11} D(e^{\tilde{\Gamma}_m}) \right]^{-1}, \tag{8.82}
$$

which is equivalent to (8.79).   ∎

Notice that the ith component of $s_m[t+1]$ in (8.80) is actually

$$
s_m^{(i)}[t+1] = \frac{e^{\tilde{\Gamma}_m^{(i)}}}{\bar{\rho}} \left[\sum_{j=1}^{N_m} H_{11}^{(j,i)} s_m^{(j)}[t] + \sum_{j=1}^{N_f} H_{21}^{(j,i)} s_f^{(j)} \right] \tag{8.83}
$$

for $i = 1, \ldots, N_m$. From (8.57) and upon recalling the partition of H, the update in (8.83) further amounts to

$$
s_m^{(i)}[t+1] = \frac{e^{\tilde{\Gamma}_m^{(i)}}}{\bar{\rho}} \left[\sum_{j \in \mathcal{N}_m \setminus \{i\}} s_m^{(j)}[t] + \sum_{j \in \mathcal{N}_f} h_{b_j,i} s_f^{(j-N_f)} \right]
$$

$$
= \frac{e^{\tilde{\Gamma}_m^{(i)}}}{\bar{\rho}} \left[\sum_{j \in \mathcal{N}_m \setminus \{i\}} s_m^{(j)}[t] + \sum_{l \neq b_0} h_{l,i} \sum_{j,b_j=l} s_f^{(j-N_f)} \right] \tag{8.84}
$$

for $i = 1, \ldots, N_m$. Clearly, $s_m^{(i)}[t+1]$ consists of the *internal* component $\sum_{j \in \mathcal{N}_m \setminus \{i\}} s_m^{(j)}[t]$ due to other macro users, and the *external* component $\sum_{l \neq b_0} h_{l,i} \sum_{j,b_j=l} s_f^{(j-N_f)}$ due to all femto users, where b_0 denotes the MBS.

With $s_f \succ 0$ known and once $s_m \succ 0$ has been determined, v_f can readily be computed. From (8.78) and (8.79), $v_f^T = s_m^T H_{12} + s_f^T H_{22}$. Then, its component can be found according to

$$
v_f^{(i)} = h_{b_0,i} \sum_{j \in \mathcal{N}_m} s_m^{(j)} + \sum_{j \neq i, b_j = b_i} s_f^{(j)} + \sum_{l \neq b_i} h_{l,i} \sum_{j,b_j=l} s_f^{(j)} \tag{8.85}
$$

for $i = 1, \ldots, N_f$. It is worth commenting that the first term of (8.85) amounts to the effects from all macro users, whereas the second term from the femto users within the same femtocell, and the third term from the femto users in all other femtocells.

8.4.2 Distributed Algorithm for Femtocell Utility Maximization and Macrocell SINR Balancing

The above parametrization $\tilde{\gamma}_f = \tilde{\gamma}_f(s_f, \tilde{\Gamma}_m, \bar{\rho})$ allows us to find all points on $\mathcal{F}_{\bar{\rho}}$. By fixing $\bar{\rho} \in [0, 1)$ and upon applying that parametrization, (8.66) can be solved via an equivalent optimization problem, albeit in the new variable s_f. The latter involves finding a direction of s_f that leads $\tilde{\gamma}_f$ and p to the optimum of the original problem. With $\varepsilon > 0$, we propose to update $s_f^{(i)}$ as follows.

$$
s_f^{(i)}[t+1] = s_f^{(i)}[t] + \varepsilon \Delta s_f^{(i)}[t+1], \tag{8.86}
$$

for $i = 1, \ldots, N_f$, where

$$\Delta s_f^{(i)}[t+1] = U_i'\!\left(\widetilde{\gamma}_f^{(i)}\right)\!\big/\!\left(\bar{\rho} q_f^{(i)}\right) - s_f^{(i)}[t]. \tag{8.87}$$

Upon recalling that $\mathbf{s}_f$ is a left eigenvector associated with eigenvalue $\bar{\rho}$ of $\mathbf{FD}(e^{\widetilde{\gamma}_f})$, it can be proved that the update of $\mathbf{s}_f$ in (8.86) and (8.87) actually represents an ascent direction for $U(\mathbf{s}_f)$ [28]. We also note that the update of macro load $s_m^{(i)}$ in (8.84) is totally different from that of femto load $s_f^{(i)}$ in (8.86).

Algorithm 8.4   PARETO-OPTIMAL POWER CONTROL ALGORITHM

Require: $\Gamma^{(m)} \succ \mathbf{0}$, $\bar{\rho} \in [0, 1)$, and $\varepsilon > 0$.
1: Initialize $\mathbf{p}_m[0] \succ \mathbf{0}$, $\mathbf{p}_f[0] \succ \mathbf{0}$, $\mathbf{s}_f[0] \succ \mathbf{0}$; $t_m = 1$, $t_f = 1$.
2: Set arbitrary $\mathbf{s}_m[0] \succ \mathbf{0}$.
3: **repeat**
4:   Macro user i computes $s_m^{(i)}[t_m]$ by (8.84).
5:   Set $t_m = t_m + 1$.
6: **until** $\mathbf{s}_m$ converges
7: Femto user j computes $v_f^{(j)}$ by (8.85) and SINR target $\widetilde{\gamma}_f^{(j)}$ by (8.77).
8: Set $t_p = 0$. Using the Foschini–Miljanic's algorithm [1],

- Macro user i measures the actual SINR $\hat{\gamma}^{(i)}$, and updates its power $p_m^{(i)}[t_p + 1] = p_m^{(i)}[t_p]\Gamma_m^{(i)}/\hat{\gamma}^{(i)}$ until $p_m^{(i)}$ converges.
- Femto user j measures the actual SINR $\hat{\gamma}^{(j)}$, and updates its power $p_f^{(j)}[t_p + 1] = p_f^{(j)}[t_p]e^{\widetilde{\gamma}_f^{(j)}}/\hat{\gamma}^{(j)}$ until $p_f^{(j)}$ converges.

9: Femto user j measures interference $q_f^{(j)}$.
10: Femto user j updates $s_f^{(j)}[t_f + 1]$ according to (8.86) and (8.87).
11: Set $t_m = 1$, $t_f = t_f + 1$; go back to step 2 and repeat until $\mathbf{s}_f$ converges.

We present the Pareto-optimal power control algorithm, which is described in the following. It is assumed that the channel gains between the DL and the UL are identical. Here, $s_m^{(i)}[t+1]$ in step 4 is computed and managed by macro user $i \in \mathcal{N}_m$. Specifically, each femto access point l is required to broadcast $\sum_{j, b_j = l} s_f^{(j)}$ at a constant power. This allows macro user i to also measure all channel gains $h_{l,i} = h_{i,l}$ required for the calculation of $\sum_{l \neq b_i} h_{l,i} \sum_{j, b_j = l} s_f^{(j)}$. On the other hand, the macro base station communicates the quantity $\sum_{j \in \mathcal{N}_m} s_m^{(j)}[t]$ to all macro users, which then permits macro user i to easily compute $\sum_{j \in \mathcal{N}_m \setminus \{i\}} s_m^{(j)}[t] = \sum_{j \in \mathcal{N}_m} s_m^{(j)}[t] - s_m^{(i)}[t]$. Finally, macro user i reports the resulting $s_m^{(i)}[t+1]$ back to macro base station for the computation of $\mathbf{s}_m$ in the next iteration. Note that each femto access point l only needs to broadcast $\sum_{j, b_j = l} s_f^{(j)}$ once. In addition, $\sum_{j \in \mathcal{N}_m} s_m^{(j)}[t]$ and $s_m^{(i)}[t+1]$ can be exchanged locally between macro base station and macro user i over the control channel of link i.

The computation of $v_f^{(j)}$ in step 7 can also be done by femto user j. Once $\mathbf{s}_m$ has been determined (i.e., its update (8.84) has converged), macro base station broadcasts the quantity $\sum_{i \in \mathcal{N}_m} s_m^{(i)}$, again at a constant power. Recall that all summations $\sum_{i, b_i = l} s_f^{(i)}$ have already been received at femto user j from all femto access points l (including the one that serves femto user j). Together with the assumption of symmetric DL–UL channel gains, $v_f^{(j)}$ can thus be computed according to (8.85). Additionally, the update of $s_f^{(j)}$ in step 10 can be accomplished in a completely distributed manner by femto user j with only local information required. Over its own control channel, femto user j then reports the new value of $s_f^{(j)}$ to its servicing femto access point b_j, to be used in the next iteration.

Theorem 8.4. For a sufficiently small $\varepsilon > 0$ and as $\bar{\rho} \to 1$, the proposed power control algorithm converges to the globally optimal solution of (8.66).

Proof: Note that the constraints $\widetilde{\gamma}_m^{(i)} \geq \widetilde{\Gamma}_m^{(i)}, \forall i \in \mathcal{N}_m$ in (8.66) are already satisfied with equality as we are operating on $\partial \mathcal{F}_{\bar{\rho}}$. With multiplier $\mu \geq 0$, the Lagrangian of problem in (8.66) is defined as

$$\mathcal{L}(\widetilde{\gamma}, \mu) := \sum_{i \in \mathcal{N}_f} U_i(\widetilde{\gamma}_f^{(i)}) - \mu[\rho(\mathbf{HD}(e^{\widetilde{\gamma}})) - \bar{\rho}]. \tag{8.88}$$

The KKT condition of (8.66) is simply $\nabla \mathcal{L}(\widetilde{\gamma}, \mu) = 0$, which can be shown equivalent to

$$U_i'(\gamma_f^{(i)}) = \mu \, v_f^{(i)} \tilde{q}_f^{(i)}, \; i = 1, \ldots, N_f \tag{8.89}$$

where $\mathbf{v}_f = [v_f^{(i)}]$ is the left eigenvector of $\mathbf{D}(e^{\widetilde{\gamma}_f})\mathbf{F}$ and $\tilde{\mathbf{q}}_f = [\tilde{q}_f^{(i)}]$ is the right eigenvector of $\mathbf{FD}(e^{\widetilde{\gamma}_f})$, both associated with eigenvalue $\bar{\rho}$.

At the point of convergence $\mathbf{s}_f^*$, we have that $\Delta \mathbf{s}_f^* = 0$. It follows from (8.87) that

$$U_i'\left(\widetilde{\gamma}_f^{(i)*}\right) = v_f^{(i)*} q_f^{(i)*}, \; i = 1, \ldots, N_f. \tag{8.90}$$

Manipulating (8.61) and (8.62) and using the matrix/vector partitions specified at the beginning of Section 8.4.1 give $\mathbf{q}_f = \mathbf{FD}(e^{\widetilde{\gamma}_t})\mathbf{q}_f + \tilde{\boldsymbol{\varphi}}_f$, where $\tilde{\boldsymbol{\varphi}}_f = \mathbf{H}_{21}\mathbf{D}(e^{\widetilde{\Gamma}_m})\boldsymbol{\sigma}_m + \boldsymbol{\sigma}_f$. It can be shown that

$$q_f^{(i)*} \to \tilde{q}_f^{(i)} / [(1 - \bar{\rho})\mathbf{s}_f^{*T} \tilde{\boldsymbol{\varphi}}_f] \tag{8.91}$$

as $\bar{\rho} \to 1$. Therefore, (8.90) is exactly (8.89) for $\widetilde{\gamma}_\mathbf{f} = \widetilde{\gamma}_\mathbf{f}^*$ and $\mu = 1/[(1 - \bar{\rho})\mathbf{s}_f^{*T} \tilde{\boldsymbol{\varphi}}_f]$ in the limit $\bar{\rho} \to 1$. Since (8.66) is a convex optimization problem, any point that satisfies the KKT conditions is also the global optimum [37].

With $\mathbf{s}_f^*$ known and upon recalling that $\mathbf{v}_f^* = \mathbf{F}^T \mathbf{s}_f^*$, the optimal SINR assignments $\widetilde{\gamma}_\mathbf{f}^*$ of all femto users are determined according to (8.77). Also recollect that the optimal macrocell SINR is indeed $\widetilde{\gamma}_\mathbf{m}^* = \widetilde{\Gamma}_\mathbf{m} \in \mathbb{R}^{N_m}$. By Foschini–Miljanic's algorithm [1], the power allocation $\mathbf{p}^*$ that achieves these SINR targets can be found. Together with $\widetilde{\gamma}^* = [\widetilde{\Gamma}_\mathbf{m}; \widetilde{\gamma}_f^*]$, $\mathbf{p}^*$ gives the global optimum of problem (8.66). ∎

It can be verified that the distributed Pareto-optimal power control algorithm proposed in this section requires much higher signaling overhead than those developed by the game theory approach in Sections 8.2 and 8.3. In fact, the employment of the Foschini–Miljanic power control updates in step 8 given temporary target SINRs $\hat{\gamma}^{(i)}$ for macro and femto users may require a similar level of signaling as the main BSA and/or power control updates in the proposed algorithms in Sections 8.2 and 8.3. However, the proposed Pareto-optimal power control algorithm in this section indeed demands to search for $s_m^{(i)}$, $s_f^{(i)}$, and $v_f^{(i)}$ (in steps 4, 10, and 7, respectively) based on which the the temporary target SINRs $\hat{\gamma}^{(i)}$ can be determined and achieved by the distributed Foschini–Miljanic power control updates. These two main operations require nested loops and the signaling overhead required by the updates in steps 4, 10, and 7 is quite heavy. Therefore, the Pareto-optimality achieved by the proposed algorithm in this section is indeed penalized by the heavy signaling overhead, which would consume large backhaul capacity. Another distributed power control algorithm for joint utility maximization of both network tiers subject to SINR constraints for macro users can be found in [38].

8.4.3 Numerical Examples

The network setting and user placement in our examples are shown in Figure 8.3, where macro and femto users are randomly deployed inside circles of radii of 1000 m and 50 m, respectively. In particular, we assume that there are $N_m = 10$ macro users, whereas $N_f = 20$ femto users are divided equally among 4 femtocells (i.e., 5 femto users per femtocell). The UL case is considered in all simulations. The absolute channel gain from the transmitter of user j to BS b_i that serves user i is calculated as follows.

$$h_{\sigma_i, j} = \begin{cases} d_{b_i, j}^{-3}, & \text{if } b_j = b_i, \\ d_{b_i, j}^{-3}/(10^{\kappa/10}), & \text{if } b_j \neq b_i, \end{cases} \tag{8.92}$$

where $d_{b_i, j}$ is their corresponding geographical distance, and $\kappa = 10\text{dB}$ is used to represent the extra cross-cell signal loss due to penetration through walls (as femto users are typically deployed indoors).

For simplicity, we consider unit bandwidth. The throughput, normalized over the total bandwidth, is thus expressed in terms of b/s/Hz. Gaussian noise power is taken as $\sigma_i = 10^{-6}$, $\forall i \in \mathcal{N}$. Normalized target SINRs $\mathbf{\Gamma}^{(m)} = [\Gamma_1, \ldots, \Gamma_{N_m}]^T$, are assumed equal for all macro users, which are chosen such that $\rho\left(\mathbf{HD}([\mathbf{\Gamma}^{(m)}; \mathbf{0}_K])\right) \leq \bar{\rho} < 1$. We choose processing gain equal 32 and 3-fair utility function is used, that is, $\alpha = 3$ in (8.67) unless stated otherwise. We set the error tolerance for the convergence of the proposed schemes and Foschini–Miljanic's algorithm as $\epsilon_m = 10^{-4}$ and $\epsilon_p = 10^{-10}$, respectively.

Figure 8.11 shows that the presented algorithm converges to an equilibrium where the exact target SINR $\bar{\gamma}^{\min} = 8.4$ dB obtained for all the macro users. Additionally, femto users are able to achieve quite a large total throughput, which confirms the efficacy of the proposed algorithm. The issue of fairly utilizing the available radio resources can be effectively resolved by regulating α in the utility function. Figure 8.12 shows the minimum and maximum throughput of all the femtocells for

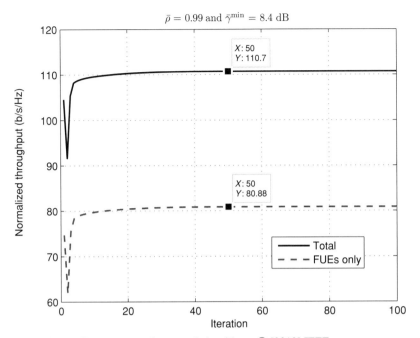

Figure 8.11   Convergence of proposed algorithms. © [2012] IEEE.

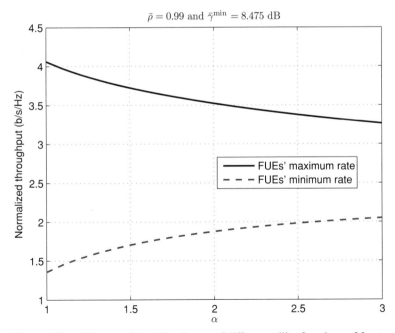

Figure 8.12   Fairness achieved by the use of different utility functions of femto users.
© [2012] IEEE.

different values of α and target SINRs of macro users equal $\overline{\gamma}^{\min} = 8.475$ dB. Apparently, as α increases, the femto user whose data rate is the highest (likely due to its advantageous link conditions) sees a decline in its throughput, whereas the femto user with the lowest throughput has its data rate gradually enhanced. For a large value of α, the femto users' minimum throughput is further expected to reach a plateau, meaning that max–min fairness is realized at that point.

8.5   SUMMARY AND OPEN RESEARCH ISSUES

In this chapter, we have exemplified the design of distributed resource allocation algorithms for CDMA multi-tier HetNets by using game theory and optimization approaches. We demonstrated how the convergence of these algorithms can be established, and discussed their different signaling overhead requirements. There are still various research issues that deserve more research efforts.

- For multi-tier HetNets with multiple antennae, various advanced communication techniques can be applied for efficient interference management and dynamic spectrum sharing. Specifically, if the BSs of each network tier are equipped with multiple antennae, then we can employ the beamforming technique to mitigate both co-tier interference and CTI [39, 40]. However, distributed and low-complexity beamforming algorithms are strongly desired to alleviate heavy capacity requirement of the wired backhaul channel due to coordination signaling. Such design can be achieved by adapting available techniques on limited feedback and opportunistic beamforming design originally developed for single-tier networks to the multi-tier HetNet setting. Finally, low-complexity algorithms based on interference alignment idea can be devised for interference control in the wireless HetNets [41–43].

- All models considered in this chapter assume fixed network topology with unvarying number of static users. In practice, users dynamically arrive and leave the network and users may move over space, which result in changes in network topologies. Therefore, design of adaptive and robust resource allocation algorithms that can provide QoS support for dynamic HetNets is an important but challenging research problem to tackle. In addition, admission control at the call level must be jointly designed with resource allocation algorithms to efficiently utilize network resources and to avoid network congestion and QoS deterioration. Rich literature on these research issues available for homogeneous single-tier networks must be revisited and extended to resolve the corresponding issues in the heterogeneous setting.

REFERENCES

1. G. J. Foschini and Z. Miljanic, "A simple distributed autonomous power control algorithm and its convergence," *IEEE Transactions on Vehicular Technology*, vol. 42, no. 4, pp. 641–646, November 1993.

2. J. Zander, "Performance of optimum transmitter power control in cellular radio systems," *IEEE Transactions on Vehicular Technology*, vol. 41, no. 1, pp. 57–62, February 1992.

3. R. D. Yates, "A framework for uplink power control in cellular radio systems," *IEEE Journal on Selected Areas in Communications*, vol. 13, no. 7, pp. 1341–1347, September 1995.

4. M. Xiao, N. B. Shroff, and E. K. P. Chong, "A utility-based power control scheme in wireless cellular systems," *IEEE/ACM Transactions on Networking*, vol. 11, no. 2, pp. 210–221, April 2003.

5. S. Koskie and Z. Gajic, "A Nash game algorithm for SIR-based power control in 3G wireless CDMA networks," *IEEE/ACM Transactions on Networking*, vol. 13, no. 5, pp. 1017–1026, October 2005.

6. M. Andersin, Z. Rosberg, and J. Zander, "Gradual removals in cellular PCS with constrained power control and noise," *Wireless Networks*, vol. 2, pp. 27–43, 1996.

7. N. Bambos, S. C. Chen, and G. J. Pottie, "Channel access algorithms with active link protection for wireless communication networks with power control," *IEEE/ACM Transactions on Networking*, vol. 8, no. 5, pp. 583–597, October 2000.

8. M. Rasti and A. R. Sharafat, "Distributed uplink power control with soft removal for wireless networks," *IEEE Transactions on Communications*, vol. 59, no. 3, pp. 833–843, March 2011.

9. C. W. Sung and W. S. Wong, "A noncooperative power control game for multirate CDMA data networks," *IEEE Transactions on Wireless Communications*, vol. 2, no. 1, pp. 186–194, January 2003.

10. M. Rasti, A. R. Sharafat, and B. Seyfe, "Pareto-efficient and goal-driven power control in wireless networks: A game-theoretic approach with a novel pricing scheme," *IEEE/ACM Transactions on Networking*, vol. 17, no. 2, pp. 556–569, April 2009.

11. C. U. Saraydar, N. B. Mandayam, and D. J. Goodman, "Efficient power control via pricing in wireless data networks," *IEEE Transactions on Communications*, vol. 50, no. 2, pp. 291–303, February 2002.

12. C. U. Saraydar, N. B. Mandayam, and D. J. Goodman, "Pricing and power control in a multicell wireless data network," *IEEE Journal on Selected Areas in Communications*, vol. 19, no. 10, pp. 1883–1892, October 2001.

13. T. Alpcan, T. Baar, R. Srikant, and E. Altman, "CDMA uplink power control as a noncooperative game," *Wireless Networks*, vol. 8, no. 6, pp. 659–670, November 2002.

14. W. Yu, G. Ginis, J. M. Cioffi, "Distributed multiuser power control for digital subscriber lines," *IEEE Journal on Selected Areas in Communications*, vol. 20, no. 5, pp. 1105–1115, June 2002.

15. T. Alpcan, T. Basar, and S. Dey, "A power control game based on outage probabilities for multicell wireless data networks," *IEEE Transactions on Wireless Communications*, vol. 5, no. 4, pp. 890–899, April 2006.

16. C. W. Sung and K. K. Leung, "A generalized framework for distributed power control in wireless networks," *IEEE Transactions on Information Theory*, vol. 51, no. 7, pp. 2625–2635, July 2005.

17. K. K. Leung and C. W. Sung, "An opportunistic power control algorithm for cellular network," *IEEE/ACM Transactions on Networking*, vol. 14, no. 3, pp. 470–478, June 2006.

18. R. D. Yates and C. Y. Huang, "Integrated power control and base station assignment," *IEEE Transactions on Vehicular Technology*, vol. 44, no. 3, pp. 638–644, August 1995.

19. S. V. Hanly, "An algorithm for combined cell-site selection and power control to maximize cellular spread spectrum capacity," *IEEE Journal on Selected Areas in Communications*, vol. 13, no. 7, pp. 1332–1340, September 1995.

20. H. S. Dhillon, R. K. Ganti, F. Baccelli, and J. G. Andrews, "Modeling and analysis of K-tier downlink heterogeneous cellular networks," *IEEE Journal on Selected Areas in Communications*, vol. 30, no. 3, pp. 550–560, April 2012.

21. P. Xia, V. Chandrasekhar, and J. G. Andrews, "Open vs. closed access femtocells in the uplink," *IEEE Transactions on Wireless Communications*, vol. 9, no. 10, pp. 3798–3809, December 2010.

22. V. Chandrasekhar and J. G. Andrews, "Spectrum allocation in tiered cellular networks," *IEEE Transactions on Communications*, vol. 57, no. 10, pp. 3059–3068, October 2009.

23. V. Chandrasekhar and J. G. Andrews, "Uplink capacity and interference avoidance for two-tier femtocell networks," *IEEE Transactions on Wireless Communications*, vol. 8, no. 7, pp. 3498–3509, July 2009.

24. H. S. Jo, C. Mun, J. Moon, and J.-G. Yook, "Interference mitigation using uplink power control for two-tier femtocell networks," *IEEE Transactions on Wireless Communications*, vol. 8, no. 10, pp. 4906–4910, October 2009.

25. H. S. Jo, C. Mun, J. Moon, and J.-G. Yook, "Self-optimized coverage coordination in femtocell networks," *IEEE Transactions on Wireless Communications*, vol. 9, no. 10, pp. 2977–2982, October 2010.

26. M. Rasti and A. R. Sharafat, "Pareto and energy-efficient distributed power control with feasibility check in wireless networks," *IEEE Transactions on Information Theory*, vol. 57, no. 1, pp. 245–255, January 2011.

27. S. Ulukus and R. D. Yates, "Stochastic power control for cellular radio systems," *IEEE Transactions on Communications*, vol. 46, no. 6, pp. 784–798, June 1998.

28. P. Hande, S. Rangan, M. Chiang, and X. Wu, "Distributed uplink power control for optimal SIR assignment in cellular data networks," *IEEE/ACM Transactions on Networking*, vol. 16, no. 6, pp. 1420–1433, December 2008.

29. D. Palomar and M. Chiang, "A tutorial on decomposition methods and distributed network resource allocation," *IEEE Journal on Selected Areas in Communications*, vol. 24, no. 8, pp. 1439–1451, August 2006.

30. M. Chiang, S. H. Low, A. R. Calderbank, and J. C. Doyle, "Layering as optimization decomposition: A mathematical theory of network architectures," *Proceedings of the IEEE*, vol. 95, no. 1, pp. 255–312, January 2007.

31. D. Ngo, L. B. Le, T. Le-Ngoc, E. Hossain, and D. I. Kim, "Distributed interference management in two-tier CDMA femtocell networks," *IEEE Transactions on Wireless Communications*, vol. 11, no. 3, pp. 979–989, March 2012.

32. A. A. D'Amico, U. Mengali, and M. Morelli, "Channel estimation for the uplink of a DS-CDMA system," *IEEE Transactions on Wireless Communications*, vol. 2, no. 6, pp. 1132–1137, November 2003.

33. V. N. Ha and L. B. Le, "Distributed base station association and power control for heterogeneous cellular networks," *IEEE Transactions on Vehicular Technology*, to appear.

34. J. Mo and J. Walrand, "Fair end-to-end window-based congestion control," *IEEE/ACM Transactions on Networking*, vol. 8, no. 5, pp. 556–567, October 2000.

35. A. Horn and A. Johnson, *Matrix Analysis*, 1st edition, Cambridge University Press, 1985.

36. H. Boche and S. Stanczak, "Convexity of some feasible QoS regions and asymptotic behavior of the minimum total power in CDMA systems," *IEEE Transactions on Communications*, vol. 52, no. 12, pp. 2190–2197, December 2004.

37. D. P. Bertsekas, *Nonlinear Programming*, 2nd edition, Boston, MA: Athena Scientific, 1999.

38. D. Ngo, L. B. Le, and T. Le-Ngoc, "Distributed Pareto-optimal power control for utility maximization in femtocell networks," *IEEE Transactions on Wireless Communications*, vol. 11, no. 10, pp. 3434–3446, October 2012.

39. K. Huang, J. G. Andrews, and R. W. Heath, "Performance of orthogonal beamforming for SDMA with limited feedback," *IEEE Transactions on Vehicular Technology*, vol. 58, no. 1, pp. 152–164, January 2009.

40. W. Choi, A. Forenza, J. G. Andrews, and R. W. Heath, "Opportunistic space-division multiple access with beam selection," *IEEE Transactions on Communications*, vol. 55, no. 12, pp. 2371–2380, December 2007.

41. M. A. Maddah-Ali, A. S. Motahari, and A. K. Khandani, "Communication over MIMO X channels: Interference alignment, decomposition, and performance analysis," *IEEE Transactions on Information Theory*, vol. 54, no. 8, pp. 3457–3470, August 2008.

42. T. Gou and S. A. Jafar, "Degrees of freedom of the K user MxN MIMO interference channel," *IEEE Transactions on Information Theory*, vol. 56, no. 12, pp. 6040–6057, December 2010.

43. C. Suh, M. Ho, and D. N. C. Tse, "Downlink interference alignment," *IEEE Transactions on Communications*, vol. 59, no. 9, pp. 2616–2626, September 2011.

CHAPTER 9

SELF-ORGANIZING SMALL CELL NETWORKS

Small cell networks such as femtocell networks have to be designed not only to support users' QoS requirements, but also to achieve cost effectiveness, deployment flexibility, and scalability. These capabilities are needed to meet the requirements of both customers and service providers. The femto access points are usually deployed by the users without prior planning. To reduce capital expenditure (CAPEX) and operation expenditure (OPEX), the concept of self-organizing network (SON) has been introduced in the different wireless standards (e.g., 3GPP Long Term Evolution [LTE]) [1]. This concept can also be adopted in the femtocell networks. The SONs are expected to automatically organize, operate, optimize, and maintain themselves with minimum control from external entity and intervention of users. In general, the SON includes three functionalities, that is, self-configuration, self-optimization, and self-healing. While the self-configuration functionality is related to the initialization and installation (i.e., pre-operational) stage of the network, self-optimization and self-healing functionalities are required in the operational stage, which will impact the performance and availability of the services.

In this chapter, the concept of self-organization in small cell networks is presented. First, the motivations behind self-organization are discussed, which naturally arise due to a large number of small cells and the complex network structure in a multi-tier deployment scenario. The desirable behaviors of the self-organizing small cell networks in terms of scalability, stability, robustness, and agility are introduced. Then, the self-x (self-configuration, self-optimization, and self-healing) concept and the interference management issues are summarized. Specifically, the concept of cognitive radio can be adopted in the small cell networks to make the system capable of observing, learning, optimizing, and adapting to environment changes. Next, a detailed review of the selected works in the literature on self-configuration, self-optimization, and self-healing is presented. The reviews clearly show that there are a number of challenging issues in designing self-organizing small cell networks. Different techniques can be adopted (e.g., from control theory, game theory, and reinforcement algorithm) to achieve the optimal network performance while meeting the QoS requirements of both macro and small cell users.

Radio Resource Management in Multi-Tier Cellular Wireless Networks, First Edition.
Ekram Hossain, Long Bao Le, and Dusit Niyato
© 2014 John Wiley & Sons, Inc. Published 2014 by John Wiley & Sons, Inc.

9.1   SELF-ORGANIZING NETWORKS

9.1.1   Motivations of Self-Organization

The concept of self-organization has been used in the ad hoc/sensor networks for more than a decade [2]. However, self-organization has emerged as a significant issue only recently in the context of cellular network due to the introduction of the concept of small cells. The needs for self-organization are as follows.

- Since the number of small cells deployed in the network is expected to be large, it is infeasible to configure, operate, control, and maintain them manually. Therefore, automated tools and intelligence in the network devices (e.g., SBSs) would be required for scalability and efficient network operation.

- The network must be cost-effective to deploy and operate. Due to the high competition, the wireless service providers seek for a solution that minimizes their capital expenditure (CAPEX) and operation expenditure (OPEX). Also, the users' experience and satisfaction must be maximized.

The self-organization concept is built based on the intelligence and autonomous adaptability to allow different entities in the network to observe, learn, optimize, decide, and adapt to environmental changes. This is also known as the cognitive cycle [3]. The cognitive cycle can be adopted for the self-organization since it makes the configuration, optimization, and healing of femtocell network autonomous (i.e., perform without or with minimum human intervention).

Nevertheless, in addition to the intelligence and autonomous adaptability, the self-organization should also encompass distributed control, local interaction, and emergent behavior capabilities. The distributed control [4] provides the capability for each network device (e.g., femto access point) to operate without an external control. For example, in a two-tier macrocell–femtocell network, the femto access points can select the sub-bands by themselves to communicate with the femto users. The distributed control can be implemented together with local interaction [5]. With the local interaction, multiple network devices can form a group, exchange information, and interact in order to achieve the group or individual objective [6]. For example, multiple nearby femto access points can form a cluster. Inside the cluster, the femto access point can coordinate and schedule the spectrum access to avoid co-tier interference. The emergent behavior [4] is the capability of a network device to act and respond to the environmental changes dynamically even without being programmed in advance. For example, the failure of a femtocell can be detected and the femto users of the failed femtocell can be handed over to nearby femtocells automatically. The benefits of implementing self-organization are scalability, stability, robustness, and agility [2].

Scalability refers to the ability of the network to accommodate an increasing number of network devices. Therefore, the complexity of the network operation must not increase sharply given the increase in the number of network devices. To achieve scalability, the network must have the following properties, that is, bounded complexity, minimum overhead, and local interaction. The bounded complexity means that the installation, deployment, operation, and maintenance of the network will require only a small amount of resources (e.g., time, space, and communication). For example, the resource allocation algorithm for a scalable network should be solvable within a reasonable time even though the number of network devices increases by an order of magnitude. Also, the network overhead (e.g., for exchanging control information) should not grow significantly as the network size grows. Local interaction is required to avoid the requirement of global information and a centralized control.

Because of learning and adaptation functions, the optimal solution of an algorithm for a SON is usually obtained in an iterative fashion. The stability of such an algorithm is its ability to converge to the final solution within an acceptable amount of time. When the network parameters change, the algorithm must be able to reach the steady state quickly so that the network can operate efficiently. For example, when a new user joins a small cell, the transmission powers of all ongoing and new users must be adjusted within a few iterations to satisfy the users' QoS requirements. Also, the stability implies that the network state should not oscillate when one or more network parameters change or when there is more than one solutions that the network can operate in. In the latter case, the self-organization algorithm must be able to determine the efficient operating point.

Robustness is the ability of a SON to handle randomness and failure of network components. Randomness is common in wireless networks. A channel can experience fading and, hence, channel quality fluctuates. The number of users may increase or decrease rapidly due to high mobility of the users. Likewise, failures of the network components (e.g., SBSs) can occur. The SBSs such as the femtocells can fail more frequently than the expensive macro base stations. The broadband link between the SBSs and the core network can also fail occasionally. The small cell networks must be able to detect such unexpected events and perform the necessary recovery action in a timely manner.

In addition to stability and robustness, agility is needed to ensure that the network can adapt quickly to environmental changes. To achieve agility, the network must have a responsive feedback and decision making infrastructure. Especially in small cell networks, where communications among network devices could be intermittent, delay and loss of feedback information may happen. Therefore, the network must be able to operate efficiently even with limited, lossy, and delayed information. It is also important to note that the responsiveness of the network should be controlled. If the network responds to the changes too fast, then the optimal operating point may not be reached and the network state could oscillate. A simple but effective approach to address this issue is to set a threshold. The network will start the adaptation if the system parameter exceeds the threshold. For example, subchannel reallocation in an OFDMA-based multi-tier network can be performed only when the macro users experience interference higher than an acceptable level or when the QoS performance of the small cell users falls below an acceptable level.

9.1.2 Use Cases of Self-Organizing Small Cell Networks

[2] discusses the use cases of the self-organizing femtocell networks defined in [7] as follows.

- *Coverage and capacity optimization:* Coverage of femtocells have to be adjusted to avoid service hole and low spectrum utilization due to overlapping [8].

- *Energy saving:* The transmission power has to be optimized and power management needs to be implemented (e.g., to put radio transceivers into the sleep and wakeup mode) [9], [10].

- *Interference reduction:* To avoid performance degradation of macro users, the interference caused by the femtocells has to be limited.

- *Automated configuration of physical cell identity:* When a femto access point is newly deployed, its identity (e.g., IP address) must be determined to avoid conflict in the network.

- *Mobility management:* Handoff management of users among macro- and femtocells has to be optimized. Mobility management and access control can also be jointly designed.

- *Random access channel (RACH) optimization:* The mobile devices with sporadic transmissions (e.g., devices for machine-type communications) could use RACH to avoid the overhead (e.g., signaling to reserve radio resource) incurred in a fixed allocation protocol.

- *Automatic neighbor discovery and interaction:* To avoid interference and recover a failed femtocell, neighboring femtocells need to be discovered and may need to be cooperated with. Nearby femtocells can be connected with each other through the FMS [11].

- *Intercell Interference (ICIC):* By using adaptive and intelligent sub-band allocation, the IC among femtocells can efficiently be managed to avoid performance degradation of femto users [12].

9.1.3   Classification of Self-Organizing Small Cell Networks

A classification of the SON based on the time-scale or deployment phase is presented in [2]. In small cell networks, different system parameters may change at different time scales, that is,

- *Very short time scale (millisecond–seconds):* The channel variation can happen within a transmission frame or after a few transmission frames. Adaptive modulation and coding, packet scheduling, and power control will be performed and adapted to such channel variations.

- *Short time scale (minutes–hour):* The channel could experience shadowing. The user mobility changes the number of ongoing connections. Also, some existing femto access points may fail (e.g., due to hardware failure) and need to be fixed. Spectrum allocation and mobility management will be performed. Also, failure detection will be executed to inform the users or service provider of a faulty network device.

- *Medium time scale (hours–day):* A new femto access point could be installed. The load balancing and coverage optimization will be invoked.

- *Long time scale (days–months):* A new wireless system can be introduced or a new spectrum access policy can be enforced. Some new applications and new QoS requirements can emerge. The resource allocation and interference management algorithm must be re-executed.

Based on the deployment phase, self-organization can be classified as follows.

- *Self-configuration [13]:* When new network components (e.g., femto access points) are deployed, they must be self-configured. Therefore, minimum or no human intervention should be required to provide the setting and adjust the operation. Self-configuration can be achieved by using the plug-and-play concept. In this case, the network components must be able to retrieve the basic configuration and the parameter settings so that they can be integrated and can work with the existing components. For example, when new femto access points are installed, they should be able to connect with a core network and provide service to users in closed, open, or hybrid access mode with some basic functions (e.g., admission control and scheduling). Frequency selection is also an important issue in self-configuration.

- *Self-optimization [14]:* Network components must work efficiently and optimally to achieve the network objectives while meeting the requirements of the service provider and users. Therefore, the network parameters should be automatically optimized and adjusted. For example, femto access points must be able to perform a power control algorithm to maximize the throughput of femto users while meeting the interference requirement of macro users. Also, an admission control of the femto access point must make a decision to maximize the spectrum utilization.

- *Self-healing [1]:* When faults and failures happen, the network components must be able to detect such events, analyze the causes, and recover to operate in a normal state automatically and quickly. A recovery plan and policy must be executed and optimized to minimize the user interruption and performance degradation. For example, when one SBS fails and stops working, the nearby BSs should be able to detect or be informed such that they can provide alternative service to the ongoing small cell users.

RRM in small cell networks can be designed and implemented using either a centralized or a distribution architecture. In a centralized SON, the components and algorithms are performed at the operations, administration, and management (OAM) system or server, located in selected locations. On the other hand, in a distributed SON, the components and algorithms are performed at the network element level (e.g., macro base station and femto access point). A self-organizing small cell network can also be a hybrid one. In this case, some components and algorithms are performed in the OAM system, while others are executed locally at the network.

9.1.4 Self-x Concept

Self-configuration, self-optimization, and self-healing are referred to as the "self-x functionality" [15]. [15] also introduces a high-level framework to support the self-x functionality. The major components of this framework are as follows.

- *Reconfiguration resource control module (RCM)* is the hardware with firmware to adjust and modify the physical transmission parameters (e.g., operating channel, transmission power, and broadband connection). The RCM utilizes the command and control from the dynamic SON planning and management.

- *Dynamic spectrum management (DSM)* provides the spectrum assignment function. DSM can be designed based on the cognitive radio technology, which will be able to learn and adapt spectrum access according to the environment change (e.g., interference level). The regulatory guideline, service provider policy and preference, and the negotiation among service providers will be taken into account in DSM.

- *Dynamic self-organizing network planning and management (DSNPM)* is responsible for controlling the overall operation and maintaining the desired performance of the small cell network. The decision making process is implemented to obtain optimal transmission parameters and mobility management strategy of the self-x functionality.

- *Radio resource management (RRM)* optimizes the available radio resources to achieve the desired network objectives. If there are multiple radio interfaces, a joint RRM must be designed to coordinate the transmission parameters across multiple radios.

The interactions among the different components to realize the self-x functionality are shown in Figure 9.1.

There are some projects that develop and implement the complete system of the self-organizing femtocell networks. Broadband evolved femto network (BeFEMTO) [16] is one of them. The BeFEMTO project aims at solving various issues and developing advanced femtocell technology based on a long-term evolution (LTE) system, providing efficient and cost-effective broadband services. The BeFEMTO project focuses on the self-organizing femtocell network for both residential and enterprise environments. In the residential environment, the femto access point is a standalone device and must operate independently, without or with minimum external control. On the other hand, in the enterprise environment, multiple femtocells will operate as a network. The femto access points of femtocells will be able to exchange information among each other to make decisions based on the objective and the constraints of the network. In addition, other typical scenarios (i.e., outdoor mobile and fixed relay) are also considered in the BeFEMTO project [16].

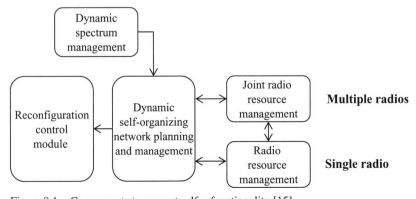

Figure 9.1   Components to support self-x functionality [15].

9.1.5 Interference Management for Self-Organizing Networks

Interference management plays an important role in small cell networks, since it affects the performances of both the macro users and small cell users. Different approaches for interference management have been proposed (e.g., power control, cross-tier resource partitioning, and co-tier resource reuse partitioning [17]) in the literature. [17] provides a survey of the interference management techniques for self-organizing small cell networks. The techniques can be categorized into two approaches, that is, system-centric and user-centric approaches.

- In the system-centric approach, the focus is at the system performance (e.g., sum of throughput of all users) rather than the individual performance. Therefore, this approach will implement the optimization to obtain optimal solutions for the transmission and QoS parameters (e.g., transmission power and/or operating sub-band). Two different scenarios are typically considered in the optimization, that is, to maximize the sum of throughput given that the transmission power is maintained below the threshold, and to minimize the sum of transmission power given that the throughput of the users is maintained above the threshold. In most cases, the system-centric approach is based on the centralized femtocell network.

- The user-centric approach focuses at optimizing the individual utility of a user. The utility represents the satisfaction of users on the received performance (e.g., throughput) and system behavior (e.g., energy consumption). An optimization can be used to achieve a fair solution (e.g., proportional fairness) for the users according to their individual utility. On the other hand, the game theory can be applied if the users adapt their transmission parameters rationally and non-cooperatively. The objective of the rational user is to maximize the individual utility given the transmission strategies of other users. An equilibrium, which ensures that none of the users will change the strategy unilaterally, is commonly considered as a solution. In most cases, the user-centric approach is applied in a distributed femtocell network.

One of the major approaches for interference management in SONs, which is discussed in [17], is based on the cognitive radio concept. In the cognitive radio, the users can be divided into licensed (i.e., primary) users and unlicensed (secondary) users. A radio resource is allocated to the primary users, but can also be opportunistically accessed by the secondary users. Therefore, the secondary users must be capable of observing, learning, optimizing, and adapting their transmission according to the environment (e.g., primary users' activities) to achieve the objective and meet the constraints. In the femtocell networks, the macro and femto users can be considered as the primary and secondary users in the cognitive radio [18], respectively. Therefore, two spectrum access models from the cognitive radio can be adopted in small cell networks.

- *Overlay spectrum access:* In an overlay spectrum access, the macro users considered as the primary users are allocated with sub-bands (i.e., radio resource

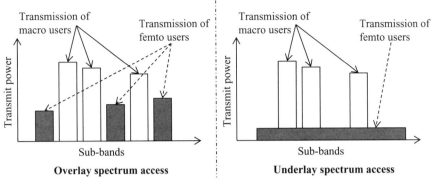

Figure 9.2   Overlay and underlay spectrum access in interference management based on cognitive radio concept.

or spectrum). The small cell users considered as the secondary users opportunistically locate and access the free sub-bands. Therefore, the transmissions of macro and femto users are on the orthogonal spectrum, fully avoiding CTI (Figure 9.2).

- *Underlay spectrum access:* In an underlay spectrum access, the macro and small cell users use the same sub-bands at the same time. However, the transmission of the femto users is constrained by the interference at the macro users (Figure 9.2). Specifically, the small cell users will use low transmission power. Using the low transmission power can not only avoid strong leaking interference but also achieve good performance due to favorable channel characteristics (e.g., short-range transmission in indoor femtocells).

Few issues (e.g., frequency scheduling, access control, and pricing) related to interference management for femtocell networks are addressed based on that done in the cognitive radio (e.g., [19], [20], and [21]). The frequency scheduling proposed in [19] uses spectrum sensing (i.e., energy detection) to obtain and analyze sub-band usage information. Then, this information is used by the scheduling algorithm aiming at avoiding interference to macro users. The access control proposed in [20] considers the femtocells in an open or close access mode. The access control algorithm allows macro users to access the femtocell to improve the overall network performance (i.e., open access mode). However, if the QoS performance needs to be guaranteed, then the close access mode will reserve the femtocell for femto users only. The economic approach can also be used for interference management. In [21], pricing is used to improve the efficiency of the distributed power control algorithm. Typically, the power control among femtocells is formulated as the noncooperative game [22], [23] and its solution in terms of the Nash equilibrium is usually inefficient (i.e., price of anarchy). The price is charged to the femtocells creating interference. As a result, the femtocells are motivated to use lower transmission power and reduce interference, and thus improve the network capacity. The detailed reviews of different approaches are provided in [17].

9.2 SELF-CONFIGURATION

The self-configuration functionality is used to provide a configuration and setting for the newly installed network devices in small cell networks automatically, so that the network devices can be in operation quickly. This self-configuration is performed in a pre-operational state, that is, from the instant the network device (e.g., femto access point) is turned on and the broadband connection is obtained until the transceiver is turned on. The self-configuration is composed of the functions including basic setup and initial radio configuration. In the typical basic setup, first the configuration of the IP address must be set to communicate with a core network through a broadband link. This can be performed through the dynamic host configuration protocol (DHCP). Then, the femto access point will search and connect to the OAM server. Next, the femto access point will perform an authentication process to verify its identity so that the new femto access point can access and transfer data to the core network. The femto access point also needs to connect and access the gateway (e.g., access gateway [AGW] in LTE). Finally, if necessary, the femto access point will download the software and other necessary configuration parameters to preset the femtocell. In the initial radio configuration, the femtocell has to obtain information about other nearby femtocells. Therefore, each femtocell has to inform its status and download the list and parameters of other femtocells. The femtocell also needs to set the operating frequency (e.g., through sub-channel allocation) and determine the coverage and capacity (e.g., through power control).

In addition to the above functions to be implemented in the SBSs, to achieve the optimal performance and the most suitable setting, self-configuration techniques can be developed to make the small cells dynamic, intelligent, and efficient. Therefore, in the literature, different issues in self-configuration have been considered and efficient schemes and algorithms have been developed. The operation of small cells (e.g., femtocells) in a densely deployed environment can be coordinated to avoid strong interference [24]. In this case, the SBSs will contend for the transmission time. The winning BS will be able to communicate with its users. Since only the winning BS can operate, the interference from other small cells can be completely avoided, thus improving the network performance. A similar contention-based sub-channel allocation among small cells is also considered in [25]. However, in [25], the contention is based on the IC link. Also, a two-tier scheduling (i.e., femtocell tier and femto user tier) method is proposed for self-optimization. In [26], the preamble allocation is proposed for an IEEE 802.16e femtocell network. The aim of the preamble allocation is to avoid collision among nearby femtocells so that the macro users are able to detect signals from different cells. The self-initialization method is proposed together with the preamble allocation, which can be performed without a centralized controller. In [27], the interference mitigation based on switched multi-element antenna (SMEA) is proposed. Specifically, a reinforcement learning algorithm is developed to adjust the optimal antenna configuration dynamically. The transmission power calibration method is proposed [28], which is divided into two stages. In the first stage, the transmission power is initialized after a femto access point is switched on. Then, based on the interference measurement, a fine tuning of the power calibration is performed to precisely avoid the performance degradation of nearby macro users.

[29] introduces the dynamic traffic off-loading scheme. The key idea is to control the coverage of a femtocell by limiting the transmission power of a pilot signal. Then, given the achievable coverage, the transmission power control is optimized to maximize the minimum transmission rate of femto users in the femtocell (i.e., max–min problem). Alternatively, [8] applies a genetic algorithm to optimize the transmission power of the pilot signal to maximize the fitness function defined as the weighted performance metrics (i.e., coverage overlap and coverage hole). In [30], a method for dynamic frequency allocation to femtocells is proposed. Frequency reuse is applied among adjacent femtocells to avoid co-tier interference. The outage probability (i.e., probability that the SINR of a femto user falls below a threshold) is analyzed. In [31], a coordinated spectrum assignment method is proposed for the densely deployed femtocell networks in an enterprise environment. The assignment to a femtocell is based on the channel quality of the femto users and the interference degree with other femtocells. However, the assignment is based on a heuristic algorithm.

In the following, a detailed discussion on the different approaches for self-configuration is presented.

9.2.1 Dynamic Traffic Off-loading

Small cells in an open access mode can provide the off-loading function for macro users. In [29], a network capacity enhancement scheme is proposed to support traffic off-loading. In particular, the femtocell can be self-configured to control the pilot power (i.e., transmission power of a pilot signal). The pilot power will determine the coverage area that the macro users can be off-loaded from the macrocell to the femtocell. For example, when the pilot power of the femtocell is increased, the coverage area will increase and the femtocell can serve more off-loaded macro users (e.g., femto access point 1 shown in Figure 9.3). On the other hand, when the pilot power is decreased, the coverage area will decrease (e.g., femto access point 2 shown in Figure 9.3). Also, a self-optimizing ICIC scheme is proposed to adjust the transmission power. The objective of this scheme is to improve the capacity of the entire network.

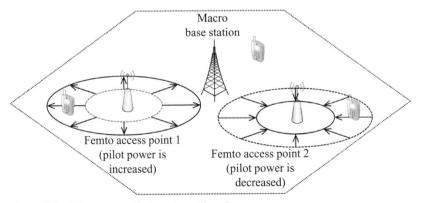

Figure 9.3   Pilot power adjustment to adjust the coverage area.

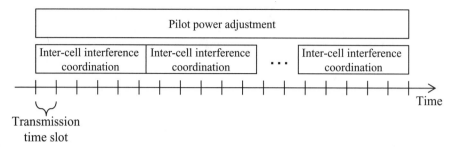

Figure 9.4   Multi time-scale structure in dynamic off-loading.

The pilot power adjustment and the ICIC work on different time scales (i.e., multi time-scale structure as shown in Figure 9.4). The smallest time scale is a frame or time slot for packet transmission. The larger time scale is an interval of ICIC for transmission power allocation. This interval is composed of multiple packet transmission frames. The largest time scale is a period of the pilot power adjustment (i.e., a period for a cell coverage update).

The pilot power adjustment requires the femtocell density information, which can be provided by the macro base station. Then, the femto access point adjusts the pilot power to achieve the optimal network performance. The network performance is derived based on the following system model. There are N stations or cells (i.e., macro base stations and femto access points). $\mathbb{A}$ is the service area of the network and $\mathbb{A}_s \subset \mathbb{A}$ is the subarea (i.e., coverage area) of the station s. l denotes a location in the service area. $h_{s,l}$ is the channel gain between station s and location l. There are R physical resource blocks (PRBs) with equal size W available for data transmission of users. The transmission power of the station s on RB k is denoted by $p_{s,k}$ and $p_{s,0}$ is the pilot power (i.e., transmission power of a pilot signal). The user will connect to the station with the strongest pilot signal. The mean SINR of a user at location l using RB k is given by

$$r_{l,k}(\mathbf{p}) = \frac{p_{s,k}h_{s,l}}{\sigma^2 + \sum_{s' \neq s} p_{s',k}h_{s',l}} \tag{9.1}$$

where $\mathbf{p}$ is composed of the transmission powers of all stations including the pilot power $p_{s,0}$, σ^2 is the noise power, and $h_{s',l}$ is the channel gain between transmitter of station s' to the receiver station s. Given the stochastic fading process, the ergodic rate of the user at location l can be obtained from

$$R(l, \mathbf{p}) = \sum_{k=1}^{K} \Psi(r_{l,k}(\mathbf{p})) \tag{9.2}$$

where

$$\Psi(r_{l,k}(\mathbf{p})) = W \int_{0}^{+\infty} \phi(r_{l,k}(\mathbf{p})x) f_X(x) \mathrm{d}x, \tag{9.3}$$

where fading of a channel is determined by a multiplicative random variable. For $\phi(\gamma) \leq \log_2(1 + \gamma)$, where γ is SINR, and $f_X(x)$ is the marginal pdf of a fading process.

While (9.2) provides the ergodic rate of a user, it is important to consider other users' performance metrics. Therefore, a queueing model is applied to obtain the network capacity, network blocking rate, and mean user throughput. [29] uses an M/G/1/PS queueing model (i.e., Poisson arrival, general distribution, one server with processor sharing). In this case, the processor sharing (i.e., "PS") is based on the round-robin scheduling, in which the users share the same RB equally. With the M/G/1/PS queueing model, the load of station s given the transmission powers $\mathbf{p}$ can be expressed as

$$\rho_s(\mathbf{p}) = \int_{\mathbb{A}_s(\mathbf{p})} \frac{\lambda}{R(l, \mathbf{p})} dl \tag{9.4}$$

where, in this case, the area $\mathbb{A}_s(\mathbf{p})$ is also defined as a function of the pilot powers, and λ is the maximum traffic density. The capacity of station s can be obtained from

$$C_s(\mathbf{p}) = \left(\int_{\mathbb{A}_s(\mathbf{p})} \frac{1}{R(l, \mathbf{p})} dl \right)^{-1}. \tag{9.5}$$

The station blocking rate is obtained from

$$B_s(\mathbf{p}) = \frac{\rho_s^N}{1 + \rho_s + \cdots + \rho_s^N} \tag{9.6}$$

where N is the total number of active users associated with the station s. The flow throughput of the user at location l is obtained from

$$\tau(l, \mathbf{p}) = R(l, \mathbf{p})(1 - \rho_s(\mathbf{p})). \tag{9.7}$$

For the entire network, the global performance metrics can be obtained as

$$C(\mathbf{p}) = \min_s C_s(\mathbf{p}) \tag{9.8}$$

$$B(\mathbf{p}) = \sum_{s=1}^{N} \left(\frac{\int_{\mathbb{A}_s(\mathbf{p})} dl}{\int_{\mathbb{A}} dl} \right) B_s(\mathbf{p}) \tag{9.9}$$

$$\tau(\mathbf{p}) = \frac{1}{\int_{\mathbb{A}} dl} \sum_{s=1}^{N} \left(\int_{\mathbb{A}_s(\mathbf{p})} \tau(l, \mathbf{p}) dl \right) \tag{9.10}$$

where the network capacity $C(\mathbf{p})$ is obtained to be the minimum capacity of all stations, the network blocking rate is obtained as the weighted sum of the station blocking rate (i.e., weight depends on the size of the coverage area of the station), and the mean user throughput $\tau(\mathbf{p})$ is the average of flow throughput of all stations weighted by the size of the coverage area of the station. These global performance metrics are obtained based on the assumption that traffic arrival to the stations is uniformly distributed. Given these global performance metrics, the pilot powers of the stations (i.e., $p_{s,0}$ in $\mathbf{p}$) can be numerically adjusted to achieve the optimal performance. For example, the network capacity can be maximized given the constraint on the network blocking rate.

Then, the ICIC is proposed by optimizing the transmission powers of all the users, given that the pilot power is fixed. The objective is to maximize the utility of the stations. In a distributed environment, the transmission power of each station is updated as follows.

$$p_{s,k} \leftarrow p_{s,k} + \alpha \frac{\partial U(\mathbf{p})}{\partial p_{s,k}} \qquad (9.11)$$

subject to $p_{s,k} \in \mathcal{H}_s$, where $\mathcal{H}_s$ is a feasible set of transmission powers to be used by station s (e.g., derived from minimum and maximum transmission power requirements) and α is the adjustment speed. The utility can be chosen based on the objective and users' preference. In [29], the proportional fair utility is considered, and its derivative $\frac{\partial U(\mathbf{p})}{\partial p_{s,k}}$ is given by [29, Eq. (14)], which can be calculated in a closed form.

In summary, the multi time-scale structure allows the pilot power adjustment and ICIC to configure the transmission parameters with flexibility. While the pilot power is adjusted in a longer time period to achieve the network performance target, the ICIC can optimize the transmission power accordingly in a short-term period. In the performance evaluation, the proposed pilot power adjustment and ICIC are shown to improve the blocking rate and transfer time significantly compared with the static scheme (i.e., transmission power is statically optimized regardless of time varying channel fading).

9.2.2   Coverage Optimization of Small Cells

Coverage area of a small cell has to be optimally planned. For example, in residential usage scenarios, the femtocell coverage has to be adjusted to avoid interference to macro users in public area and to provide high capacity [32], [33]. For business and commercial usage (i.e., enterprise environment), the small cell coverage has to be adjusted to minimize service holes in a building or campus. The traditional approaches to adjust the coverage of macrocells are not suitable for the small cell networks due to the unplanned deployment of a large number of small cells and their plug-and-play mode of operation. For self-organizing distributed coverage optimization, a genetic programming approach is adopted in [8].

[8] considers the enterprise environment to optimize the femtocell coverage in a certain service area. Therefore, the goal is to minimize the coverage holes (e.g., as shown in Fig. 9.5) in the service area, to balance traffic load among femtocells, and to minimize the interference to macro users. While the coverage can be controlled by adjusting the transmission power of the pilot signal, there is a trade-off that needs to be taken into account. Firstly, to minimize the coverage holes, the transmission power should be increased and, hence, enlarge the coverage. However, increasing the transmission power will create more interference to macro users (if the same sub-channel is used) or create more overlapping area (if different sub-channels are used), resulting in low utilization of the reused spectrum.

The performance metrics that are used to optimize the femtocell coverage are as follows [8].

- The highest experienced load L is measured based on the voice traffic in Erlang.
- The estimated coverage overlap O is the ratio of the coverage area of a femtocell overlapping with that of a nearby femtocell to the total coverage area of that

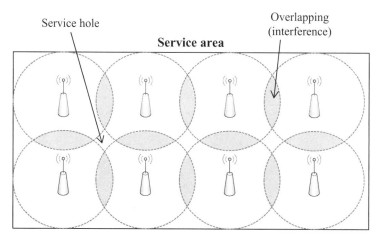

Figure 9.5   Coverage area of small cell networks, overlapping, and service hole problems.

femtocell. This ratio is estimated by dividing the frequency that the users can receive the pilot signals, from more than one femtocells at the same time, by the total number of times that the pilot signal is received by the users. Specifically, if the user is in an overlapping area, this user will be able to receive the pilot signals from multiple femtocells, each of which has a signal strength higher than the threshold.

- The probability of users entering a femtocell coverage hole H is the ratio of the time that the users are in the coverage hole to the total time that the users are in the coverage area of the femtocell. The time duration for which the users are in the coverage hole is estimated by the events that the users are handed over to the underlaid macrocell (e.g., from the instant the user is handed over to the macrocell to the instant when the user is handed over back to the femtocell). Alternatively, if the macrocell is not available, the users can measure the time span during which the received pilot signal strength falls below the threshold.

A genetic algorithm based on tree representation is applied to find the best pilot power adjustment strategy. This algorithm is composed of the component building blocks, that is, function and terminal. The function is the branch node of the tree. The branch node defines the condition to evaluate and branches of the node define the choices. The terminal is the leaf node of the tree, in which the terminal defines the action. The major steps of the tree-based genetic algorithm are as follows. Firstly, the functions and terminals are defined. Secondly, the objective is defined and the corresponding fitness function is derived. Given the tree, the tree-based genetic algorithm performs a search to achieve the objective by maximizing the fitness. The algorithm works as follows. The initial population of multiple trees is created by randomly generating the combinations of functions and terminals. Then, each tree is tested in a simulated scenario and the fitness is calculated based on the defined fitness function. Then, some trees are selected as the parents based on the calculated fitness (e.g., some with the highest fitness). The standard mutation and crossover processes [34] are performed. The crossover process selects the branch of the function in the parent tree and then swaps this branch with that of another parent tree. The mutation process allows some branches of the parent tree to be replaced by the randomly generated branches.

Following the standard tree-based genetic algorithm, the functions and terminals are defined for the femtocell coverage optimization. The major functions used in the algorithm are defined based on the performance metrics compared with the thresholds.

- Function "if $L > L_{th}$": If the load L is larger than the threshold L_{th}, then the action of branch 1 will be executed. Otherwise, branch 2 will be executed.
- Function "if $O > O_{th}$": If the coverage overlap O is larger than the threshold O_{th}, then the action of branch 1 will be executed. Otherwise, branch 2 will be executed.
- Function "if $H > H_{th}$": If the probability of users entering a femtocell coverage hole H is larger than the threshold H_{th}, then the action of branch 1 will be executed. Otherwise, branch 2 will be executed.
- Function "Execute all": All branches are executed.

Then, three actions are defined, that is, to increase pilot power, to decrease pilot power, and to do nothing. An example of a tree composed of functions and terminals is shown in Figure 9.6. The pseudocode of this tree can be expressed as follows.

```
IF H is larger than the threshold THEN
 IF L is larger than the threshold THEN
  Do nothing
 ELSE
  If O is larger than the threshold THEN
```

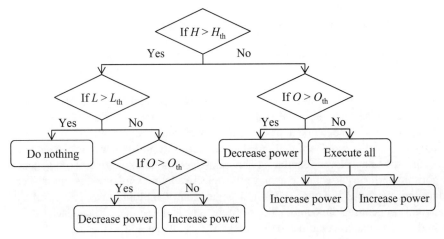

Figure 9.6   Example of tree.

```
  Decrease power
 ELSE
  Increase power
 ENDIF
 ENDIF
ELSE
 IF O is larger than the threshold THEN
  Decrease power
 ELSE
  Increase power
 ENDIF
ENDIF.
```

Note that as in a genetic algorithm, this tree will be revised based on the mutation and crossover processes. The fitness function is then defined to quantify the suitability of the generated tree as follows.

$$f_{\text{femto}} = \frac{w_1(1 - \overline{H}) + w_2(1 - \overline{O}) + w_3(1 - \overline{L})}{w_1 + w_2 + w_3} \tag{9.12}$$

where $\overline{H}$ and $\overline{O}$ are the mean of the probability of users entering a femtocell coverage hole H and the mean of estimated coverage overlap O, respectively. w_1, w_2, and w_3 are the weights of the performance metrics. $\overline{L}$ is defined as

$$\overline{L} = \begin{cases} \frac{1}{F} \sum_{f=1}^{F} \frac{L_f}{L_{\text{th}}}, & L_f \leq L_{\text{th}} \\ 0, & \text{otherwise} \end{cases} \tag{9.13}$$

where F is the total number of femtocells and L_f is the load of femtocell f. $\overline{L}$ is the indication if the load in any femtocell is lower than or equal to the threshold. It is important to note that the above tree-based genetic algorithm uses a centralized approach, where the global information of the network is needed to evolve to the globally optimal solution.

The performance of the proposed tree-based genetic algorithm is evaluated based on simulations of realistic femtocell networks in an enterprise environment. It is shown that the algorithm converges to the optimal tree within about 8 hours and the evolved tree is shown in [8, Figure 5]. Using this evolved tree, the femtocell capacity is found to be approximately even. However, how the thresholds for the traffic load L_{th}, coverage overlap O_{th}, and the probability of users entering a femtocell coverage hole H_{th} can be obtained is not discussed.

9.2.3 Dynamic Frequency Allocation

One of the approaches for interference mitigation in small cell networks is dynamic frequency allocation. In dynamic frequency allocation, limited available frequencies must be efficiently allocated to macrocells and small cells to maximize the spectrum utilization, while maintaining the QoS performances of macro and femto users at the

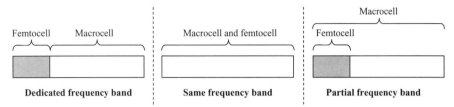

Figure 9.7 Traditional frequency allocation schemes, that is, dedicated, same, and partial frequency band.

target level. [30] introduces a dynamic frequency reuse scheme for dense femtocell networks. In such a network, due to a large number of femto access points deployed overlaid by the macrocells, interference will be a critical problem. The traditional frequency allocation schemes (Figure 9.7) as discussed below may not be suitable for dense femtocell networks.

- *Dedicated frequency band:* A macrocell and femtocells use different non-overlapping frequency bands, but all femtocells use the same frequency. This allocation is suitable for a moderately dense femtocell deployment scenario, but will incur heavy interference in a dense femtocell deployment scenario. This allocation is also not suitable for a sparse femtocell deployment scenario due to the low utilization of the frequency.

- *Same frequency band:* A macrocell and all the femtocells use the same frequency band. This allocation is suitable only for the sparse femtocell deployment scenario, but will incur high interference when the density of the femtocells increases.

- *Overlapping frequency band:* A macrocell uses the entire frequency band, while the femtocells share part of it. All the femtocells will use the same band. Again, this allocation is not suitable for a dense femtocell deployment scenario.

Instead, the dynamic frequency reuse scheme proposed in [30] allows different femtocells to use different frequencies. The entire frequency band is divided into three sub-bands, each of which is allocated to a sector of a macrocell. The femtocell in that sector can use different frequency sub-bands. In addition, to avoid IC, the femtocell uses a different frequency sub-bands to communicate with femto users in the center and edge of the femtocell.

Figure 9.8 shows an example of the frequency reuse scheme for dense femtocell networks. The total frequency band is divided into sub-bands k_1, k_2, and k_3 and they are allocated to sectors 1, 2, and 3 of the macrocell, respectively. Let us consider sector 1, which has two femtocells 1 and 2. The frequency sub-band k_2 is allocated to the center of these femtocells. Then, the frequency sub-band k_3 is divided into 3 parts, that is, $k_{3,1}$, $k_{3,2}$, and $k_{3,3}$. In this case, frequency parts $k_{3,1}$ and $k_{3,3}$ are allocated to the edge of the femtocells 1 and 2, respectively. Although the edges of these femtocells are overlapping, using different frequency parts can avoid the interference.

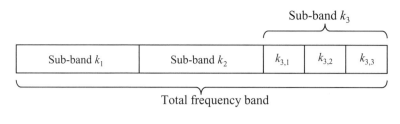

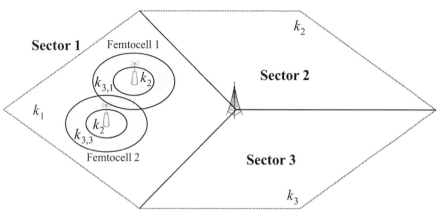

Figure 9.8   Frequency reuse scheme for dense small cell networks.

To obtain the optimal frequency band assignment, the outage probability of a reference femtocell is used as a performance metric. The outage probability is defined as

$$P_{\text{out}} = \Pr\left(I_{\text{f}} > \left(\frac{S_o}{\Gamma} - I_{\text{m}}\right)\right) \tag{9.14}$$

where S_o is the received signal from the reference femtocell, Γ is the target SINR, I_{f} and I_{m} are the interference from all F neighboring femtocells and the macrocell to the reference femtocell, respectively. They are obtained from

$$I_{\text{f}} = \sum_{f=1}^{F} I_{\text{f}}(f) = \sum_{f=1}^{F} p_f h_f X_f \tag{9.15}$$

$$I_{\text{m}} = p_{\text{m}} h_{\text{m}} Y \tag{9.16}$$

where $I_{\text{f}}(f)$ is the interference from neighboring femtocell f, p_f and h_f are the transmission power and channel gain from neighboring femtocell f to the reference femtocell, respectively. Similarly, p_{m} and h_{m} are the transmission power and channel gain from the macrocell to the reference femtocell, respectively. X_f is a binary variable, whose value is 1 if the neighboring femtocell f uses the same frequency band as that of the reference femtocell. Likewise, the value of a binary variable Y is 1 if the macrocell uses the same frequency band as that of the reference femtocell.

Otherwise, their values are set to 0. Finally, the expression for the outage probability is given by

$$P_{\text{out}} = 1 - \left(\prod_{f=1}^{F} \exp\left(-\frac{\Gamma}{\overline{S}} I_{\text{f}}(f) X_f \right) \right) \exp\left(-\frac{\Gamma}{\overline{S}} I_{\text{m}} Y \right) \qquad (9.17)$$

where $\overline{S}$ is the average received signal. To minimize the outage probability, the frequency band allocation for the dynamic frequency reuse scheme can be performed through setting the values of variables X_f and Y. The simulation results confirm that the proposed dynamic frequency reuse scheme can significantly reduce the outage probability compared with the dedicated and same frequency band allocations.

9.2.4   Coordinated Spectrum Assignment

The distributed and localized approach for coordinated spectrum assignment for densely deployed femtocells is introduced in [31]. An enterprise environment (i.e., a number of femtocells are deployed in a building or campus) similar to that in [8] is considered. The coordinated spectrum assignment is divided into two steps, that is, the autonomous dedicated sub-band selection process for basic connectivity and the cooperative shared sub-band allocation process for high data capacity.

To support the coordinated spectrum assignment of a particular femtocell, the interference cell list and interference degree are constructed to maintain the basic information of interference sources from the reference signal received power (RSRP) of all other nearby femtocells.

- The interference cell list of one femtocell contains the details of the other interfering femtocells. In this case, any nearby femtocell f' is an interfering femtocell of the femtocell f if the signal strength from femtocell f' received by femtocell f is higher than the threshold.
- The interference degree is defined as the number of entries in the interference cell list. Therefore, if femtocell f has many interfering femtocells, the interference degree of femtocell f will be high.

The interference cell list contains basic information including the cell identity and IP address and measurement data including the RSRP and interference degree of the interfering femtocells. The femtocells will periodically exchange the information in the interference cell list among themselves, which will allow the femtocells to cooperate for dynamic sub-band allocation. The sub-bands (or sub-channels) are divided into two groups, that is, dedicated and shared sub-bands. The dedicated sub-bands will be allocated to the femtocell to be used on a long-term basis (e.g., several hours) to meet the transmission requirements of the users. On the other hand, the shared sub-bands will be used by the femtocells dynamically on a short-term basis to improve the performance and to meet the instantaneous demand (e.g., from handed over macro users). In this case, the shared sub-bands will be allocated to different femtocells based on the fairness scheduling rule.

In the first step (i.e., the autonomous dedicated sub-band selection process), each femtocell chooses and proposes to allocate one dedicated sub-band for providing basic cell connectivity. This sub-band selection is based on the channel quality of the sub-band for communication with the corresponding femto users. Specifically, the femto access point measures and collects a channel quality report from femto users for all sub-bands. Then, the femto access point will choose the sub-band with the highest transmission rate. However, this sub-band selection needs global information of the network and is not trivial to solve for the optimal solution (e.g., it is NP-hard if the sub-band selection is formulated as a graph coloring problem). Therefore, the self-organizing method for the dedicated sub-band selection is proposed [31] as follows.

1. At the beginning of the selection period, each femto access point maintains a pool of available sub-bands. The femto access point listens to the sub-bands and waits for the reallocation signaling messages from other femtocells. Upon receiving this message, the femto access point removes the corresponding sub-band from the pool.

2. The femto access point checks its interference degree. If its interference degree is larger than that of other femtocells, the femto access point will select the sub-band with the best channel quality to the users. Then, the femto access point broadcasts the sub-band selection (reallocation) signaling message to all the femtocells in the interference cell list. The information to be included is the index of the selected sub-band and its usage efficiency. The usage efficiency is defined as the ratio of transmission rate to the allocated bandwidth. Similar to Step 1, upon receiving the reallocated signaling message, other femtocells will update their interference cell lists accordingly.

3. However, if there are more than one femtocells with the same interference degree without selected sub-band, these femtocells will compare the usage efficiency. The femtocell with the highest usage efficiency will select the target sub-band. Nevertheless, if the usage efficiency is the same for both the femtocells, the femtocell with the smaller cell identity will select the sub-band. Then, similar to Step 2, the reallocation signaling message is broadcast to all the femtocells in the interference cell list.

4. If the pool of sub-bands is empty in the above Steps 2 and 3, the femto access point will try to reuse some sub-band that has the lowest IC (i.e., the least overlapping). Specifically, the femto access point will select the sub-band k with the minimum value of $\exp(Ef(k))\gamma(k)/G_k$, where $Ef(k)$ is the usage efficiency, $\gamma(k)$ is the measured RSRP of the femtocell that uses sub-band k, and G_k is the channel quality of sub-band k.

5. When the new femto access point is switched on, all interfering femtocells will send the allocation signaling messages once their interference cell lists are updated. Then, Step 2 or 4 is performed depending on whether the pool of sub-bands is empty or not.

The above algorithm is performed for the selection period that is relatively long (e.g., several hours). However, there could be the case that there are some unused sub-bands remaining in the pool. These sub-bands can be dynamically allocated among femtocells in an on-demand basis, for example, when the femtocell cannot support QoS requirements of the users due to instantaneous demand. The cooperative shared sub-band allocation process [31] works as follows.

1. If the pool of sub-bands is not empty, the femto access point selects the sub-band with the best channel quality. In this case, the femto access point sends the request messages to all interfering femtocells in its interference cell list. In the request message, there is the index of the sub-band to be selected, the QoS priority, and a proportional fairness factor defined as $\frac{G_k}{\tau_f + w}$ where G_k is the channel quality, τ_f is the average transmission rate of femtocell f, and w is a constant.

2. If the interfering femtocell accepts the request, it will reply with an acknowledgement message. Then, upon receiving the acknowledgement, the femto access point which has sent the request message will allocate the sub-band. In this case, all interfering femtocells will update their interference cell lists and interference degrees accordingly.

3. However, if the pool of sub-bands is empty, the femto access point will search for the sub-band by checking the QoS priority. If the current QoS priority is higher than the lowest QoS priority of the allocated sub-band, this femto access point will select the corresponding sub-band. Then, the request message will be broadcast. However, if the QoS priority of the femto access point is equal to the lowest QoS priority of the allocated sub-bands, this femto access point will check the proportional fairness factor. If the proportional fairness factor of the requesting femto access point is higher than the lowest proportional fairness factor of the allocated sub-bands, the femto access point will select the corresponding sub-band. The request message will be broadcast.

4. If the pool of sub-bands is empty, and the QoS priority and proportional fairness factors are lower than those of the lowest ones among the allocated sub-bands, then the femto access point will not select any sub-band.

To avoid conflict of the sub-band selection, the acknowledgement message is used to confirm the success of selection. If there are request messages sent after the earlier request, the request messages sent later will be blocked.

The performance evaluation is performed through simulations and a comparison with random sub-band assignment is given. Clearly, the proposed dedicated and shared sub-bands assignment algorithm, which can work in a distribution fashion (i.e., no centralized controller is needed) can achieve much higher throughput.

9.3 SELF-OPTIMIZATION

The main functions of the self-optimization process are to measure, collect, and exchange information about the operating network. Then, the information is used to optimize the performance of the small cell networks automatically to achieve the objective and meet the constraints. The self-optimization process is performed in an operational state, that is, after the transceiver is turned on. The self-optimization is similar to the initial radio configuration. In particular, the small cell obtains information about other nearby small cells and this information is used to adjust the coverage and capacity parameters of the small cell.

In [35], a coalitional game framework is used to analyze the group (i.e., coalition) formation among femtocells to cooperate and coordinate their transmissions. The objective is to manage the interference among femtocells. This objective is achieved through cooperative and optimal power control such that the individual utility of the femtocell is maximized while the interference experienced by macro users is limited. In [36], a bio-inspired algorithm (i.e., biomimetics) is proposed for optimizing system-wide antenna tilts used in femtocell networks. The objective is to maximize the aggregate throughput of the femtocells. The algorithm is based on the flock of common cranes. Specifically, the flock-wide flight parameters can be optimized to maximize the group flight efficiency in the flock. This can be achieved by minimizing the air drag. In the femtocell context, the flock is a femtocell network. The flight parameters are the antenna tilts. The flight efficiency is the spectral efficiency and the air drag is the interference. The performance evaluation through simulation shows that the proposed bio-inspired algorithm can improve the network performance noticeably. In [37], a Q-learning algorithm is applied to solve the power control problem of multiple femto access points. The distributed Q-learning algorithm used by femtocells utilizes the information in the macrocell networks to adapt the transmission power decision. Using the 3GPP LTE as a test case, the proposed algorithm can quickly react to the environmental changes (e.g., transmission of nearby macro users). A similar design of using Q-learning and reinforcement learning algorithms to address an interference mitigation issue is also discussed in [38], [39], and [40].

[41] presents an admission control mechanism designed jointly with the interference avoidance for self-organizing femtocell networks. Each femto access point makes a decision to accept or reject an incoming connection independently. In addition, the femto access point also allocates the sub-channel to the new connection based on the required QoS performance (i.e., outage probability). In the admission control and sub-channel allocation, a counter is used to keep track of the available sub-channels. A similar problem is also considered in [42]. In [11], a joint self-configuration and self-optimization method is proposed to provide service differentiation resource allocation for femtocell networks. The handshaking method to exchange information among femtocells through FMS is introduced. Then, the RB allocation is optimized for the femtocell. The power control is focused in [43]. Specifically, the control theory and game theory are applied to meet SINR requirements of femto users. Firstly, the maximum transmission power is optimized. Then, based on the optimized maximum transmission power, the femtocells adjust their instantaneous power to meet the SINR requirements. In [38], evolutionary and reinforcement learning methods are adopted for the power control problem. With these methods, the femtocells can learn and adapt the transmission power strategies to achieve the SINR

requirements. In [44], coordinated groups of femtocells are formed. The femtocells in the same group can adjust their transmission powers to reduce interference to a macro user. An analytical model based on stochastic geometry is also developed.

In the following, details of the selected issues and approaches in self-optimization are presented.

9.3.1 Resource Allocation for Different Service Classes

While most of the works treat the femto users equally (i.e., all users are in the same class), service differentiation for users in different classes with different QoS requirements is an important issue in femtocell networks. For example, the 3GPP LTE specifications have 9 QoS class identifiers (QCIs) to support 9 service classes. The QoS requirement for each QCI is defined by the L2 packet delay constraint, L2 packet loss probability, and guarantee bit rate (GBR).

[11] introduces a self-configuration and self-optimization framework to provide the service differentiation function to support different service classes. The objective of the framework is to maximize the efficiency of PRB; while the QoS requirements of different service classes can be guaranteed, co-channel and co-tier interferences can be avoided. A centralized femtocell network is considered, where the femto access points of all femtocells are connected with the FMS. The FMS works as a controller and gateway for the cellular system. The femtocells in the nearby geographical location connected with each FMS are considered to be in the same group. The FMS is responsible for collecting network information from all femtocells and performing interference avoidance.

With global network information, the resource allocation proposed in [11] can be performed into two steps, that is, self-configuration and self-optimization steps. First, in an initialization phase (i.e., self-configuration functionality), the configuration of each femtocell has to be determined. Therefore, the femtocell will exchange the neighbor-informing message using a handshaking signal with the transmission power that is two times higher than that of the normal signal for data transmission. This high transmission power of the neighbor-informing message is to avoid the hidden terminal problem. Then, the femtocell waits for the response. The information in the response messages is used to create a list of neighboring femtocells. Then, once the configuration is finished, the femtocell network composed of femto users, femto access point, and FMS can perform resource allocation. In addition, the neighbor-informing messages are also periodically transmitted to monitor the network environment continuously.

For resource allocation, the QoS constraint is defined based on the required number of PRBs. Each femto user belongs to one service class which is identified by the QoS class identifier. An example considered in [11] has nine applications for nine classes, where the identifiers 1–9 are for conventional voice connection, conventional video, buffered streaming, real-time gaming, IP multimedia signaling, live streaming, file sharing, e-mail, P2P, and Web, respectively. These service classes are determined by the L2 packet delay constraint and GBR. Specifically, the service classes 1–4 have the requirements for the guaranteed bit rate while the service classes 5–9 do not have such requirements such restrictions.

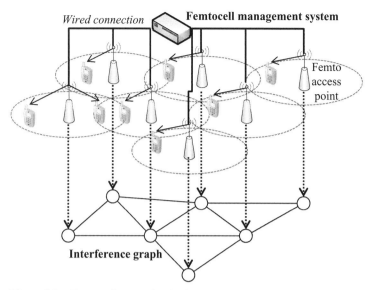

Figure 9.9   Femtocell network with femtocell management system and the interference graph.

The PRB allocation is performed based on graph optimization. First, an inter-ference graph is constructed based on the structure of the femtocell network. An example of the interference graph is shown in Figure 9.9. Then an optimization problem is formulated. Let $\mathcal{F}$ denote a set of femtocells. A PRB is identified by the time slot t and sub-channel k. The PRB allocation variable is obtained from $a_{f,t,k} = x_f$, if RB at time slot t on sub-channel k is allocated to femtocell $f \in \mathcal{F}$, and $a_{f,t,k} = 0$, otherwise. In this case, $x_f = 1$ if femtocell f is in the group (i.e., connected to the same FMS), and $x_f = 0$ otherwise. The accumulative utilization of the PRB can be defined as $\eta_{t,k} = \sum_{f \in \mathcal{F}} a_{f,t,k}$, which is used as the objective function in the optimization formulation.

Based on the interference graph, there are some constraints that need to be imposed. Firstly, the same PRB cannot be assigned to multiple femtocells that have a direct interference link among them. The variable $n_{f,f'}$ indicates the status of the interference link. That is, $n_{f,f'} = 1$ if there is an interference link between femtocells f and f' (i.e., femtocells f and f' overlap) and $n_{f,f'} = 0$ otherwise. Secondly, the utilization ratio for the transmission frame must be less than one. The utilization ratio is defined as

$$\beta = \sum_{t=1}^{T} \sum_{k=1}^{K} \frac{u_{t,k}}{TK} \tag{9.18}$$

where T is the total number of time slots in a frame and K is the total number of sub-channels. $u_{t,k}$ is the indicator variable, where $u_{t,k} = 1$ if $\sum_{f \in \mathcal{F}} a_{f,t,k} \neq 0$ (i.e., the PRB is assigned to any femtocell), and $u_{t,k} = 0$ otherwise.

The formulation for the optimization of femtocell-to-femtocell interference (OPT-FF) introduced in [11] can be expressed as

$$\max_{a_{f,t,k}} \sum_{t=1}^{T} \sum_{k=1}^{K} \eta_{t,k} \tag{9.19}$$

$$\text{subject to } n_{f,f'} \sum_{t=1}^{T} \sum_{k=1}^{K} a_{f,t,k} a_{f',t,k} = 0 \tag{9.20}$$

$$\beta \leq 1 \tag{9.21}$$

$$\sum_{t=1}^{T} \sum_{k=1}^{K} a_{f,t,k} \leq a_{f,\max} \tag{9.22}$$

where the last constraint is to ensure that the femtocell is not allocated with the PRBs more than the maximum threshold $a_{f,\max}$. In addition, to avoid co-tier interference as stated in the above formulation, the QoS requirement of the user has to be satisfied. The constraints are as follows. Firstly, the transmission rate of the femto user must be larger than or equal to the guaranteed bit rate. Secondly, the delay must be smaller than or equal to the L2 packet delay constraint. It is noted that the formulation of the OPT-FF problem is hard to solve due to the NP-complete nature of the problem.

To address the complexity issue, the resource allocation scheme of femtocell-to-femtocell interference (RAFF) based on a greedy algorithm is also proposed [11]. The resource allocations scheme is divided into three major steps as follows.

1. *Initialization:* Similar to the configuration phase, in the initialization step, the femtocell transmits and collects information about its neighboring femtocells through the FMS.

2. *Admission control:* The femto user submits the L2 packet delay constraint, the required number of PRBs, and the GBR to the femto access point. This information is forwarded to the FMS afterward. The resource allocation is performed to check whether the QoS requirements can be met or not. If the QoS requirements can be satisfied, the femto users will be accepted, and rejected otherwise.

3. *Allocation:* The algorithm, performed by the FMS, allocates the PRBs to achieve the objective and constraints defined in (9.19–9.22). Also, the information of the femtocells is updated periodically to perform the allocation continuously.

In summary, using an interference graph to optimize the PRB can achieve the nearly optimal performance, by using the greedy algorithm. However, there are some issues for extension. Firstly, the interference level in the graph is binary (i.e., interfered or not interfered). However, in reality, the interference could be avoided if the transmission power can be adjusted. The interference level can be defined as a continuous value. Secondly, the algorithm is based on the FMS, which makes it a centralized approach and requires global information of the network. If the global information is not available, the resource allocation cannot be performed efficiently.

A distributed approach needs to be devised. Thirdly, the impact of interference due to the macro users must be taken into account.

9.3.2 Self-Organizing Small Cell Management Architecture

[43] focuses on interference management for UL transmissions in a femtocell. Specifically, a self-organizing femtocell management architecture, called the complementary tri-control loops, is proposed. Due to random deployment of femtocells, the UL transmission of a macrocell can be interfered with if femtocells are located near the macro base station. Similarly, UL transmissions in a femtocell can be interfered with if macro users using the same sub-band as the femto users are located near the femto access point. In addition, the interference can be bursty due to unpredictable sources of interference. Therefore, the objectives of the architecture are to protect the UL transmission of the macrocell from the interference from femtocells, to provide efficient radio resource allocation coordination among femtocells, and to protect the UL transmission of the femtocell from the bursty interference.

The design principles of the architecture are as follows. First, the architecture does not require any change and modification of the RRM framework at the macrocell. Also, the interruption and performance degradation to the macro users can be avoided. Second, the architecture supports a distributed environment and can work independently. In addition, the architecture does not require global information to optimize the operation of the femtocell to adapt quickly to the environmental changes. Third, the architecture requires no changes in the user devices and no additional costly hardware.

The complementary tri-control loops architecture is composed of three control loop components, that is, maximum transmission power control loop, target SINR control loop, and instantaneous transmission power control loop (Figure 9.10).

- The maximum transmission power control loop is used to determine the maximum transmission power of a femto user based on the load margin of the macrocell UL transmission. The load is a monotonically increasing function of the total received power at the macro base station. The load margin is the

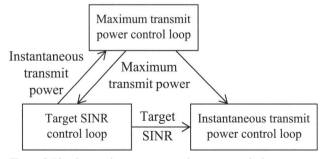

Figure 9.10   Interactions among maximum transmission power control loop, target SINR control loop, and instantaneous transmission power control loop in the complementary tri-control loops architecture [43].

difference between the load and the threshold of macrocell load. The load margin indicates the capability of the macrocell to accommodate more load or not. The positive and negative values of the load margin represent under- and overload conditions, respectively. The objective of this component is to protect the macro users by controlling the maximum transmission power of femto users. Although the maximum transmission power is limited, the femto access points can adjust their instantaneous transmission powers to meet the SINR requirements of the femto users.

- The target SINR control loop is used to provide the coordination among femto-cells with minimum signaling to achieve the optimal utility of resource usage. The target SINR control loop determines the required SINR for femto access points using implicit feedback (e.g., interference level and achieved SINR). The target SINR control loop uses the maximum transmission power to determine the Nash equilibrium solution in the noncooperative game setting.

- The instantaneous transmission power control loop is used to allocate the actual transmission power to the femto users to achieve the target SINR under the constraint of maximum transmission power. This component can adapt the transmission power quickly within a frame to handle the random and bursty interference from other macro and femto users.

In the complementary tri-control loops architecture, the load margin has to be exchanged among macro base stations and femto access points. [43] considers two options, that is, using a wired or wireless links. In the case of wired link, there are two possible approaches. In the first approach, the load margin can be exchanged directly among macro base stations and femto access points through backhaul connections, without the centralized controller. However, the signaling in this approach will be complex and difficult to manage. Alternatively, the exchange of the load margin can be performed on the existing interface in a centralized fashion. Specifically, OAM server in the core network can be used for this. The macro base stations and femto access points will be connected with the OAM server. The macro base stations periodically transmit the load margin to the server and the server will forward this load margin to the femtocell manager, which is another server in the femtocell core network. The femtocell manager then distributes the load margin to the corresponding femto access points. It is noted that using a wired link to transfer the load margin does not require additional hardware for the femto access point. However, it could incur a large delay, especially when the transfer is through the server.

The load margin can be exchanged over a wireless link. The macro base station broadcasts the load margin to the femto access points directly without using backhaul connections. However, many wireless standards do not support this functionality. For example, an additional receiver would be required for the femto access point if the load margin is transmitted using a different frequency from that of the data transmission. Also, an additional interference cancellation method is needed if the load margin is transmitted using the same frequency as that of the data transmission.

The maximum transmission power control loop is formulated as the steady-state tracking problem. The simplified and abstracted model of the control block diagram is shown in Figure 9.11. The objective of this control loop is to limit the upper bound

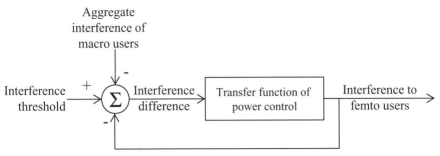

Figure 9.11   Simplified and abstracted control block diagram of the maximum transmission power control loop.

of the transmission power, especially from the femto users, such that the load of the macrocell is maintained close to the threshold as much as possible, that is,

$$\mathbf{P}^* = \min_{\mathbf{P}} \lim_{t \to \infty} \left| e_{m,k}(t) \right| \tag{9.23}$$

for the macrocell $m \in \mathcal{M}$ and resource block (RB) $k \in \mathcal{K}$, where $\mathbf{P}$ is composed of the maximum transmission powers $P_{i,k}$ for user i. The maximum transmission power $P_{i,k}$ is the decision variable of the maximum transmission power control loop. $\mathcal{M}$ is a set of macro users and $\mathcal{K}$ is a set of RBs. $e_{m,k}$ is the load margin of the macrocell m, defined as

$$e_{m,k} = L_{m,k}^{\text{th}} - L_{m,k} \tag{9.24}$$

where $L_{m,k}$ and $L_{m,k}^{\text{th}}$ are the load and load threshold, respectively. The load is defined as $L_{m,k} = L_{m,k}^{\text{m}} + L_{m,k}^{\text{f}}$, where $L_{m,k}^{\text{m}}$ and $L_{m,k}^{\text{f}}$ are the loads from macro and femto users, respectively. The load from femto users can be obtained from

$$L_{m,k}^{\text{f}} = \Gamma_m(I_{m,k}^{\text{f}}(\mathbf{p})) \quad \text{where } I_{m,k}^{\text{f}}(\mathbf{p}) = \sum_{f \in \mathcal{F}} \sum_{i \in \mathcal{N}_f} a_{i,k} h_{i,m,k} p_{i,k}, \tag{9.25}$$

where $a_{i,k}$ is the ratio of the slots that the user i transmits on resource block (RB) k, $h_{i,m,k}$ and $p_{i,k}$ are the channel gain and transmission power from user i to macro base station m, respectively. $\Gamma_m(\cdot)$ is an interference-load function, which is assumed to be monotonically increasing in the aggregate interference $I_{m,k}^{\text{f}}(\mathbf{p})$. $\mathbf{p}$ is composed of the transmission powers of all the users $p_{i,k}$. $\mathcal{F}$ is the set of femtocells and $\mathcal{N}_f$ is the set of femto users associated with femtocell f. It is proved that there exists a feasible solution for the maximum transmission power $\mathbf{P}$ and transmission power $\mathbf{p}$ such that the load margin is zero (i.e., $e_{m,k} = 0$).

The target SINR control loop is formulated as a noncooperative game to obtain the target SINR for the femtocell. The *players* are the femto access points. The *strategy* is the transmission power denoted by $p_{i,k}$ for user i on resource block (RB) k and the time sharing $b_{i,k}$ (i.e., the normalized duration that the user i transmits on the resource block (RB) k). The *payoff* is the utility of femto access point f defined as

$$U_f(\mathbf{p}, \mathbf{b}) = \sum_{i \in \mathcal{N}_f} \sum_{k \in \mathcal{K}} b_{i,k} W \log_2 \left(1 + \gamma_{i,k}/c\right) - \sum_{i \in \mathcal{N}_f} \sum_{k \in \mathcal{K}} \mu_{i,k} b_{i,k} p_{i,k} \quad (9.26)$$

where $\mathbf{p}$ and $\mathbf{b}$ are composed of the transmission powers and time sharing of all the femto access points, respectively. W is the size of RB and c is a constant to achieve the target BER given by $c = -\ln(5\text{BER})/1.6$. $\mu_{i,k}$ is the weight (e.g., price paid for the interference caused). In this case, the transmission power (i.e., strategy) must be less than or equal to the maximum transmission power $P_{i,k}$ obtained from (9.23), that is, maximum transmission power control loop. It is proved that the noncooperative game for the target SINR control loop has a Nash equilibrium in terms of transmission power and time sharing if the weight $\mu_{i,k}$ is large enough. The proof is based on the fact that the utility function $U_f(\cdot)$ defined in (9.26) is a nonempty, convex, and compact subset of some Euclidean space. Also, the utility function is continuous and quasi-concave. The Nash equilibrium of the target SINR is given by

$$\gamma_{i,k}^* = \max \left(\left[\frac{W h_{i,k}}{(\ln 2) I_{i,k}(\mathbf{p}_{-i}) \mu_{i,k}} - c \right]^+, \frac{h_{i,k} P_{i,j}}{I_{i,k}(\mathbf{p}_{-i})} \right) \quad (9.27)$$

where $I_{i,k}(\mathbf{p}_{-i})$ is the interference to user i on the RB k given the transmission powers of other users $\mathbf{p}_{-i}$. This interference can be obtained similar to that in (9.25). $h_{i,k}$ is the channel gain of the femto user i to the corresponding femto access point on the RB k.

The instantaneous transmission power control loop is formulated as a control system with the objective to achieve the target SINR. The difference between the target SINR and received SINR is $e_{i,k} = \gamma_{i,k}^* - \gamma_{i,k}$, where

$$\gamma_{i,k} = \frac{h_{i,k} p_{i,k}}{I_{i,k}(\mathbf{p}_{-i})}. \quad (9.28)$$

However, this function is nonlinear. Therefore, to simplify the analysis, the control function is linearized as

$$\tilde{\gamma}_{i,k} = \log \gamma_{i,k} = h_{i,k} + \tilde{p}_{i,k} - \tilde{I}_{i,k}(\mathbf{p}_{-i}) \quad (9.29)$$

where $\tilde{p}_{i,k} = \log p_{i,k}$ and $\tilde{I}_{i,k}(\mathbf{p}_{-i}) = \log I_{i,k}(\mathbf{p}_{-i})$. Then, the transmission power is adjusted iteratively as

$$p_{i,k}(t+1) = 10^{\tilde{p}_{i,k}(t)} \quad (9.30)$$

where t is the iteration index.

The performance evaluation is performed for both the cases when the load margin is transferred over the wired and wireless links. The results show that when the load margin is transferred over the wireless link, the received interference of the macro users converges to the target level without oscillation and faster than that of the wired link. This is due to the smaller delay of a wireless link (i.e., broadcasting). Also, when the users' traffic is generated randomly, the interference to the macro users fluctuates. However, the received interference is still not much different from the target level, confirming the effectiveness of the proposed received instantaneous transmission power control loop.

9.3.3   Evolutionary and Learning-Based Power Control

As an alternative to the control theory, the techniques in game theory and artificial intelligence can be applied to solve the power control problem. These techniques are suitable for small cell networks such as the femtocell networks, since the femto access points may belong to different owners. Also, due to the limited capacity of a broadband link, the femto access points may not be able to communicate and exchange information for resource management and interference mitigation (e.g., through power control). Therefore, the femto access point has to make a decision (e.g., transmission power) independently. However, the decision of one femto access point could impact the performances and decisions of other femto access points (e.g., interference). Therefore, the interaction among femto access points in the femtocell network can be modeled using evolutionary game [38] [39]. Then, to implement the algorithm for resource management in the femto access points, the reinforcement learning method is applied. The Q-learning algorithm, which is one of the reinforcement learning methods [45], allows the self-organizing femto access points to learn and adapt their decisions gradually without global information of the network.

System Model: In the system model under consideration in [39], there is one macro base station using K sub-channels whose set is denoted by $\mathcal{K}$. Let $\Gamma_0 = (\Gamma_0^{(1)}, \ldots, \Gamma_0^{(N)})$ denote the minimum time-average signal-to-interference-plus-noise ratio (SINR) offered by the macro base station to its macro users. There are F femtocells (i.e., F femto access points) underlaying the macrocell, in which the set of femto access points is denoted by $\mathcal{F}$. Each femto access point can use any of the available sub-channels for the transmission to its femto users. However, the transmission of a femto access point must not violate the minimum time-average SINR required by the macro users. A time-division multiple access (TDMA) method is applied for the transmission of multiple femto users at each time interval.

Let $t \in \{1, \ldots, \infty\}$ be a discrete time index. The transmission power of a macro user on a given sub-channel k is denoted by $p_0^{(k)}$. Let $h_{0,0}^{(k)}$ denote the channel gain between the macro base station and its associated macro user on sub-channel $k \in \mathcal{K}$. Likewise, $h_{i,j}^{(k)}$ denotes the channel gain between transmitter i and receiver j on sub-channel k. The transmission power of femto access point $f \in \mathcal{F}$ on sub-channel k is denoted by $p_f^{(k)}$.

The signal to interference plus noise ratio of macro base station at the macro user (assuming Gaussian signaling) is given as

$$\gamma_0^{(k)} = \frac{h_{0,0}^{(k)} p_0^{(k)}}{\sigma^2 + \underbrace{\sum_{f \in \mathcal{F}} h_{f,0}^{(k)} p_f^{(k)}}_{\text{femtocells}}} \tag{9.31}$$

where σ^2 is the variance of additive white Gaussian noise.

The signal to interference plus noise ratio for femto access point $f \in \mathcal{F}$ serving its femto user is given as

$$\gamma_f^{(k)} = \frac{h_{f,f}^{(k)} p_f^{(k)}}{\sigma^2 + \underbrace{h_{0,f}^{(k)} p_0^{(k)}}_{\text{macrocell}} + \underbrace{\sum_{f' \in \mathcal{F} \backslash \{f\}} h_{f',f}^{(k)} p_{f'}^{(k)}}_{\text{femtocells}}}. \qquad (9.32)$$

Macro-Femtocell Coexistence: Evolutionary Game Approach. We first consider the case that femto access points of femtocells are able to exchange information among each other through a controller. Then, the case without the controller is considered, where the fictitious play is used (i.e., one femtocell imitates strategies of another femtocell).

Evolutionary Game and Replicator Dynamics: In the evolutionary game approach, a femto access point observes the behavior of other femto access points (i.e., interfering femto access points). Then, the femto access point learns from observation and adapts to the best decision based on their instantaneous payoff compared with the average payoff of all femto access points. The players of evolutionary game are the femto access points. The strategy is the set of power allocation vectors denoted by $\mathcal{A}_f = \{q_f^{(l,k)} : l \in \mathcal{L} = \{0, \ldots, L_f\}, f \in \mathcal{F}, k \in \mathcal{K}\}$, where L_f is the total number of power levels usable at femto access point f. u_f is the payoff or interference mitigation metric of femto access point $k \in \mathcal{K}$. The *average* payoff of the entire femtocell population (i.e., all femto access points) at time t is defined as $\bar{u}(t) = \frac{\sum_{f \in \mathcal{F}} u_f(t)}{|\mathcal{F}|}$.

In this scenario, there is a gateway (e.g., OAM server) to collect the payoffs for all femto access points and calculate the average payoff (Figure 9.12). The payoff $u_f(t)$ of femto access point f is then compared with the average payoffs $\bar{u}(t)$. If the payoff of femto access point f is less than the average payoff, then a random strategy is chosen and the whole process is repeated. Let $\zeta_a^{(l,k)}(t) = \sum_{f=1}^{F} \mathbf{1}_{p_f(t)=q_f^{(l,k)}=a}$ represent the total number of femto access points choosing strategy $q_f^{(l,k)}$, and $x_a(t) = \frac{\zeta_a^{(l,k)}(t)}{\sum_a \zeta_a^{(l,k)}(t)}$

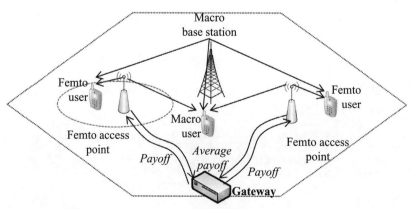

Figure 9.12 System model with gateway for evolutionary game model of power control.

be the proportion of femto access points using strategy $a \in \mathcal{A}$. Then, the replicator dynamics can be defined as $\forall (l, k) \in (\mathcal{L} \times \mathcal{K})$:

$$\dot{x}_a(t) = x_a(t)\big(u_a(t) - \bar{u}(t)\big) \tag{9.33}$$

where $u_a(t)$ is the payoff at time t of femto access point f when using action a, and $\bar{u}(t)$ is the average payoff over all strategies $a \in \mathcal{A}$, where $\bar{u}(t) = \sum_{a \in \mathcal{A}} x_a u_a(t)$.

Replication by Imitation: To implement the replicator dynamics, we first consider the replication by imitation approach. In this approach, a femto access point uses some strategy and reviews the payoff from that strategy, and changes the strategy if needed. The time rate that the femto access point reviews its strategy depends on the current payoff of the strategy used. In this case, $r_f^i(x)$ denotes the average review rate of femto access point $f \in \mathcal{F}$ using strategy $i \in \mathcal{A}_f$. The probability that femto access point f in population x will change its strategy from i to j is denoted by $pr_i^j(x)$. Let $pr_i(x) = \big(pr_i^{(1)}(x), \ldots, pr_i^{(K)}(x)\big)$ denote the probability distribution vector over the set of $\mathcal{A}_f$. $pr_i^i(x)$ means that femto access point f does not change its strategy i. The dynamics of the replication by imitation are given as

$$\dot{x}_i = \sum_{j \in \mathcal{A}_f} (1 - x_i) r_i(x) pr_j^i(x) - \sum_{j \in \mathcal{A}_f} x_i \, pr_i^j r_i(x) \tag{9.34}$$

where the first term on the right-hand side of (9.34) regards the proportion of femto access points that randomly meet another femto access point and change their strategies from j to i. The second term is the number of remaining femto access points that change their strategy i to other strategies. In this replication by imitation approach, there is no need to know the average payoff as in the replicator dynamics.

Fictitious Play with Complete Information: In a fictitious play-based algorithm, we first assume that femto access points have complete and perfect information. In particular, they know the structure of the game and observe at each time t the power allocation vector taken by all other femto access points. Each femto access point $f \in \mathcal{F}$ assumes that all other femto access points play independent and stationary (time-invariant) mixed strategies $\pi_{f'}$, $\forall f' \in \mathcal{F} \setminus \{f\}$, where $\pi_{f'} = \big(\pi_{f', q_{f'}^{(1,1)}}, \ldots, \pi_{f', q_{f'}^{(L_{f'}, K)}}\big)$ and $\pi_{f', q_{f'}^{(l,k)}} = Pr\big(p_{f'}(t) = q_{f'}^{(l,k)}\big)$. Under these conditions, the femto access point f is able to build an empirical probability distribution over each set $\mathcal{A}_{f'}$, for all $f' \in \mathcal{F} \setminus \{f\}$. Let $e_{f, p_f(t)} = \frac{1}{t} \sum_{s=1}^{t} \mathbf{1}_{p_f(t) = q_f^{(l,k)}}$ be the empirical probability with which femto access points $f' \in \mathcal{F} \setminus \{f\}$ observe that femto access point f plays action $q_f^{(l,k)} \in \mathcal{A}_f$. Hence, $\forall f \in \mathcal{F}$ and $\forall p_f \in \mathcal{A}_f$; the following recursive expression holds:

$$e_{f, p_f}(t+1) = e_{f, p_f}(t) + \frac{1}{t+1}\big(\mathbf{1}_{p_f(t) = q_f^{(l,k)}} - e_{f, p_f}(t)\big). \tag{9.35}$$

Let $\bar{e}_{f, p_{-f}}(t) = \prod_{f' \neq f} e_{f', p_{f'}}(t)$ be the probability with which femto access point f observes the action profile p_{-f} of all other femto access points except femto access point f at time $t > 0$, for all $f \in \mathcal{F}$. Let the vector $e_f(t)$ be regarded as the beliefs of femto access point f over the strategies of all its corresponding counterparts. Hence, at each time instant t and based on its own beliefs $e_f(t)$, each femto access point f chooses its action $p_f(t) = q_f^{(l,k)}$, that is,

$$(l, k) \in \arg \max_{(l,k)\in(\mathcal{L}\times\mathcal{K})} \bar{u}_f\big(p_f(t), e_f(t)\big) \tag{9.36}$$

where for all $f \in \mathcal{F}$, $\bar{u}_f(\pi) = \mathbb{E}_\pi\big[u_f(p_f, p_{-f})\big]$. From (9.35), it can be implied that by playing fictitious play, femto access points become myopic. In particular, the femto access points build beliefs on the strategies being used by all other femto access points. At each time t, the femto access points choose the actions that maximize their instantaneous expected utility.

Macro-Femtocell Coexistence: Reinforcement Learning Approach. Then, we consider the case that femto access points cannot exchange information among themselves. Therefore, femto access points must learn about their strategies from instantaneous payoffs. The Q-learning aims at finding a policy that maximizes the observed rewards (i.e., payoffs) over a certain time period. Every femto access point $f \in \mathcal{F}$ explores the environment, observes its current state s, and takes a subsequent action a, according to a decision policy $\pi : s \rightarrow a$, which is a mapping from state s to action a. With the ability to learn, the knowledge about other femto access points' strategies is not needed. Instead, a Q-function maintains the knowledge about other femto access points in the network. Let $\mathcal{G}^Q = \big(\mathcal{F}, \{\mathcal{P}_f\}_{f\in\mathcal{F}}, \{u_f\}_{f\in\mathcal{F}}\big)$ denote the Q-learning game. Here, the players are the femto access points $f \in \mathcal{F}$; $s_f(t) = \big(s_f^{(1)}(t), \ldots, s_f^{(K)}(t)\big)$ is the composite state of femto access point f at time t in sub-channel $k \in \mathcal{K}$. $s_f^{(k)}(t) \in \{0, 1\}$ indicates whether femto access point f generates interference to a macro user or not, on sub-channel k. $\boldsymbol{a}_f(t) = \big(a_f^{(1)}(t), \ldots, a_f^{(K)}(t)\big)$ is the action vector of femto access point f, where $a_f^{(k)}(t) \in \{0, 1\}, \forall k \in \mathcal{K}$. Finally, $\mathbf{u}_f(t) = \big(u_f^{(1)}(t), \ldots, u_f^{(K)}(t)\big)$ is the utility function or payoff vector of femto access point f at time t. The expected discounted reward (i.e., payoff) over an infinite horizon is given by

$$V^\pi(s) = \mathbb{E}\Big[\gamma^t \times r(s_t, \pi^*(s_t))|s_0 = s\Big] \tag{9.37}$$

where $0 \leq \gamma \leq 1$ is a discount factor and r is the femto access point's reward at time t. (9.37) can be rewritten as

$$V^\pi(s) = R(s, \pi^*(s)) + \gamma \sum_{v\in S} P_{s,v}(\pi(s))V^\pi(v) \tag{9.38}$$

where $R(s, \pi^*(s)) = \mathbb{E}\{r(s, \pi(s))\}$ is the mean value of reward $r(s, \pi(s))$, and $P_{s,v}$ is the transition probability from state s to v. Moreover, the optimal policy π^* satisfies the optimality criterion

$$V^*(s) = V^{\pi^*}(s) = \max_{a\in\mathcal{A}} \left(R(s, a) + \gamma \sum_{v\in S} P_{s,v}(a)V^*(v) \right). \tag{9.39}$$

Through Q-learning, the knowledge of reward $R(s, a)$ and transition probability $P_{s,v}(a)$ can be gradually learnt and reinforced with time. For a given policy π, define a Q-value as

$$Q^*(s, a) = R(s, a) + \gamma \sum_{v \in S} P_{s,v}(a) V^\pi(v), \tag{9.40}$$

which is the expected discounted reward when executing action a at state s, following policy π.

Here, the Q-learning algorithm is used to estimate the state-action value function $Q(s, a)$. The femto access point keeps trying all actions in all states with non-zero probability. Also, the femto access point occasionally explores by choosing a random action with probability $\epsilon \in (0, 1)$, and the greedy action with probability $(1 - \epsilon)$. This is referred to as ϵ-greedy exploration. Another approach is to use the Boltzmann exploration strategy with temperature parameter κ, where the action a in state s is taken with a probability $P(a|s)$. Then, the femto access point receives a reinforcement r. The actions are chosen according to their Q-values as

$$p(a|s) = \frac{e^{Q(s^f, a)/\kappa}}{\sum_{a' \neq a} e^{Q(s^f, a')/\kappa}}. \tag{9.41}$$

The Q-learning process aims at finding $Q(\mathbf{s}, \mathbf{a})$ recursively where the update is given as

$$Q_t(\mathbf{s}, \mathbf{a}) = (1 - \alpha)Q_{t-1}(\mathbf{s}, \mathbf{a}) + \alpha\left[r_t(\mathbf{s}, \mathbf{a}) + \gamma V_{t-1}(v_t)\right] \tag{9.42}$$

where $V_{t-1}(v_t) = \max_{b \neq a} Q_{t-1}(v, b)$ and α is the learning rate.

Cooperative Q-learning: Instead of learning by themselves, femto access points can adopt the learning from an expert. In such a scheme, the knowledge is built based on the accumulated rewards over time gained by a given femto access point f. The cooperation between femto access points is allowed based on temporal difference of their utility function $\Delta_f(t) = u\left(p_f(t), p_{-f}(t)\right) - u\left(p_f(t-1), p_{-f}(t-1)\right)$. Therefore, if femto access point f finds a femtocell f' to have a similar difference such that $|\Delta_f(t) - \Delta_{f'}(t)| \leq \zeta$, then femto access point f considers femto access point f' as an expert to which an appropriate weight is assigned. As a result, a given femto access point modifies its Q-values and learns from a small group of other femto access points that it considers as the expert. This exchange of information between femto access points is performed periodically.

Simulation Results: The following parameter setting is used for the performance evaluation. There is one macrocell with radius 500 m underlaid with F femtocells each of radius 20 m, transmitting over $K = 8$ sub-channels. Femto access points have $L = 3$ transmission power levels. The minimum SINR of the macro users is given by $\Gamma_0^{(1)} = \cdots = \Gamma_0^{(K)} = 3$ dB. The transmission power of the macro BS is set to 43 dBm, whereas the femto access point adjusts its power through the various learning schemes to a value of maximum 10 dBm. For the learning through replication by imitation, we set the average review rate to be $r = 0.01$ for all femto access points. For Q-learning and cooperative Q-learning approaches, we set the discount factor $\gamma = 0.95$, exploration probability $\epsilon = 0.1$, and impressionability $\beta = 0.3$. For cooperative Q-learning, cooperation was set to occur after every four intervals of time. The following interference mitigation metric,

$$u\left(p_f(t), p_{-f}(t)\right) = \sum_{n=1}^{N} \log_2\left(1 + \gamma_f^{(k)}\right).\mathbf{1}_{\gamma_0^{(k)} > \Gamma_0^{(k)}}, \tag{9.43}$$

is considered. This metric at a given instant t is different from zero only if the macrocell satisfies at time t, the minimum SINR level required for their own macro users.

Figure 9.13 shows the average femtocell sum-rate for $F = 50$ femto access points underlaying the macrocell with $K = 8$ sub-channels. We observe that replicator dynamics, fictitious play, Q-learning, and cooperative Q-learning schemes eventually converge to some steady state, whereas the replication by imitation converges to zero. This is due to the fact that the reviewing femto access point imitates a random femto access point based on the population distribution of femto access points over the strategy space, and not necessarily the femto access point with better payoff. As a result, the reviewing femto access point has a higher probability of choosing the more "popular" strategy, instead of a performance enhancing strategy. Also, Figure 9.13 shows the plot for average network sum-rate in a dashed line. Even though the performance of the macrocell is degraded over time by the femtocells, the number of femtocells operating within the macrocell area tends to increase the overall network performance.

Figures 9.14 and 9.15 show the effect of femtocell density on the average femtocell sum-rate and average system sum-rate for different learning algorithms,

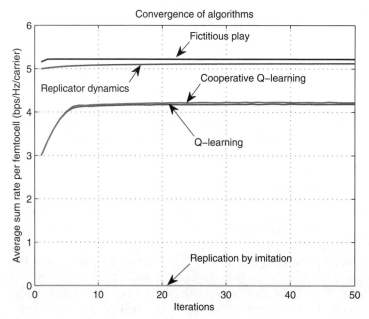

Figure 9.13   Convergence of various learning algorithms and their impact on the average femtocell sum-rate and total average system sum-rate.

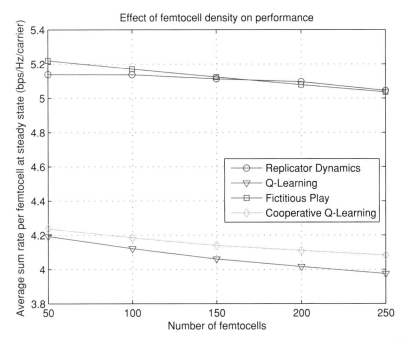

Figure 9.14   Effect of femtocell density on the average femtocell sum-rate (for different learning algorithms).

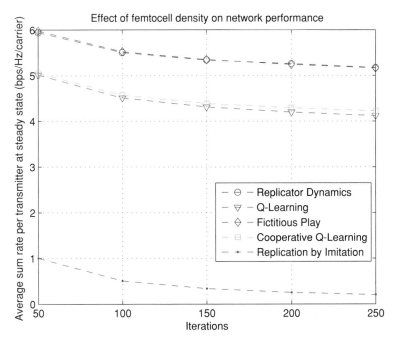

Figure 9.15   Effect of femtocell density on the average femtocell sum-rate (for different learning algorithms).

respectively. The steady state performance is shown for $F = \{50, 100, 150, 200, 250\}$ femtocells. From Figure 9.14, we observe a general decline in performance as the number of femtocells increases. For $F = 250$, a femtocell average sum-rate of around 5 bps/Hz is achieved via replicator dynamics and fictitious play. However, the sum-rate of around 4 bps/Hz is achieved via Q-learning and cooperative Q-learning. This discrepancy might have resulted from the fact that the Q-learning algorithms allow for only two levels of transmission power, that is, on or off. We also observe from Figure 9.14 that cooperative Q-learning clearly outperforms the traditional Q-learning method.

Based on above results, different learning algorithms can be applied to the distributed interference mitigation. The results show that when the information exchange is possible among femto access points, the overall performance and the speed of convergence of the learning algorithms improve. However, there are some open research issues that are worth investigating. For example, channel selection methods based on learning algorithm can be developed. The effect of delay due to information exchange among femto access points on the system performance has to be analyzed.

9.3.4 Coordination Mechanism and Stochastic Geometry Analysis

In a multi-tier small cell network, a coordination mechanism among small cells can be developed so that the power control and sub-channel allocation can be performed effectively. For example, in a femtocell network, femto access points creating co-channel interference to a macro user can be identified and then a coordination mechanism can be applied to such a group of femto access points. This approach is proposed in [44] and analyzed using stochastic geometry [46].

The system model considered in [44] follows the spectrum arrangement as defined in the 3GPP standards. In particular, there are dedicated and shared sub-bands similar to that in [31]. Traditionally, the dedicated sub-bands are used by a macrocell, while the shared sub-bands are used by both femtocells and a macrocell (i.e., "partial frequency band" scheme in Figure 9.7). However, such a spectrum arrangement may result in low spectrum utilization if the macro users are few. Therefore, [44] extends the traditional arrangement by allowing femtocells to access all sub-bands as long as the interference to macro users is acceptable (i.e., "same frequency band" scheme in Figure 9.7). This can be achieved by reducing the transmission power. With this approach, the femto users will enjoy better transmission performance due to more sub-bands for their transmissions.

[44] proposes a method to identify whether or not there is any "victim macro user" nearby femtocells that use the same sub-bands. The macro user is identified to be a victim user if this user experiences an aggregate co-channel interference that is higher than a predefined threshold. In this case, the victim user will send a signaling message to the nearby femto access points. These femto access points can coordinate to adjust their transmission parameters (e.g., reducing transmission power or switching to a different sub-band). To avoid low spectrum utilization, the range of the coordination, which determines the number of coordinated femtocells, should be kept small. Therefore, only the femto access points with strong interference will have

to adapt their transmission parameters. One way to limit the number of femto access points to adapt their transmissions (i.e., through coordination) is that the victim user transmits the signaling message with small transmission power (e.g., 0 dBm). The femto access point will be able to receive the signaling message from the victim user if $d_f^{-\alpha} h_f \geq \Gamma_{th}$, where d_f is the distance from the victim user to femtocell f, α is the path-loss exponent, h_f is the channel fading, and Γ_{th} is the threshold.

The threshold Γ_{th} can divide the femto access points into two groups (i.e., coordinating regions) based on the indicator function defined as

$$1_{d_f^{-\alpha} h_f > \Gamma_{th}} = \begin{cases} 1, & d_f^{-\alpha} h_f > \Gamma_{th} \\ 0, & \text{otherwise} \end{cases} \tag{9.44}$$

The first coordinating region is denoted as $\mathcal{R}_1$. The femto access point is in this region if $1_{d_f^{-\alpha} h_f > \Gamma_{th}} = 1$. Otherwise, the femto access point is in the region denoted by $\mathcal{R}_2$, that is, the femto access point does not receive the signaling message from the victim user. Clearly, these regions $\mathcal{R}_1$ and $\mathcal{R}_2$ are not overlapping and, hence, are statistically independent. An example of these regions is shown in Figure 9.16.

An analytical model based on stochastic geometry is proposed [44]. The entire region of analysis is referred to as the observation region, which is composed of two coordinating regions $\mathcal{R}_1$ and $\mathcal{R}_2$. The observation region has the minimum and maximum radii denoted as D_{min} and D_{max}, respectively (e.g., $D_{min} = 1$ meter and $D_{max} = 100$ meters). The DL transmission is considered. The victim user is assumed to be at the origin and the distance to the macro base station is denoted by d_0. The femto access points are uniformly located in the observation region, where the active femto access points are modeled as the PPP denoted by Φ. The density of

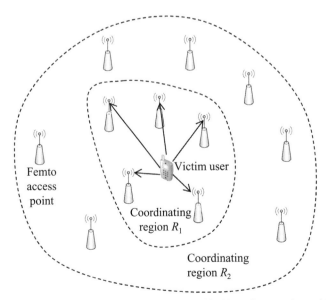

Figure 9.16   Coordinating regions classified based on co-channel interference to a victim user.

femto access points in the observation region is denoted by λ. The fading effect is assumed to follow a random mark. Therefore, the entire process can be modeled as a Marked point process (MPP). Considering the fading effect, the point (i.e., femto access point's location) in the stationary (unmarked) point process Φ can be defined as $\tilde{\Phi} = \{(\phi, h); \phi \in \Phi\}$, where ϕ is the point of the original PPP Φ and h is the channel fading.

The aggregate co-channel interference at the victim user is analyzed based on the cumulants approach [47]. In particular, the aggregate interference is approximated to be the log-normal and shifted log-normal distributions. In this case, the characteristic function (CF) of the aggregate interference is given by

$$\Psi_I(\omega) = E\{\exp(j\omega I)\} \tag{9.45}$$

where I is the total interference from the femto access points and $j = \sqrt{-1}$. The nth cumulant is obtained as follows:

$$\kappa_n = \frac{1}{j^n} \left[\frac{\partial^n}{\partial h^n} \ln \Psi_I(\omega) \right]_{\omega=0}. \tag{9.46}$$

The nth cumulant is needed for obtaining the parameters of the log-normal and shifted log-normal distributions, which are used to approximate the aggregate co-channel interference [44, Eq. (8)].

A few cases are considered in the analysis [44], which are, full interference, OPC, and spectrum reallocation. The full interference case refers to when all femto access points use the same sub-band as that of the victim user. Therefore, this case corresponds to the worst-case scenario. Firstly, in the full interference case, the femto access point transmits with a fixed power denoted by p. The CF of the co-channel interference to the victim user can be expressed as

$$\Psi_I(\omega) = \exp\left(2\pi \int_{\mathcal{H}} \int_{D_{\min}}^{D_{\max}} \left(\exp(jpd^{-\alpha}h) - 1 \right) \lambda f_H(h) d d d h \right) \tag{9.47}$$

where $\mathcal{H}$ is a feasible set of channel fading and $f_H(h)$ is the pdf of channel fading. The nth cumulant is given by

$$\kappa_n = 2\pi\lambda \int_{\mathcal{H}} \int_{D_{\min}}^{D_{\max}} p^n d^{1-n\alpha} h^n f_H(h) d d d h \tag{9.48}$$

$$= \frac{2\pi\lambda p^n}{n\alpha - 2} \left(D_{\min}^{2-\alpha n} - D_{\max}^{2-\alpha n} \right) E_H(h^n) \tag{9.49}$$

where $E_H(h^n)$ is the nth moment of the random variable H for the channel fading.

The next case is the OPC. The transmission power of the femto access points in the coordinating region $\mathcal{R}_1$ will be reduced to $p - 3\text{dB}$, and the new transmission power is denoted by p'. The CF for the co-channel interference from the femto access points in the coordinating region $\mathcal{R}_1$ can be expressed as

$$\Psi_I(\omega) = \exp\left(2\pi \int_{-\infty}^{\infty} \int_{D_{\min}}^{D_{\max}} \left(\exp(jp'd^{-\alpha}h) - 1 \right) \lambda f_H(h) 1_{d^{-\alpha}h > \Gamma_{\text{th}}} d d d d h \right).$$

$$\tag{9.50}$$

The nth cumulant is given by

$$\kappa_n = 2\pi\lambda \int_{\Gamma_{\max}}^{\infty} \int_{D_{\min}}^{D_{\max}} (p')^n d^{1-n\alpha} h^n f_H(h) dd\,dh$$

$$+2\pi\lambda \int_{\Gamma_{\min}}^{\Gamma_{\max}} \int_{D_{\min}}^{(h/\Gamma_{\text{th}})^{1/\alpha}} (p')^n d^{1-n\alpha} h^n f_H(h) dd\,dh \qquad (9.51)$$

where $\Gamma_{\min} = \Gamma_{\text{th}} D_{\min}^{\alpha}$ and $\Gamma_{\max} = \Gamma_{\text{th}} D_{\max}^{\alpha}$. On the other hand, the transmission power of the femto access points in the coordinating region $\mathcal{R}_2$ will remain the same. The CF for the co-channel interference from the femto access points in the coordinating region $\mathcal{R}_2$ can be expressed as

$$\Psi_I(\omega) = \exp\left(2\pi \int_{-\infty}^{\infty} \int_{D_{\min}}^{D_{\max}} \left(\exp(jpd^{-\alpha}h) - 1\right) \lambda f_H(h) 1_{d^{-\alpha}h < \Gamma_{\text{th}}} dd\,dh\right)$$

$$(9.52)$$

where the original transmission power p and the indicator function of the femto access points in the region $\mathcal{R}_2$ (i.e., $1_{d^{-\alpha}h < \Gamma_{\text{th}}}$) are used. Similarly, the nth cumulant is obtained from

$$\kappa_n = 2\pi\lambda \int_{0}^{\Gamma_{\min}} \int_{D_{\min}}^{D_{\max}} p^n d^{1-n\alpha} h^n f_H(h) dd\,dh$$

$$+2\pi\lambda \int_{\Gamma_{\min}}^{\Gamma_{\max}} \int_{(h/\Gamma_{\text{th}})^{1/\alpha}}^{D_{\max}} p^n d^{1-n\alpha} h^n f_H(h) dd\,dh. \qquad (9.53)$$

Then, the aggregate co-channel interference from the femto access points in regions $\mathcal{R}_1$ and $\mathcal{R}_2$ is combined. Since these regions are independently marked, the additivity property of the cumulants can be applied. The resulting cumulant of the total aggregate co-channel interference can be obtained from $\kappa_n(\tilde{\Phi}_1 + \tilde{\Phi}_2) = \kappa_n(\tilde{\Phi}_1) + \kappa_n(\tilde{\Phi}_2)$.

Then, the concept of dynamic exclusive regions is introduced. In particular, the femto access points in the coordinating region $\mathcal{R}_1$ will stop using the dedicated sub-bands and use only the shared sub-bands to reduce the interference to the victim user. On the other hand, the femto access points in the region $\mathcal{R}_2$ do not switch the sub-band. As a result, the aggregate co-channel interference at the victim user is from the femto access points in the region $\mathcal{R}_2$ only. The cumulant of the total aggregate co-channel interference is then given by $\kappa_n(\tilde{\Phi}_1 + \tilde{\Phi}_2) = \kappa_n(\tilde{\Phi}_2)$. In addition, the power control can be used by letting the femto access points in region $\mathcal{R}_2$ reduce their transmission powers. The CF and the nth cumulant can be obtained similar to those in (9.50) and (9.51), respectively.

Simulations are used to verify the correctness of the stochastic geometry-based analytical model. The shifted log-normal distribution is shown to have better accuracy than the log-normal distribution. The coordination mechanism among the femto access points reduce the co-channel interference of the macro users significantly.

9.4 SELF-HEALING

To cope with the failures of network devices (e.g., faulty hardware or software), a small cell network must have the self-healing functionality to provide continuous and efficient services. The traditional approaches to fix the problem in wireless networks are not feasible and cost-effective due to a large number of deployed small cells and complex network structure. Three major steps in self-healing are monitoring, diagnosis, and recovery. A small cell network must be able to continuously monitor and observe the system parameters and performances. These observed parameters and performances are then evaluated to check whether there are or there will be any problems or not. To analyze the system parameters and performances, the conditions can be defined and evaluated. Upon evaluation of the conditions, a trigger will be sent to the related entity, if necessary. For example, the interference of the nearby macro users is measured and reported to the small cell management system if the interference is higher than a threshold. Given the trigger, the system has to diagnose the cause of the problem and optimize the best action. The additional information could be collected, measured, and used in the diagnosis. A recovery step is taken to resolve the problem based on the solution from the diagnosis. The results are observed and used to analyze and adjust the recovery action until the desired network operation state is reached.

In the following, the details of selected issues and approaches related to self-healing are presented.

9.4.1 Collaborative Resource Allocation

When a SBS does not work properly or stops working (i.e., called a faulty small cell), the service availability and performance of the users will be affected. One of the solutions is to let the small cell users switch to a macrocell. However, the quality of the connections to the macro base station could be poor if the users of the faulty small cell are located at the cell edge of the macrocell. Also, the congestion can happen if the macrocell has to accommodate a number of macro users and the users from the faulty small cells. One simple solution is to allow some nearby small cells to serve the users from the faulty small cell. [48] considers this approach and proposes a joint self-healing and self-optimization scheme to provide smooth and seamless service for small cell networks.

Figure 9.17 shows the system model considered for the joint self-healing and self-optimization scheme proposed in [48]. Specifically, in the centralized architecture, femto access points of multiple femtocells are connected with an OAM server. The connection can be through Itf-N [49]. The OAM server can perform an optimization for channel allocation and power control by collecting required parameters from the macro base stations and femto access points. Also, the femto access points can communicate directly with each other through a wireless link (e.g., X2 air interference). Let $\mathcal{K}$ denote a set of available sub-channels and $\mathcal{F}$ denote a set of femtocells. The femtocell $f \in \mathcal{F}$ has N_f users. All femtocells are required to periodically transmit a message (e.g., using a network management protocol) to the server. The faulty femtocell can be detected if such a message is not received by the server. In this case, the users in the faulty femtocell will detect a preamble or pilot signal of nearby femtocells. The nearby femtocells that cooperate with the faulty femtocell will allocate the sub-channels to support the users from the faulty femtocell. Therefore, the sub-channels are divided into two sets denoted by $\mathcal{K}_N$ and $\mathcal{K}_H$ (i.e., $\mathcal{K}_N \cup \mathcal{K}_H = \mathcal{K}$) for the

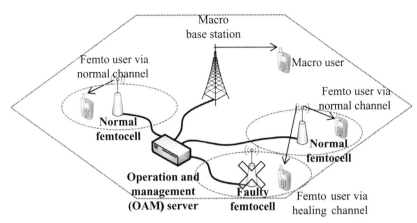

Figure 9.17   Use of healing channel for femto users of faulty femtocell [48].

users in the normal femtocells (i.e., the femtocells that work normally) and in faulty femtocells, respectively. The sub-channels in $\mathcal{K}_H$ are called the healing channels.

When a faulty femtocell is detected, the nearby collaborative femtocells will reallocate the available resources (i.e., sub-channels) to serve both their own users and new users from the faulty femtocell. An optimization problem is formulated for resource allocation in the self-healing and self-optimization scheme [48] with the objective to maximizing the total network throughput, that is,

$$\max \sum_{f\in\mathcal{F}_N}\sum_{k\in\mathcal{K}_N} w_{x(f,k)} R^{(k)}_{f,x(f,k)}(\mathbf{p}^{(k)})$$
$$+ \sum_{f'\in\mathcal{F}_F}\sum_{k'\in\mathcal{K}_H} w_{x(f',k')} R^{(k')}_{f',x(f',k')}(\mathbf{p}^{(k')}) \tag{9.54}$$

$$\text{subject to } \sum_{k\in\mathcal{K}} p_{f,k} \leq p_{\max} \tag{9.55}$$

for $p_{f,k} \geq 0$, where $p_{f,k}$ is the transmission power used by femtocell f on sub-channel k. $\mathcal{K}$ is the set of all sub-channels. $\mathcal{F}_N$ and $\mathcal{F}_F$ are the sets of normal femtocells and faulty femtocells, respectively. $x(f,k)$ denotes the user in femtocell f assigned with sub-channel k. $w_{x(f,k)}$ is the weight representing the priority of the user from femtocell f to use sub-channel k. $R^{(k)}_{f,x(f,k)}(\mathbf{p}^{(k)})$ is the transmission rate of the user in femtocell f transmitting on sub-channel k. $\mathbf{p}^{(k)}$ is composed of the transmission powers of all femtocells on sub-channel k. $p_{\max}$ is the maximum power.

Due to the decision variable $x(f,k)$ (i.e., the user in femtocell f to transmit on sub-channel k), the above optimization problem is complicated to solve (i.e., NP-hard and non-convex). An algorithm is proposed for allocating the sub-channels to users in the normal and faulty femtocells. The major steps of the algorithm are first to identify sub-channels to be the healing channel and second to allocate the transmission power. In the first step, the healing channel is determined to be the worst sub-channels for the users in the normal femtocells (i.e., sub-channel which yields the lowest throughput). These worst sub-channels should be allocated to the faulty femtocells without significantly degrading the performance of the users in the normal femtocells.

The sub-channel allocation algorithm (i.e., for a healing channel) works as follows.

1. The sets of sub-channels for normal and faulty femtocells are initialized as $\mathcal{K}_N = \mathcal{K}$ and $\mathcal{K}_H = \emptyset$.

2. The sub-channels in $\mathcal{K}_N$ are allocated to the users in the normal femtocells. The allocation is performed by solving

$$x(f, k) = \arg\max_y R_{f,y}^{(k)}(\mathbf{p}^{(k)}) \quad \text{for } k \in \mathcal{K}_N \tag{9.56}$$

where y is a user in a normal femtocell.

3. The worst sub-channel k^* is determined from

$$k^* = \arg\min_{k \in \mathcal{K}_N} \left(\frac{U_{x(f,k),k} - U_{x(f,-k),-k}}{U_{x(f,k),k}} \right) \tag{9.57}$$

where $U_{x(f,k),k}$ is the network throughput when the sub-channel k is allocated to a user in each normal femtocell. On the other hand, $U_{x(f,-k),-k}$ is the network throughput when sub-channel k is not allocated to any user. In other words, k^* is the sub-channel that has the smallest contribution to the network throughput.

4. Sub-channel k^* is assigned to be the healing channel. This sub-channel k^* is then allocated to the user in the faulty femtocell with the highest transmission rate, that is,

$$x(f^*, k^*) = \arg\max_{y', f' \in \mathcal{F}_F} R_{f',y'}^{(k^*)}(\mathbf{p}^{(k^*)}) \tag{9.58}$$

where y' is a user in a faulty femtocell.

5. Then, the gain due to assigning sub-channel k^* to be a healing channel and allocated to the user in a faulty femtocell is obtained from

$$g = \frac{U_{x(f^*,k^*),k^*} - U_{x(f,k),k}}{U_{x(f,k),k}} \tag{9.59}$$

where $U_{x(f^*,k^*),k^*}$ is the network throughput when sub-channel k^* is allocated to a user in the faulty femtocell with the highest transmission rate.

6. If the gain g obtained from (9.59) is less than or equal to zero, then there is no throughput surplus from moving sub-channel k^* for the user in a normal femtocell to a user in a faulty femtocell. Therefore, the algorithm terminates. However, if the gain g is larger than zero, there is a throughput surplus, and this sub-channel k^* is moved from a user in a normal femtocell to a user in a faulty femtocell, that is, $\mathcal{K}_N = \mathcal{K}_N \setminus \{k^*\}$ and $\mathcal{K}_H = \mathcal{K}_H \cup \{k^*\}$. Then, go to Step 2.

Now, given the decision variable $x(f, k)$ obtained from the above algorithm, the transmission power of the optimization problem defined in (9.54) and (9.55) can be solved. The Lagrangian (i.e., given the maximum transmission power $p_{\max}$

constraint) is defined as

$$L(\mathbf{p}, \lambda) = \sum_{f \in \mathcal{F}_N} \sum_{k \in \mathcal{K}_N} w_{x(f,k)} R^{(k)}_{f,x(f,k)}(\mathbf{p}^{(k)})$$

$$+ \sum_{f^* \in \mathcal{F}_F} \sum_{k' \in \mathcal{K}_H} w_{x(f^*,k')} R^{(k')}_{f^*,x(f^*,k')}(\mathbf{p}^{(k')})$$

$$\sum_{f \in \mathcal{F}_N} \lambda_f \left(p_{\max} - \sum_{k \in \mathcal{K}} p_{f,k} \right) \tag{9.60}$$

$$\text{subject to } p_{f,k} \geq 0, \tag{9.61}$$

where λ_f is a nonnegative Lagrangian multiplier. By using KKT condition, the optimal transmission power can be obtained in a closed form [48, Eq. (11)], as in a standard power control problem.

Simulations are performed to evaluate the performance of the self-healing and self-optimization scheme. It is shown that when the number of sub-channels as healing channels increases, the fairness among the users can be affected. Specifically, the users in the normal femtocells could receive a lower transmission rate. In this case, the proposed algorithm can dynamically adjust the collaboration among femtocells and, hence, the sub-channels can be assigned and allocated efficiently as healing channels. As a result, the fairness is marginally affected while the network throughput is maximized.

9.4.2 Transfer Learning-Based Diagnosis for Configuration Troubleshooting

In small cell networks such as the femtocell networks, femto access point deployment is usually not as well planned as that of macro base stations. Also, the operation of femto access points can be random (e.g., users can turn on/off and change the operating frequency of the femto access point arbitrarily), making the femtocell network highly dynamic. As a result, inappropriate configuration can cause performance degradation and service disruption problems. This is referred to as the misconfiguration [50]. The misconfiguration happens when the parameter setting of the femto access points is inappropriate and the operation functions are not optimally tuned. The typical misconfiguration problems are usually based on the following factors [50].

- *Frequency:* Femto access points will be deployed to provide femtocell services underlaid to the macrocell. However, the available frequency spectrum is limited and in most cases femtocells and the macrocell will operate on the same frequency. If the frequencies used by the femtocells and the macrocell are not properly chosen, CTI will be severe. Likewise, the co-tier interference can be significant if multiple femtocells use the same frequency. In this case, the femto access points near the macro base station should use different frequencies to avoid performance degradation in the UL transmission for macro users. Also, a group of femtocells should coordinate on the operating frequencies to avoid strong co-tier interference.

- *Transmission power:* The QoS performance in femtocell networks is measured based on the signal-to-interference-plus-noise ratio (SINR). SINR is determined by the transmission powers. The transmission powers need to be chosen optimally. If the transmission power is too low, there will be service outage or service gap between femtocells. However, if the transmission power is too high, there will be strong interference.

- *Mobility management:* Handoff must be performed efficiently. If the mobility management parameter is not properly set, strong interference can occur. For example, when a femto user moves out of the coverage area of any femtocell, this user should be handed over to a macrocell. Otherwise, high transmission power would be required to maintain the connection with the femto access point, creating strong interference to other users.

Some approaches to detect and diagnose the misconfiguration problem exist, for example, using Bayesian networks [51], [52]. However, such approaches require full historical data, which may not be available in the femtocell networks. An alternative approach is considered in [50], the details of which are presented in the following.

[50] introduces the transfer learning-based diagnosis for configuration troubleshooting for femtocell networks. The diagnosis is designed to work with minimum historical data of the target femtocell. Instead, the diagnosis framework presented in [50] can utilize the historical data from other femtocells to identify the problem by using the transfer learning techniques.

Figure 9.18 shows a framework that is a combination of the traditional learning and transfer learning approaches. In the traditional approach, the historical data collected from the target femtocell (i.e., the femtocell to be diagnosed for the problem) is used to build (i.e., train) a diagnosis model. Then, the model is used to predict unknown problems of the target femtocell given the current condition and state of the network. However, this traditional approach requires full historical data for diagnosis, which may not be available due to random deployment and unpredictable operation of the femtocell networks. Therefore, the transfer learning approach is adopted and integrated with the traditional approach. In this approach, the historical data from other femtocells is also used to build the diagnosis model. An example of how the

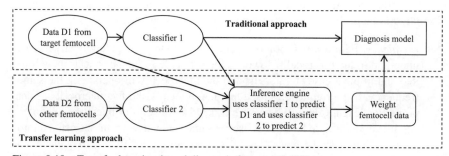

Figure 9.18   Transfer learning-based diagnosis framework [50].

diagnosis framework shown in Figure 9.18 works is as follows. The historical data of the target femtocell D1 and the other femtocell D2 are collected. Then, the classifiers 1 and 2 use data D1 and D2, respectively, to identify the condition and network state. This result is used by the inference engine to predict the events related to data D1 and D2. The result from the inference engine is weighted and used by the diagnosis model to identify the misconfiguration problem.

One example of a diagnosis scenario of the framework shown in Figure 9.18 is to identify whether a call is dropped or blocked. The call is dropped if the interference is too high and the call cannot maintain the SINR at the target level. On the other hand, the call is blocked if there is a service gap (i.e., receive signal is too weak in a particular location). Clearly, only the SINR information collected at the target femtocell cannot be used to identify the problem and to determine the solution. The other information from other femtocells would be required. In this case, the data D1 and D2 are the transmission powers in the target femtocell and the other femtocells, respectively. The classifiers 1 and 2 are used to determine the level of transmission powers. In this case, if the transmission power in the target femtocell is low, while those of other femtocells are high, the call is likely to be dropped (due to high interference). On the other hand, if the transmission powers in the target and other femtocells are low, it is likely that the call is blocked due to insufficient SINR.

An important function of the framework is to extract useful information from other femtocells for diagnosis. A cell-aware transfer scheme is proposed based on the following observations. First, the different misclassification (i.e., when one misconfiguration problem is classified to be another different problem) events have different impacts on the diagnosis model. Second, the network condition and state of the different femtocells could be substantially different. Given these two observations, the misclassification cost and degree of femtocell dissimilarity are defined. In this case, the measurement data (e.g., received signal strength and SINR) has a different distribution and each misconfiguration problem is associated with a particular distribution. As a result, the weight of the measurement data is determined based on the decisiveness. The weights are the divergence associated with a misconfiguration problem. The Kullback–Leibler (K–L) divergence method [53] is adopted. The divergence represents the difference between two distributions defined as

$$D(\omega_{F_1}, \omega_{F_2}) = \sum_{i \in \Theta} \omega_{F_1}(i) \log \left(\frac{\omega_{F_1}(i)}{\omega_{F_2}(i)} \right) + \sum_{i \in \Theta} \omega_{F_2}(i) \log \left(\frac{\omega_{F_2}(i)}{\omega_{F_1}(i)} \right) \quad (9.62)$$

where ω_{F_1} and ω_{F_2} are the distributions of measurement data of two different misconfiguration problems and Θ is the sample space. The weight of each measurement data is defined as the information gain as

$$G(A_i) = H(I) - H(I|A_i) \quad (9.63)$$

where $H(I)$ is the entropy of the misconfiguration instance in the historical data, and $H(I|A_i)$ is the condition entropy of the misconfiguration instance given the measurement data A_i. The misclassification cost between two misconfiguration problems F_1 and F_2 is expressed as

$$C_{F_1,F_2} = \sum_{i \in \Theta} G(A_i) D(A_i^{F_1}, A_i^{F_2}) \tag{9.64}$$

where $A_i^{F_j}$ is the distribution of A_i in the misconfiguration F_j. Then, the dissimilarity for the measurement data of two femtocells f_1 and f_2 is defined as

$$S_{f_1,f_2} = \sum_{i \in \Theta} G(A_i) D(A_i^{f_1}, A_i^{f_2}) \tag{9.65}$$

where $D(A_i^{f_1}, A_i^{f_2})$ is the divergence of the two femtocells f_1 and f_2. This divergence is calculated similar to that in (9.62), except that the distributions are for the measurement data from different femtocells. The misclassification cost and degree of femtocell dissimilarity are used to construct the classifiers. The weights are assigned to the outcomes of the classifiers, which are used to vote and identify the misconfiguration problem given the measurement data. In particular, the misconfiguration problem is identified as that with the largest number of votes from the classifiers.

In the performance evaluation, the following misconfiguration problems are considered, that is, too strong transmission power, too weak transmission power, inappropriate operating frequency, too high handoff threshold, and too small handoff threshold. The diagnosis framework is used to identify the problem whether it is from the strong interference or from the service gap. The support vector machine (SVM) is adopted as the traditional approach (Figure 9.18). The simulation results show that the proposed diagnosis framework based on the transfer learning techniques achieves higher diagnosis accuracy than that of using the traditional approach alone.

9.5 FUTURE RESEARCH DIRECTIONS

In this chapter, we have presented an overview of the self-organizing small cell networks. We have discussed a classification of the self-organization functionalities. Then, we have reviewed the related work on self-configuration, self-optimization, and self-healing. A number of significant research challenges still exist and there are exciting research opportunities, some of which are outlined below.

- *Self-organization in presence of multi-radio access technology (Multi-RAT):* In a Multi-RAT environment, different wireless technologies will be integrated to provide seamless and high performance mobile services. Methods for joint self-configuration of multiple radio access technologies therefore need to be designed. For self-configuration, for example, the coverage of Wi-Fi hotspots and small cells operating in the licensed spectrum can be optimized when the mobile devices are equipped with both the radio interfaces. Also, mobility management and access control across multiple networks need to be designed to achieve an optimal network performance.

- *Cognitive self-organization:* In the existing works, the spectrum is assumed to be always available for macrocells and femtocells to use. However, some licensed spectrum bands can be used opportunistically by macrocells and femtocells to improve the utilization and network performance. For example, TV bands can be opportunistically used by femtocells to avoid interference to macro users using the licensed band. The self-configuration to identify the available licensed spectrum bands and the self-optimization to determine the transmission parameters (e.g., transmission power) can be developed in such a cognitive self-organizing femtocell network.

- *Reliability analysis of small cell networks:* Reliability analysis can be applied to identify the cause, chance, and impact of the failures in small cell networks. By using reliability analysis, the risk of service disruption can be determined, which will be useful for developing self-healing strategies and policies. For a mission-critical small cell deployment in an enterprise environment, redundant SBSs can be deployed to avoid costly service interruption. Optimal redundancy design for small cell networks is worth investigating.

- *Security:* Most of existing work in the literature do not take the security concerns in the self-organization process into account. The attacker can take advantages of the vulnerable self-organizing functionality to damage and harm femtocell networks and femto users. For example, the SBSs must be equipped with a strong authentication method to protect a small cell configuration from being changed arbitrarily by attackers. Impact of jamming and falsified network information collection and monitoring must be taken into account in self-optimization. The self-healing methods must be designed to handle the security attacks in addition to typical failures.

REFERENCES

1. 3GPP TS 32.541, "Telecommunication management; self-organizing networks (SON); Self-healing concepts and requirements," July 2010.
2. O. G. Aliu, A. Imran, M. A. Imran, and B. Evans, "A survey of self organisation in future cellular networks," *IEEE Communications Surveys & Tutorials*, vol. 15, no. 1, pp. 336–361, First Quarter 2013.
3. S. Haykin, "Cognitive radio: Brain-empowered wireless communications," *IEEE Journal on Selected Areas in Communications*, vol. 23, no. 2, pp. 201–220, February 2005.
4. C. Prehofer and C. Bettstetter, "Self-organization in communication networks: principles and design paradigms," *IEEE Communications Magazine*, vol. 43, no. 7, pp. 78–85, July 2005.
5. W. Elmenreich and H. de Meer, "Self-organizing networked systems for technical applications: A discussion on open issues," in *Proceedings of International Workshop on Self-Organizing Systems (IWSOS)*, pp. 1–9, 2008.
6. M. M. S. Marwangi, N. Fisal, S. K. S. Yusof, R. A. Rashid, A. S. A. Ghafar, F. A. Saparudin, and N. Katiran, "Challenges and practical implementation of self-organizing networks

in LTE/LTE-Advanced systems," in *Proceedings of IEEE International Conference on Information Technology and Multimedia (ICIM)*, November 2011.

7. 3GPP TR 36.902 "Evolved universal terrestrial radio access network (E-UTRAN); self-configuring and self-optimizing network use cases and solutions," release 9, December 2009.

8. L. T. W. Ho, I. Ashraf, and H. Claussen, "Evolving femtocell coverage optimization algorithms using genetic programming," in *Proceedings of IEEE International Symposium on Personal, Indoor and Mobile Radio Communications (PIMRC)*, pp. 2132–2136, September 2009.

9. C. H. M. de Lima, M. Bennis, K. Ghaboosi, M. Latva-aho, "Interference management for self-organized femtocells towards green networks," in *Proceedings of IEEE International Symposium on Personal, Indoor and Mobile Radio Communications Workshops (PIMRC Workshops)*, pp. 352–356, September 2010.

10. L. M. del Apio, E. Mino, L. Cucala, O. Moreno, I. Berberana, and E. Torrecilla, "Energy efficiency and performance in mobile networks deployments with femtocells," in *Proceedings of IEEE International Symposium on Personal Indoor and Mobile Radio Communications (PIMRC)*, pp. 107–111, September 2011.

11. Y.-S. Liang, W.-H. Chung, G.-K. Ni, I.-Y. Chen, H. Zhang, and S.-Y. Kuo, "Resource allocation with interference avoidance in OFDMA femtocell networks," *IEEE Transactions on Vehicular Technology*, vol. 61, no. 5, pp. 2243–2255, Jun 2012.

12. J. Kim, H. Kim, K. Cho, and N. Park, "SON and femtocell technology for LTE-Advanced system," in *Proceedings of IEEE International Conference on Wireless and Mobile Communications (ICWMC)*, pp. 286–290, September 2010.

13. 3GPP TS 32.501, "Telecommunication management; self-configuration of network elements; concepts and requirements," March 2010.

14. 3GPP TS 32.521, "Telecommunication management; self-organizing networks (SON) policy network resource model (NRM) integration reference point (IRP); requirements," March 2010.

15. J. Belschner, P. Arnold, H. Eckhardt, E. Kuhn, E. Patouni, A. Kousaridas, N. Alonistioti, A. Saatsakis, K. Tsagkaris, and P. Demestichas, "Optimisation of radio access network operation introducing self-x functions: Use cases, algorithms, expected efficiency gains," in *Proceedings of IEEE Vehicular Technology Conference (VTC Spring)*, April 2009.

16. M. Bennis, L. Giupponi, E. M. Diaz, M. Lalam, M. Maqbool, E. C. Strinati, A. De Domenico, and M. Latva-aho, "Interference management in self-organized femtocell networks: The BeFEMTO approach," in *Proceedings of International Conference on Wireless Communication, Vehicular Technology, Information Theory and Aerospace & Electronic Systems Technology (Wireless VITAE)*, February–March 2011.

17. F. Mhiri, K. Sethom, and R. Bouallegue, "A survey on interference management techniques in femtocell self-organizing networks," *Journal of Network and Computer Applications*, vol. 36, no. 1, pp. 58–65, January 2013.

18. M. Bennis and S. M. Perlaza, "Decentralized cross-tier interference mitigation in cognitive femtocell networks," in *Proceedings of IEEE International Conference on Communications (ICC)*, June 2011.

19. M. E. Sahin, I. Guvenc, M.-R. Jeong, and H. Arslan, "Handling CCI and ICI in OFDMA femtocell networks through frequency scheduling," *IEEE Transactions on Consumer Electronics*, vol. 55, no. 4, pp. 1936–1944, November 2009.

20. P. Xia, V. Chandrasekhar, and J. G. Andrews, "Open vs. closed access femtocells in the uplink," *IEEE Transactions on Wireless Communications*, vol. 9, no. 12, pp. 3798–3809, December 2010.
21. A. Agustin, J. Vidal, and O. Munoz-Medina, "Interference pricing for self-organisation in OFDMA femtocell networks," in *Proceedings of Future Network & Mobile Summit (FutureNetw)*, June 2011.
22. I. W. Mustika, K. Yamamoto, H. Murata, and S. Yoshida, "Potential game approach for self-organized interference management in closed access femtocell networks," in *Proceedings of IEEE Vehicular Technology Conference (VTC Spring)*, May 2011.
23. W. Cheng, L. Hongyan, L. Jiandong, and M. Yinghong, "A self-configuration scheme for power and bandwidth assignment in femtocell networks," in *Proceedings of IEEE International Conference on Signal Processing, Communications and Computing (ICSPCC)*, September 2011.
24. J. Yun, S.-G. Yoon, J.-G. Choi, and S. Bahk, "Contention based scheduling for femtocell access points in a densely deployed network environment," *Computer Networks*, vol. 56, no. 4, pp. 1236–1248, March 2012.
25. S. Park and S. Bahk, "Dynamic inter-cell interference avoidance in self-organizing femtocell networks," in *Proceedings of IEEE International Conference on Communications (ICC)*, June 2011.
26. Y. J. Sang, H. G. Hwang, and K. S. Kim, "A self-organized femtocell for IEEE 802.16e system," in *Proceedings of IEEE Global Telecommunications Conference (GLOBECOM)*, November– December 2009.
27. R. Razavi and H. Claussen, "Self-configuring switched multi-element antenna system for interference mitigation in femtocell networks," in *Proceedings of IEEE International Symposium on Personal Indoor and Mobile Radio Communications (PIMRC)*, pp. 237–242, September 2011.
28. C. Patel, V. Khaitan, S. Nagaraja, F. Meshkati, Y. Tokgoz, and M. Yavuz, "Downlink interference management techniques for residential femtocells," in *Proceedings of IEEE International Symposium on Personal Indoor and Mobile Radio Communications (PIMRC)*, pp. 117–121, September 2011.
29. S. Akbarzadeh, R. Combes, and Z. Altman, "Network capacity enhancement of OFDMA system using self-organized femtocell off-load," in *Proceedings of IEEE Wireless Communications and Networking Conference (WCNC)*, pp. 1234–1238, April 2012.
30. M. Z. Chowdhury, Y. M. Jang, and Z. J. Haas, "Interference mitigation using dynamic frequency re-use for dense femtocell network architectures," in *Proceedings of IEEE International Conference on Ubiquitous and Future Networks (ICUFN)*, pp. 256–261, June 2010.
31. Y. Wu, H. Jiang, and D. Zhang, "A novel coordinated spectrum assignment scheme for densely deployed enterprise LTE femtocells," in *Proceedings of IEEE Vehicular Technology Conference (VTC Spring)*, May 2012.
32. L. T. W. Ho and H. Claussen, "Effects of user-deployed, co-channel femtocells on the call drop probability in a residential scenario," in *Proceedings of IEEE International Symposium on Personal, Indoor and Mobile Radio Communications (PIMRC)*, September 2007.
33. H. Claussen, L. T. W. Ho, and L. G. Samuel, "Self-optimization of coverage for femtocell deployments," in *Proceedings of IEEE Wireless Telecommunications Symposium (WTS)*, pp. 278–285, April 2008.

34. W. B. Langdon and R. Poli, *Foundations of Genetic Programming*, Springer, December 2010.

35. F. Pantisano, M. Bennis, W. Saad, M. Latva-aho, and R. Verdone, "Enabling macrocell–femtocell coexistence through interference draining," in *Proceedings of IEEE Wireless Communications and Networking Conference Workshops (WCNCW)*, pp. 81–86, April 2012.

36. A. Imran, M. Bennis, and L. Giupponi, "Use of learning, game theory and optimization as biomimetic approaches for self-organization in macro-femtocell coexistence," in *Proceedings of IEEE Wireless Communications and Networking Conference Workshops (WCNCW)*, pp. 103–108, April 2012.

37. A. Galindo-Serrano, L. Giupponi, and G. Auer, "Distributed learning in multiuser OFDMA femtocell networks," in *Proceedings of IEEE Vehicular Technology Conference (VTC Spring)*, May 2011.

38. M. Nazir, M. Bennis, K. Ghaboosi, A. B. MacKenzie, and M. Latva-aho, "Learning based mechanisms for interference mitigation in self-organized femtocell networks," in *Proceedings of Asilomar Conference on Signals, Systems and Computers (ASILOMAR)*, pp. 1886–1890, November 2010.

39. M. Bennis, S. Guruacharya, and D. Niyato, "Distributed learning strategies for interference mitigation in femtocell networks," in *Proceedings of IEEE Global Telecommunications Conference (GLOBECOM)*, December 2011.

40. A. L. Stefan, M. Ramkumar, R. H. Nielsen, N. R. Prasad, and R. Prasad, "A QoS aware reinforcement learning algorithm for macro– femto interference in dynamic environments," in *Proceedings of IEEE International Congress on Ultra Modern Telecommunications and Control Systems and Workshops (ICUMT)*, October 2011.

41. S. Namal, K. Ghaboosi, M. Bennis, A. B. MacKenzie, and M. Latva-aho, "Joint admission control & interference avoidance in self-organized femtocells," in *Conference Record of Asilomar Conference on Signals, Systems and Computers (ASILOMAR)*, pp. 1067–1071, November 2010.

42. C. H. M. de Lima, K. Ghaboosi, M. Bennis, A. B. MacKenzie, and M. Latva-aho, "A stochastic association mechanism for macro-to-femtocell handover," in *Conference Record of the Forty Fourth Asilomar Conference on Signals, Systems and Computers (ASILOMAR)*, pp. 1570–1574, November 2010.

43. J.-H. Yun and K. G. Shin, "Adaptive interference management of OFDMA femtocells for co-channel deployment," *IEEE Journal on Selected Areas in Communications*, vol. 29, no. 6, pp. 1225–1241, June 2011.

44. C. H. M. de Lima, M. Bennis, and M. Latva-aho, "Coordination mechanisms for stand-alone femtocells in self-organizing deployments," in *Proceedings of IEEE Global Telecommunications Conference (GLOBECOM)*, December 2011.

45. R. S. Sutton and A. G. Barto, *Reinforcement Learning: An Introduction*, A Bradford Book, March 1998.

46. J. F. C. Kingman, *Poisson Processes*, Oxford University Press, January 1993.

47. A. Ghasemi and E. S. Sousa, "Interference aggregation in spectrum-sensing cognitive wireless networks," *IEEE Journal on Selected Areas in Communications*, vol. 2, no. 1, pp. 205–230, February 2008.

48. K. Lee, H. Lee, and D.-H. Cho, "Collaborative resource allocation for self-healing in self-organizing networks," in *Proceedings of IEEE International Conference on Communications (ICC)*, June 2011.

49. 3GPP TS 32.101, "Telecommunication management; Principles and high level requirements," version 5.5.0, September 2003.

50. W. Wang, J. Zhang, and Q. Zhang, "Transfer learning based diagnosis for configuration troubleshooting in self-organizing femtocell networks," in *Proceedings of IEEE Global Telecommunications Conference (GLOBECOM)*, December 2011.

51. R. Barco, V. Wille, L. Diez, and M. Toril, "Learning of model parameters for fault diagnosis in wireless networks," *Wireless Networks*, vol. 16, no. 1, pp. 255–271, January 2010.

52. R. M. Khanafer, B. Solana, J. Triola, R. Barco, L. Moltsen, Z. Altman, and P. Lazaro, "Automated diagnosis for UMTS networks using Bayesian network approach," *IEEE Transactions on Vehicular Technology*, vol. 57, no. 4, pp. 2451–2461, July 2008.

53. S. Kullback and R. A. Leibler, "On information and sufficiency," *The Annals of Mathematical Statistics*, vol. 22, no. 1, pp. 79–86, March 1951.

CHAPTER 10

RESOURCE ALLOCATION IN MULTI-TIER NETWORKS WITH COGNITIVE SMALL CELLS

10.1 INTRODUCTION

For small cells to succeed, cognition is expected to be a necessary feature that should be implemented in all types of small cells for their efficient operation with limited centralized control. This cognition will enable the small cells to have the SON capabilities (e.g., self-configuration, self-optimization, and self-healing) which will result in low capital expenditure (CAPEX) and operation expenditure (OPEX) for the network operators. For cognitive small cells, the corresponding BSs, that is, SBSs, will be capable of monitoring the surrounding environment, locating major interference sources and avoiding them by opportunistically accessing the orthogonal channels. The concepts of dynamic spectrum access used in cognitive radio can be applied to multi-tier networks with small cells to tackle some of the major challenges of small cell deployment which include the following: limited centralized control, unplanned and massive deployment, limited transmission powers of small cells and huge transmission power gap between macrocells and small cells, coexistence and efficient operation, and requirement for self-organizing capability. Cognitive SBSs should be able to distributively monitor the spectrum and avoid major interference sources by opportunistically accessing the radio channels so that the interference from nearby MBSs and closed access SBSs is avoided. The closed access SBSs (e.g., closed access femtocells) can also be cognitive to avoid using the same frequency band used by nearby MBSs, so that the macro users are not affected by the interference from the nearby closed access SBSs.

The resource allocation in a multi-tier network and the network performance are closely related to the off-loading technique used to divide the traffic load among the different network tiers. In this chapter, we will discuss how to evaluate the share that each network tier serves from the complete set of users and how to optimize the design parameters to off-load users from one network tier to another. We are concerned with traffic off-loading via access network connectivity. That is, we design the multi-tier network such that the users are more oriented to connect to (i.e., associate with) the

Radio Resource Management in Multi-Tier Cellular Wireless Networks, First Edition.
Ekram Hossain, Long Bao Le, and Dusit Niyato

designated network tier, or transmit more of their traffic via the designated network tier. For instance, traffic off-loading to the small cell tier implies that users prefer to connect to the SBSs, and hence, the percentage of the traffic communicated via the small cell tier is increased. Off-loading users to the small cells has two merits, one from the access network and the other from the core network point of view. For the access network, with off-loading, we can decrease the congestion in the macro access network and achieve the optimal balance between the users served by each access network tier. In this way, the blocking probability in the macro tier due to the unavailability of radio resources can be minimized and the utilization of wireless resources can be improved. For the core network, since some small cells (such as femtocells) are IP-backhauled, traffic is off-loaded to the IP network, and therefore, the congestion in the service providers core network decreases.

We will investigate and quantify the off-loading efficiency for three major off-loading techniques in a two-tier network, namely, off-loading via power control, off-loading via increasing the intensity of SBSs, and off-loading via biasing [1]. The concept of biasing was suggested in [2] to motivate the users to associate with the SBSs (femtocells in their case) even if the strongest signal is received from the nearest MBS. The biasing can be viewed as a virtual increase in the relative transmission power of the SBSs to extend their range [3, 4]. In [2], it was assumed that the SBSs and the MBSs aggressively use the same channel. Hence, due to the severe interference from the geographically nearest MBS, biasing off-loads users to the small cell tier at the expense of increased outage probability. However, as will be discussed later, if the SBSs are cognitive, severe interference from the nearby MBS will be avoided and off-loading via biasing would not increase the outage probability of small cell users [5–7]. The off-loading efficiency of macro users to small cells can be quantified through the *tier association probability*. The tier association probability is the probability that a generic user, located at a generic location is associated with a given network tier (i.e., small cell network tier or macro network tier). This is based on the fact that each user associates (connects) with the network entity (i.e., MBS or SBS) which provides the highest instantaneous signal power.

Since the deployment of small cells brings topological randomness in the network, the traditional grid-based modeling approach used for single-tier cellular networks may not be suitable to model and analyze such multi-tier networks. Recently, a new modeling approach based on *Stochastic Geometry* has been used to model and analyze multi-tier cellular networks [8]. In the next section, basics of the stochastic geometry-based modeling will be discussed in the context of cellular wireless networks.

10.2 BACKGROUND

In wireless communications, the signal power decays with the distance between the transmitter and the receiver according to the power law, that is,

$$P_r(y) = P_t(x)Ah_{xy} \|x - y\|^{-\eta} \tag{10.1}$$

where $P_t(x)$ is the transmission power from a transmitter located at $x \in \mathbb{R}^d$, $y \in \mathbb{R}^d$ is the receiver position, h_{xy} is a random variable accounting for the random channel

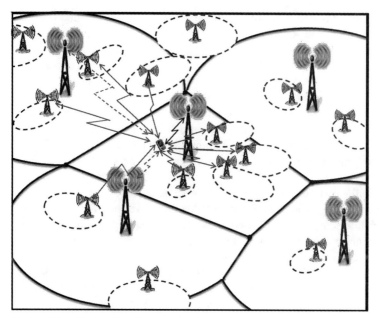

Figure 10.1   Test receiver in a two-tier network.

(power) gain between the two locations x and y, $\|.\|$ is the Euclidean norm, A is a propagation constant, and η is the path-loss exponent.[1] Although (10.1) holds for any number of dimensions, the dimensions $d = 1$, 2, and 3 are of primary interest due to their physical interpretations. Due to the distance-dependent signal power decay along with the shared nature of the wireless medium, the network geometry has a significant impact on the performance of wireless networks. That is, the position of a test receiver with respect to its serving network entity strongly affects the useful signal. On the other hand, the position of the test receiver with respect to other network entities that are simultaneously using the same channel affects the interference signal seen by the test receiver. Therefore, the network geometry has a significant impact on the SINR experienced by the receivers.

The SINR for a test receiver in the network (Figure 10.1) can be calculated as follows:

$$\text{SINR}(y) = \frac{P_t(x_0)Ah_{x_0 y}\|x_0 - y\|^{-\eta}}{W + \sum_{x \in \mathcal{I}} P_t(x)Ah_{xy}\|x - y\|^{-\eta}} \tag{10.2}$$

where y is the location of the test receiver, x_0 is the location of the test transmitter (desired transmitter), $\mathcal{I} = \{x_1, x_2, \ldots\}$ is the set of interferers (active transmitters using the same channel as that of the test transmitter), and W is the noise power. The term $\sum_{x \in \mathcal{I}} P_t(x)Ah_{xy}\|x - y\|^{-\eta} = I_{\text{agg}}$ is the aggregate interference power at the test receiver. Note that according to the network model, $\mathcal{I}$ can be finite or infinite and the locations and the intensity of the interferers depend on the network

---

[1]This model is valid for far field signal power only [9].

characteristics (e.g., network topology and number of channels) and MAC layer protocol (e.g., ALOHA, CSMA, TDMA, and CDMA). At a generic time instant, the SINR experienced by each receiver depends on its location, the positions of the interference sources as well as the instantaneous channel gains. Hence, the SINR is a random variable that strongly depends on the network geometry and significantly varies from one receiver to another and from one time instant to another.

For a generic node in the network, the aggregate interference, $I_{\text{agg}} = \sum_{x \in \mathcal{I}} P_t(x) A h_{xy} \|x - y\|^{-\eta}$ is a stochastic process that depends on the locations of the interferers captured by the point process $\mathcal{I} = \{x_i\}$ and the random channel gains h_{xy}. Note that $\mathcal{I}$ is defined by the network properties and the MAC layer. Generally, there is no known expression for the *pdf* of the aggregate interference in large-scale networks. Hence, the aggregate interference is usually characterized using the LT of the *pdf* (or equivalently its CF or moment generation function [MGF]). The LT of the aggregate interference is given by

$$\mathcal{L}_{I_{\text{agg}}}(s) = \mathbb{E}[e^{-s I_{\text{agg}}}]. \tag{10.3}$$

With the LT, CF, or MGF, we are able to generate the moment (if they exist) of the aggregate interference as $\mathbb{E}[I_{\text{agg}}^n] = (-1)^n \left. \mathcal{L}_{I_{\text{agg}}}^{(n)}(s) \right|_{s=0}$, where $\mathcal{L}_{I_{\text{agg}}}^{(n)}(s)$ is the nth derivative of $\mathcal{L}_{I_{\text{agg}}}(s)$. In the general case, it is not possible to derive the exact performance metrics (e.g., outage probability, transmission capacity, and average achievable rate) from the LT, CF, or the MGF.

Stochastic geometry is a mathematical tool that provides spatial averages, that is, averages taken over large number of nodes at different locations and over many network realizations [10], for the quantities of interest (e.g., interference, SINR, outage probability, and achieved data rate) [11]. In other words, the stochastic geometry averages over all network topologies seen from a generic node weighted by their probability of occurrence [12, 13].

Due to the variation of capacity demand across the network, both single-tier and multi-tier cellular networks are characterized by their random topologies [14, 15]. In this chapter, we will use stochastic geometry for modeling and analysis of a two-tier network. Stochastic geometry is a very powerful tool to model and analyze networks with random topologies, where point processes are used to model the spatial distribution for the network entities [11, 16]. Recently, it has been shown that stochastic geometry provides a tractable yet accurate modeling for cellular networks as well as multi-tier networks [14, 15], where point processes are used to model the spatial distribution of the network entities. The PPP is the most popular and well-understood point process in the literature due to its simplicity and tractability.

A point process in $\mathbb{R}^d$ is a PPP if and only if the number of points inside a bounded Borel set $B \in \mathbb{R}^d$ has a Poisson distribution with a mean directly proportional to the Lebesgue measure of B, and the numbers of points in disjoint Borel sets are independent [16]. That is, the PPP assumes that the positions of the points are uncorrelated. Although the assumption that the positions of the MBSs are uncorrelated is unrealistic, it was shown in [14] that the PPP assumption for the spatial location of the MBSs provides a lower bound on coverage probability (i.e., the complement of the outage probability) and the average achievable rate that is as much tight as the upper bound provided by the idealized grid-based model traditionally used for modeling cellular networks. In [15], it was shown that the PPP assumption is accurate to within 1–2 dB of the performance of an actual LTE network overlaid by heterogeneous tiers modeled as PPP.

Definition 10.1 (Definition 1). (Poisson point process (PPP)): A point process $\Pi = \{x_i; i = 1, 2, 3, \ldots\} \subset \mathbb{R}^d$ is a PPP if and only if the number of points inside any compact set $\mathcal{B} \subset \mathbb{R}^d$ is a Poisson random variable, and the numbers of points in disjoint sets are independent.

When the locations of the interferers are modeled as a PPP, the LT of the *pdf* of aggregate interference can be determined using the following lemma.

Lemma 10.1. Following (sec. 3.7.1, [17]), in a Rayleigh fading environment, the LT of the *pdf* of the aggregate interference measured at the origin from a PPP with intensity λ and existing outside $\mathbf{B}_{a_i}(r_e)$ is given by

$$\mathcal{L}_I(s) = \exp\left\{-\lambda\pi\left((Ps)^{\frac{2}{\eta}}\mathbb{E}_h\left[h^{\frac{2}{\eta}}\Gamma_L\left(1 - \frac{2}{\eta}, sPhr_e^{-\eta}\right)\right] - \frac{Psr_e^2}{Ps + \mu r_e^\eta}\right)\right\}. \quad (10.4)$$

Subsequently, the LT of the *pdf* of aggregate interference above can be used to obtain performance measures such as outage probability, network coverage, and data rate. Under Rayleigh fading, the outage probability for a test receiver can be obtained as follows. Without loss of generality, let $r = \|x_0 - y\|$ be the constant distance between the transmitter and the test receiver, $h_0 \sim \exp(\mu)$ be the channel power gain of the useful link, then we have

$$F_{\text{SINR}}(\theta) = \mathbb{P}\{\text{SINR} \leq \theta\}$$

$$= \mathbb{P}\left\{\frac{P_t A h_0 r^{-\eta}}{W + I_{\text{agg}}} \leq \theta\right\}$$

$$= 1 - \mathbb{P}\left\{h_0 > \frac{(W + I_{\text{agg}})\theta r^\eta}{P_t A}\right\}$$

$$\overset{(i)}{=} 1 - \mathbb{E}\left[\exp\left(-\frac{(W + I_{\text{agg}})\mu\theta r^\eta}{P_t A}\right)\right]$$

$$= 1 - \exp\left(-\frac{W\mu\theta r^\eta}{P_t A}\right)\mathbb{E}\left[\exp\left(-\frac{I_{\text{agg}}\mu\theta r^\eta}{P_t A}\right)\right]$$

$$= 1 - \exp\left(-\frac{W\mu\theta r^\eta}{P_t A}\right)\mathcal{L}_{I_{\text{agg}}}(s)\Big|_{s=\frac{\mu\theta r^\eta}{P_t A}}$$

$$= 1 - \exp(-Wc\theta)\mathcal{L}_{I_{\text{agg}}}(s)\Big|_{s=c\theta} \quad (10.5)$$

where the expectation in (i) is with respect to both the point process and the channel gains between the interference sources and the test receiver, and $c = \frac{\mu r^\eta}{P_t A}$ is a constant.

With channel sensing-based dynamic spectrum access capability, the spatial distribution of the simultaneously active small cells (i.e., cognitive small cells), can be modeled by a Matérn hard core point process (HCPP) [10, 16, 18–20]. A Matérn HCPP is a repulsive point process where no two points can coexist if their distance is less than the hard core radius r_h. The Matérn HCPP is derived from a PPP via dependent thinning. The dependent thinning is applied in two steps. First, an independent uniformly distributed time mark is applied to the PPP. Then, a point is chosen to be in the Matérn HCPP if and only if it has the lowest mark in its contention domain. The contention domain of a point is defined by a circle of radius r_h around that point.

Definition 10.2 (Definition 2). (HCPP): An HCPP is a repulsive point process where no two points of the process coexist with a separating distance less than a predefined hard core parameter r_h. A point process $\Pi = \{x_i; i = 1, 2, 3, \ldots\} \subset \mathbb{R}^d$ is an HCPP if and only if $\|x_i - x_j\| \geq r_h$, $\forall x_i, x_j \in \Pi$, $i \neq j$, where $r_h \geq 0$ is a predefined hard core parameter.

The LT of the *pdf* of the aggregate interference due to an HCPP has always been approximated by the LT of the *pdf* of the aggregate interference due to the PPP with the same intensity but existing outside the contention domain of the test transmitter [19–22]. The rationale behind this approximation is that the main factors affecting the aggregate interference are the number of interferers and their locations with respect to the test node. However, the locations of the interferers with respect to each other have minimal effect on the interference at the test node. The number of interferers has been captured in the calculation of the intensity of the HCPP and the locations of the interferers with respect to the test receiver have been captured by conditioning on having the PPP outside the contention domain of the test transmitter.

10.3 ANALYSIS OF TIER-ASSOCIATION PROBABILITY

10.3.1 System Model and Assumptions

In multi-tier cellular networks, the coverage of each network entity depends on its type (i.e., a macrocell base station [MBS], microcell base station [MiBS], picocell base station [PiBS], or a femto access point [FAP]) and the network geometry (i.e., its location with respect to other network entities). That is, assuming that each user will associate with (is covered by) the network entity that provides the highest signal power, the coverage of each network entity will depend on its transmission power as well as the relative positions of the neighboring network entities and their transmission powers. For instance, if two MBSs have the same transmission power, a line bisecting the distance between them will separate their coverage areas. However, for an MBS with 100 times higher transmission power than an SBS, a line dividing the distance between them with a ratio of 100:1 will separate their coverage areas, and so on. If all tiers are modeled via independent homogenous PPPs, due to the high variation of the transmission power of BSs belonging to different tiers, the multi-tier cellular network coverage will constitute a weighted Voronoi tessellation. The weighted Voronoi tessellation is the planar graph constructed by bisecting the distances between the points of a PPP according to the ratio between

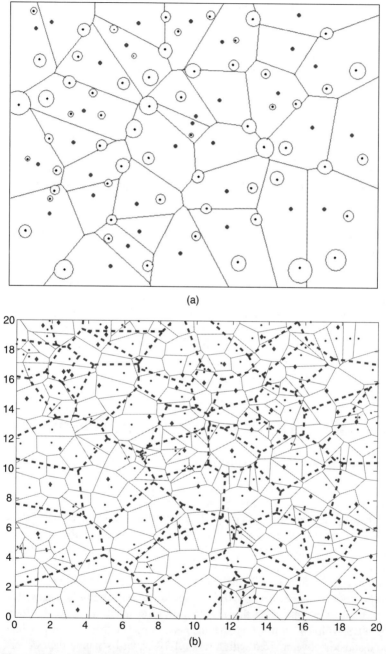

Figure 10.2 (a) The network modeled as a weighted Voronoi tessellation (b) the network modeled as a superposition of two independent Voronoi tessellations (the diamond dots with the dashed Voronoi represent the macro network tier) (© [2012] IEEE).

their weights, where the weights correspond to the transmission powers of the network entities.

Let us consider a two-tier network where MBSs overlaid by FAPs. As the FAPs are deployed randomly according to the end user requirements without any network planning from the cellular network provider. Then, we assume that the two network tiers are independent and each is represented by an independent PPP. That is, the MBSs are spatially distributed according to the PPP $\Psi_b = \{b_i ; i = 1, 2, 3, \ldots\}$ with intensity $\mathcal{B}$ where b_i is the location of the ith MBS. The FAPs are spatially distributed according to the PPP $\Psi_a = \{a_i ; i = 1, 2, 3, \ldots\}$ with intensity $\mathcal{A}$ where a_i denotes the location of the ith FAP. The coverage of each tier forms a Voronoi tessellation [16]. Due to the difference in transmission powers between the MBS and the FAPs, the network model can be represented by a weighted Voronoi tessellation as shown in Figure 10.2(a) [15]. However, since the two tiers are independent, the network model can be considered as a superposition of two independent Voronoi tessellations, one for the MBSs and the other for FAPs as shown in Figure 10.2(b). By construction, the Voronoi cells belonging to the same tier do not intersect. Hence, each user will fall in an intersection between two Voronoi cells belonging to different tiers (i.e., one of an MBS and the other of a FAP). Based on the radio signal strength (RSS) level, each user will be associated to either the MBS or the FAP of the Voronoi cells covering the user. In particular, the user will always be connected to the geographically nearest MBS or to the geographically nearest FAP.

The UEs are spatially distributed according to an independent PPP Ψ_u with intensity $\mathcal{U}$. We consider a general power-law path-loss model in which the signal power decays at the rate $r^{-\eta}$ with the distance r, where $\eta > 2$ is the path-loss exponent. To obtain general results, Nakagami-m fading environment is assumed where $m = 1$ represents Rayleigh fading and $m = \infty$ represents deterministic path-loss. In this chapter, we will use the notation $h_{a_i} \sim Gamma(m_a, \mu_a)$ to denote the channel (power) gain between the test user and FAP a_i, and the notation $h_{b_i} \sim Gamma(m_b, \mu_b)$ to denote the channel (power) gain between the test user and the MBS b_i. The channel power gains have the gamma pdf, $f_h(h) = \frac{h^{m-1} e^{-x/\mu}}{\mu^m \Gamma(m)}$ and cdf $F_h(h) = 1 - \frac{\Gamma_u(m, h/\mu)}{\Gamma(m)}$, where $\Gamma(\cdot)$ is the gamma function, $\Gamma_u(s, x) = \int_x^\infty t^{s-1} e^{-1} dt$ is the upper incomplete gamma function, m is the shape parameter and μ is the scaling parameter.

10.3.2   Tier Association Probability

Based on the instantaneous RSS level, each user will associate with either the MBS or the FAP of the Voronoi cells covering that user. Therefore, each user will only have two candidate network entities to associate with, the geographically nearest MBS and the geographically nearest FAP. Without any loss of generality, the analysis is conducted on a typical user located at the origin. According to Slivnyak's theorem [16], conditioning on having a user at the origin does not change the statistical properties of the coexisting PPPs. Hence, the analysis holds for any generic user located at a generic location [14, 15]. The association probability to the femto network tier and the macro network tier can be obtained from the following lemma.

Lemma 10.2. In a Nakagami-m fading environment, the probability that a generic user is associated with the femto network is given by

$$\mathcal{P}_a = 1 - \int_0^\infty \left(\frac{\mu_a}{\mu_b}\right)^{m_a} \frac{\mathcal{B}h^{m_a-1}}{\beta(m_a, m_b)(1 + \frac{\mu_a}{\mu_b}h)^{m_a+m_b}\left(\left(\frac{P_aT}{P_b}h\right)^{\frac{2}{\eta}}\mathcal{A} + \mathcal{B}\right)}dh \quad (10.6)$$

where $\beta(x, y) = \int_0^1 t^{x-1}(1-t)^{y-1}dt$ is the beta function, and $T \geq 1$ is a biasing factor to bias users to associate with the femto network tier. The probability that a generic user is associated with the macro network is given by

$$\mathcal{P}_b = 1 - \mathcal{P}_a = \int_0^\infty \left(\frac{\mu_a}{\mu_b}\right)^{m_a}$$

$$\times \frac{\mathcal{B}h^{m_a-1}}{\beta(m_a, m_b)(1 + \frac{\mu_a}{\mu_b}h)^{m_a+m_b}\left(\left(\frac{P_aT}{P_b}h\right)^{\frac{2}{\eta}}\mathcal{A} + \mathcal{B}\right)}dh. \quad (10.7)$$

Proof: Let $R_a = \min_i(\|a_i\|)$, $\forall a_i \in \Psi_a$ be the distance from the user located at the origin to the nearest FAP, and $R_b = \min_i(\|b_i\|)$, $\forall b_i \in \Psi_b$ be the distance from the user located at the origin to the nearest MBS. Then, the *cdf* of the distance R_a can be easily derived for the PPP null probability as follows [14, 16]:

$$F_{R_a}(r) = \mathbb{P}\{R_a < r\} = e^{-\mathcal{A}\pi r^2}, \quad r > 0. \quad (10.8)$$

Differentiating the *cdf* of R_a, the *pdf* of R_a is obtained as follows:

$$f_{R_a}(r) = \frac{dF_{R_a}(r)}{dr} = 2\pi\mathcal{A}re^{-\mathcal{A}\pi r^2}, \quad r > 0. \quad (10.9)$$

Similarly, the *pdf* of R_b is obtained as

$$f_{R_b}(r) = 2\pi\mathcal{B}re^{-\mathcal{B}\pi r^2}, \quad r > 0. \quad (10.10)$$

From the system model, we know that there are only two candidate network entities for a generic user to associate with (the nearest MBS and the nearest FAP). Hence, the probability that a test user u_i located at the origin associates with the nearest FAP is given by the probability that the nearest FAP signal power (with biasing) is greater than the received signal power from the nearest MBS.

$$\mathcal{P}_a = \mathbb{P}\{u_i \text{is connected to femto access point}\}$$
$$= \mathbb{P}\left\{P_aTh_{a_i}R_a^{-\eta} > P_bh_{b_i}R_b^{-\eta}\right\}$$
$$= \mathbb{P}\left\{\left(\frac{R_a}{R_b}\right)^\eta < \left(\frac{P_aTh_{a_i}}{P_bh_{b_i}}\right)\right\}$$
$$= \mathbb{P}\left\{R_{a/b} < \left(\frac{P_aT}{P_b}h_{a/b}\right)\right\}$$
$$= \int_h F_{R_{a/b}}\left(\frac{P_aT}{P_b}h\right)f_{h_{a/b}}(h)dh \quad (10.11)$$

where $h_{a/b} = \frac{h_{a_i}}{h_{b_i}}$ and $R_{a/b} = (\frac{R_a}{R_b})^\eta$. The *pdf* of $h_{a/b}$ is given by

$$f_{h_{a/b}}(h) = \left(\frac{\mu_a}{\mu_b}\right)^{m_a} \frac{h^{m_a-1}}{\mathcal{B}(m_a, m_b)(1 + \frac{\mu_a}{\mu_b}h)^{m_a+m_b}},$$

$$0 \leq h < \infty. \tag{10.12}$$

The *cdf* of $R_{a/b}$ is obtained as

$$F_{R_{a/b}}(r) = \mathbb{P}\left\{R_{a/b} < r\right\}$$

$$= \mathbb{P}\left\{\left(\frac{R_a}{R_b}\right)^\eta < r\right\}$$

$$= \int F_{R_a}(r^{\frac{1}{\eta}}R_b) f_{R_b}(R_b) dR_b$$

$$= 1 - \int_0^\infty 2\pi \mathcal{B} R_b e^{-\pi R_b^2 (r^{\frac{2}{\eta}}\mathcal{A} + \mathcal{B})} dR_b$$

$$= 1 - \frac{\mathcal{B}}{r^{\frac{2}{\eta}}\mathcal{A} + \mathcal{B}}, \quad 0 \leq r \leq \infty. \tag{10.13}$$

Substituting (10.12) and (10.13) in (10.11), Equation (10.6) is obtained and the lemma is proved. ∎

The tier association probability can be directly interpreted as the probability that a generic user will associate with one of the two network tiers at a generic time instant. Also, the association probability can be viewed as the percentage of time that a generic user will be connected to a given network tier, or as the percent of traffic that a user communicates through a certain network tier, or as the share that each network tier serves from the complete set of users, or as the portion of the plane which each tier is serving. The off-loading of users can be quantified by using any of these different interpretations for the tier association probability.

From (10.6), it can be shown that the femto tier association probability (and hence off-loading of the users) can be controlled via manipulating three main parameters, namely, the relative transmission power of FAPs, the relative intensity of FAPs by deploying more FAPs, and the biasing factor. The biasing factor is suggested in [2] to bias the users to associate with the FAPs even if the strongest signal is received from the nearest MBS. The biasing can be viewed as a virtual increase in the relative transmission power of FAPs. It is shown in [2] that biasing off-loads users to the femto tier, and hence increases the users' minimum achievable rate at the expense of increased outage probability. The relative channel gain of the users toward the FAPs has also a fundamental impact on the association probability. FAPs are usually deployed at indoor locations. As a result, there will be favorable channel gains toward the FAPs in comparison with the channel gains toward the MBS. The favorable channel gains toward the FAPs also result in a natural biasing of the users to associate with the femto tier as will be evidenced from the numerical results.

Another important quantity of interest is the number of users associated with each network entity. From the interpretations of the tier association probability and the fact that independently thinning a PPP produces another PPP [16], the PPP representing the complete set of users Ψ_u can be divided into two independent PPPs: Ψ_{ua} with intensity $\mathcal{U}_a = U\mathcal{P}_a$ and Ψ_{ub} with intensity $\mathcal{U}_b = U\mathcal{P}_b$ which denote, respectively, the PPP for the users associated to the FAPs and the users associated to the MBSs. As shown in (10.6), off-loading to the FAPs can be achieved via increasing the intensity of the FAPs. However, increasing the intensity of the FAPs will decrease the number of users associated with each FAP. In order to avoid having lots of idle FAPs (i.e., FAPs without any associated users) we have to derive the number of users associated with a generic FAP, which is a discrete random variable. The distribution of the number of users associated to a generic FAP is given by the following lemma.

Lemma 10.3. Given that the intensity of users associated with the femto network tier is $\mathcal{U}_a$ and that each femto user is associated with its geographically nearest FAP, the distribution of the number of users associated with each FAP is given by:

$$v(k) = \sum_{n=0}^{k} \frac{\Gamma(n+c)}{\Gamma(n+1)\Gamma(c)} \frac{(\mathcal{U}_a)^n (c\mathcal{A})^c}{(c\mathcal{A}+\mathcal{U}_a)^{n+c}}, \quad 0 \le k < \infty \quad (10.14)$$

where $c = 3.575$ is a constant.

Proof: Conditioning on a Voronoi cell area V, the number of femto users N_f in that Voronoi cell is a Poisson random variable with the probability mass function (*pmf*) given by

$$\mathbb{P}\{N_f = n\} = \frac{(\mathcal{U}_a V)^n e^{-\mathcal{U}_a V}}{n!}. \quad (10.15)$$

Note that in (10.15) we account only for femto users by considering $\mathcal{U}_a$. The Voronoi cell area V of the FAPs' Voronoi tessellation is a random variable that follows the gamma distribution $f_V(v) = \frac{(\mathcal{A}c)^c v^{c-1} e^{-c\mathcal{A}v}}{\Gamma(c)}$, $0 \le v < \infty$, where $c = 3.575$ is a constant defined for the Voronoi tessellation in the $\mathbb{R}^2$. Therefore, the unconditional *pmf* of N_f is given by

$$f_{N_v}(n) = \int_0^{\infty} \frac{(\mathcal{U}_a v)^n e^{-\mathcal{U}_a v}}{n!} f_V(v) da$$

$$f_{N_v}(n) = \frac{(\mathcal{U}_a)^n}{n!} \int_0^{\infty} v^n e^{-\mathcal{U}_a v} \frac{(\mathcal{A}c)^c v^{c-1} e^{-c\mathcal{A}v}}{\Gamma(c)} dv$$

$$= \frac{(\mathcal{U}_a)^n (\mathcal{A}c)^c}{(n)!\Gamma(c)} \int_0^{\infty} v^{n+c-1} e^{-v(c\mathcal{A}+\mathcal{U}_a)} dv$$

$$\overset{(*)}{=} \frac{(\mathcal{U}_a)^n (\mathcal{B}c)^c}{(n)!\Gamma(c)} \frac{\Gamma(n+c)}{(c\mathcal{A}+\mathcal{U}_a)^{n+c}}$$

$$\underbrace{\int_0^{\infty} \frac{(c\mathcal{A}+\mathcal{U}_a)^{n+c}}{\Gamma(n+c)} v^{n+c-1} e^{-v(c\mathcal{A}+\mathcal{U}_a)} dv}_{c}$$

$$= \frac{\Gamma(n+c)}{\Gamma(n+1)\Gamma(c)} \frac{(\mathcal{U}_a)^n (\mathcal{A}c)^c}{(c\mathcal{A}+\mathcal{U}_a)^{n+c}}$$

where in $(*)$, the integration C evaluates to 1 because it is an integration of a gamma *pdf* over its entire support domain. We will denote the *cdf* of N_v by $\upsilon(k)$ and it is given by

$$\upsilon(k) = \sum_{n=0}^{k} \frac{\Gamma(n+c)}{\Gamma(n+1)\Gamma(c)} \frac{(\mathcal{U}_a)^n (\mathcal{A}c)^c}{(c\mathcal{A}+\mathcal{U}_a)^{n+c}}, \quad 0 \le k < \infty. \qquad (10.16)$$

∎

10.3.3   Numerical Results

Figure 10.3 shows the effect of the different system parameters on the off-loading in a two-tier network. Let λ_s and λ_m denote, respectively, the intensities of SBSs and the MBSs. In Figure 10.3, we start at $\frac{\lambda_s}{\lambda_m} = 1$, $\frac{\mathbb{E}[h_s]}{\mathbb{E}[h_m]} = 1$, and $\frac{P_s T}{P_m} = 0.01$. Then according to the tested off-loading technique, we increase the corresponding off-loading parameter until it reaches 10 times of its initial value. For instance, when testing the off-loading via power control or biasing, we start at $\frac{P_s T}{P_m} = 0.01$. Then we increase $P_s T$ until we reach $\frac{P_s T}{P_m} = 0.1$ (i.e., 10 times of its original value). We do the same for off-loading via increasing the intensity. That is, we start at $\frac{\lambda_s}{\lambda_m} = 1$, then we increase λ_s until we reach $\frac{\lambda_s}{\lambda_m} = 10$. Finally, for the natural off-loading due to the favorable channel gain, we start at $\frac{\mathbb{E}[h_s]}{\mathbb{E}[h_m]} = 1$, then we increase $\mathbb{E}[h_s]$ until we reach $\frac{\mathbb{E}[h_s]}{\mathbb{E}[h_m]} = 10$.

Figure 10.3 shows that increasing the intensity of the SBSs has the greatest impact on the tier association probability. On the other hand, the power control has the least impact on the tier association probability. For instance, increasing the intensity

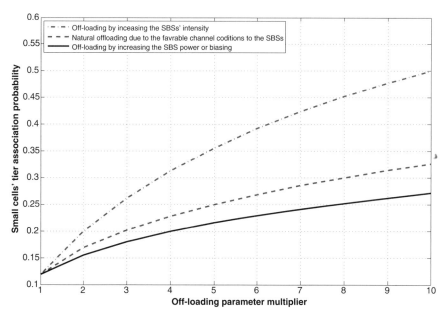

Figure 10.3   The effect of each offloading technique on the tier association probability (© [2013] IEEE).

of SBSs by 5 times increases the small cell tier association probability by 2.92 times. However, increasing the transmission power of SBSs by 5 times increases the small cell tier association probability by only 1.7 times. For more numerical results, refer to [23]. It is worth mentioning that the presented paradigm can be easily extended to evaluate the share of each network tier in a multi-tier network and quantify the off-loading achieved by manipulating different design parameters.

10.4   COGNITION TECHNIQUES FOR SMALL CELLS

The share that each network serves from the complete set of users is not the only important parameter that affects the performance of a multi-tier network. Even more important is how efficiently the users are served within each network tier. As discussed before, mutual interference is the most performance limiting parameter in multi-tier networks and it is infeasible to use traditional centralized techniques to coordinate the spectrum access by a large number of SBSs, allocate their resources, and limit the mutual interference. In this section, we discuss distributed spectrum access via cognitive radio, its design, and its effect on the network performance. A cognitive SBS will not access a channel unless the interference received on that channel from any other network entities is less than the spectrum sensing threshold.

Due to the distance-dependent signal power decay, the spectrum sensing threshold defines an area where no interference source exists (Figure 10.4). That is, a channel used by a network entity (i.e., an MBS or an SBS) located at $x \in \mathbb{R}^2$ can be reused by a cognitive SBS located at $y \in \mathbb{R}^2$ if and only if

$$\|x - y\| \geq \left(\frac{P_{tx} h(x, y)}{\gamma} \right)^{\frac{1}{\eta}} \tag{10.17}$$

where $h(x, y)$ denotes the random channel gain between the two locations x and y, P_{tx} is the transmission power of the network entity located at x, γ is the spectrum sensing threshold, $\|.\|$ is the Euclidean norm, and η is the path-loss exponent. It can be seen from (10.17) that the spectrum sensing threshold is the design parameter that controls the minimum frequency reuse distance $r_e = \left(\frac{P_{tx} h(x,y)}{\gamma} \right)^{\frac{1}{\eta}}$, and hence the spatial reuse efficiency. The higher the value of the spectrum sensing threshold, the lower is the frequency reuse distance and the more aggressive will be the cognitive SBSs in accessing the channels. Hence, the mutual interference increases leading to a higher outage probability, and vice versa. Therefore, there is a trade-off between the spatial frequency reuse efficiency and the outage probability that can be optimized by carefully tuning the spectrum sensing threshold.

Note that the spatial frequency reuse efficiency also translates to an outage probability due to the opportunistic spectrum access by the SBSs. In other words, a cognitive SBS will not access a channel which is being used by the nearby MBS and the SBSs (such a scheme is referred to as **Scheme-1** here); hence, unavailability of radio channels may lead to outage. Spectrum access for cognitive SBS can be implemented in a way similar to the traditional slotted carrier-sense multiple access (CSMA). For example, each time slot can be divided into three main parts (Figure

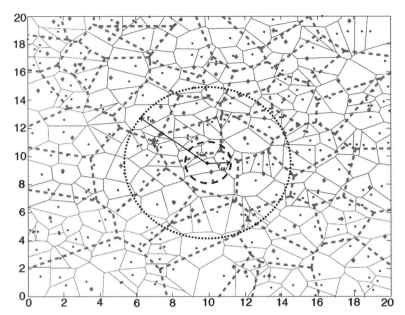

Figure 10.4   Spectrum sensing range for cognitive small cells (r_{sa}, r_{sb} are spectrum sensing range with respect to other small cells and the macrocell, respectively).

10.5(a)). In the first part, each cognitive SBS senses the available spectrum to detect the channels which are not being used by the MBS. Then, in the second part, each cognitive SBS contends to access one of the available channels (e.g., using a random backoff process while persistently sensing the channel). Finally, in the third part, if the sensed channel is available during the entire backoff duration (i.e., not used by a nearby SBS), the cognitive SBS transmits on that channel for the rest of the time slot. Otherwise, the cognitive SBS is considered to be in outage due to channel unavailability.

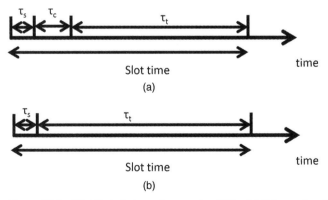

Figure 10.5   Distributed channel access by SBSs: (a) Scheme-1 and (b) Scheme-2 (© [2013] IEEE).

For DL transmission, the total outage probability for a small cell user can be expressed as follows [23]:

$$P_{out} = (1 - \mathbb{P}\{\text{opportunistic access}\}) + \mathbb{P}\{\text{SINR} \leq \beta\}$$
$$\times \mathbb{P}\{\text{opportunistic access}\} \tag{10.18}$$

where SINR is the signal-to-interference-plus-noise ratio and β is the threshold defined for correct signal reception. Note that using stochastic geometry, P_{out} can be obtained in a systematic manner. According to (10.18), the outage could be due to channel unavailability for opportunistic access and/or due to SINR violation (i.e., resulting from aggregate interference). As discussed above, both of these outages are functions of the spectrum sensing threshold. Increasing the spectrum sensing threshold decreases the frequency reuse distance and increases the opportunistic channel access. However, this increases the aggregate interference and hence the SINR outage. On the other hand, decreasing the sensing threshold will increase the frequency reuse distance and hence decrease the aggregate interference, and therefore, decrease the SINR outage. However, the outage due to opportunistic channel access may increase due to increased contention resulting from higher number of SBSs within the frequency reuse distance.

For a given spectrum sensing threshold, since the opportunistic spectrum access performance of the small cells will be deteriorated when the intensity of the deployed SBSs is high, introducing spectrum awareness at an SBS with respect to the spectrum usage at the other SBSs may not be the best solution. Instead, cognition can be introduced only with respect to the macro-tier. That is, each SBS senses the spectrum to locate the channels which are not used by the MBS and uses any of them without considering the other SBSs (**Scheme-2** in Figure 10.5(b)). In this case, for example, each time slot can be divided into two main parts. In the first part, each cognitive SBS senses the available spectrum to detect the channels which are not used by the MBS. Then, in the second part, each cognitive SBS selects one of the available channels and transmits in that channel. Hence, channels will be aggressively used in the small cell tier to increase their opportunistic spectrum access performance at the expense of higher mutual interference in the small cell tier.

Figure 10.6 shows the performance gain in the outage probability obtained by introducing cognition to the SBSs and to compare these two different cognition techniques. Figure 10.6(a) shows that there exists an optimal spectrum sensing threshold that depends on the network parameters and the cognition technique. A higher value of spectrum sensing threshold results in shorter frequency reuse distances and more spectrum opportunities. However, the aggregate interference increases and dominates the outage probability. This results in a degraded outage performance. For very high values of spectrum sensing threshold, the cognitive SBSs become very aggressive and their performance matches with that of the non-cognitive SBSs. On the other hand, lower values of spectrum sensing threshold result in higher frequency reuse distance and lower aggregate interference; however, the opportunistic spectrum access probability decreases and dominates the outage probability. Figure 10.6(b) shows the decomposed outage probability for both the cognition techniques. As shown in the figure, for **Scheme-1**, the decreased

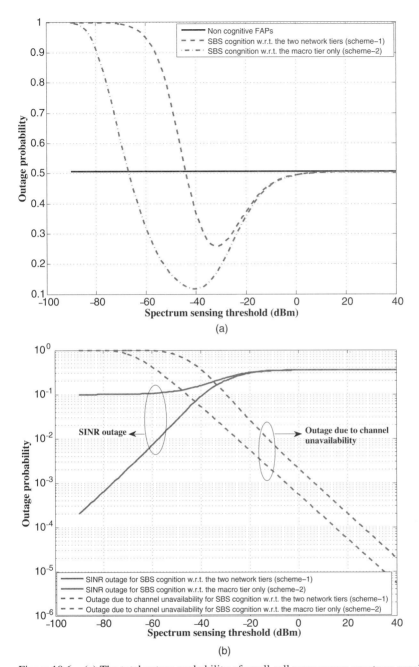

(a)

(b)

Figure 10.6   (a) The total outage probability of small cell users versus spectrum sensing threshold for cognitive techniques for $P_b = 5$ W, $P_s = 100$ dBm, $\lambda_m = 1$ MBS/km^2, $\lambda_s = 10$ SBS/km^2, $\beta = 1$, $\mathbb{E}[h_m] = \mathbb{E}[h_s] = 1$, and $\eta = 4$ and (b) for the same parameters, the decomposed outage probability of small cell users versus spectrum sensing threshold for both the cognitive techniques (© [2013] IEEE).

SINR outage probability is wasted by the outage probability due to the channel unavailability. On the other hand, the degraded SINR outage probability of **Scheme-2** is balanced by the improved spectrum access probability. Importantly, Figure 10.6 shows that cognition is an important feature that can significantly enhance the multi-tier performance and that introducing cognition with respect to the macro-tier only is more beneficial than introducing cognition with respect to both the network tiers. This is due to the massive deployment of the small cells along with the uncoordinated access (i.e., contention) among the coexisting SBSs.

10.5 DISCUSSIONS AND FUTURE RESEARCH DIRECTIONS

The main design challenge for multi-tier networks arises due to the complicated relationship between the design parameters and the performance metrics due to the coexistence of multiple network tiers. To elaborate, we give the following example. Increasing the intensity of the deployed SBSs achieves better performance from higher off-loading from the macro network tier to the small cell tier. Hence, the number of users served per MBS as well as the number of channels used per MBS decreases. Therefore, more channels are available for the SBSs to contend for, which should increase their opportunistic spectrum access performance. This might not be true, because increasing the intensity of SBSs increases the contention in the small cell tier and hence decreases the probability of their successful spectrum access. The same applies for increasing the transmission power for SBSs. Increasing the power of the SBSs increases off-loading of users to the small cell tier. However, it will also increase the contention region for the SBSs and decrease the opportunistic spectrum access probability. From these simple examples, we can see how one design parameter can have contradictory effects on the same performance metric. The same applies for other design parameters and performance metrics. This shows the importance to have a good modeling paradigm that is capable of capturing the interactions among the different design parameters and the performance metrics. It is worth mentioning that the presented concept applies to multi-tier networks including relay nodes as presented in [15]. However, only a two-tier network has been considered here for simplicity.

The two cognition techniques presented in this chapter introduces two extremes for the cognition of SBSs, namely, the contention resolution-based channel access (**Scheme-1**) and the uncoordinated aggressive channel access (**Scheme-2**). The concept of clustering may be used to optimize the trade-off between the outage due to opportunistic spectrum access and outage due to the aggregate interference. In clustering, adjacent SBSs group together and elect a cluster head to coordinate the spectrum access within the cluster. However, many challenges need to be addressed to implement clustering. For instance, what is the optimal cluster size, what information is needed to be exchanged between the cluster members, how to elect the cluster head, and what is the allocation strategy that maximizes the throughput in the small cells while maintaining fairness among the cluster members.

We would like to emphasize that the spectrum sensing threshold is not the only design aspect for cognitive small cells. For instance, designing the sensing duration/strategy is also an important design parameter for cognitive small cells. That is, how long should be the channel sensing duration for an SBS, which channel should it sense first, and should the SBS perform the spectrum sensing and assign channels for its users or should sensing be done cooperatively (i.e., between the SBS and its associated users).

In multi-tier networks, sensing duration and channel access strategy should be determined and updated via learning techniques which are adaptive to topological changes and account for the topology randomness to achieve the SON feature. Another aspect that should also be considered is the resource allocation within each SBS. Mobility also introduces a nontrivial challenge for multi-tier networks due to the lack of coordination and signaling among the MBSs and the coexisting SBSs. Note that, in this chapter, for simplicity, we have assumed a very simple association technique (i.e., based on the received signal strength [RSS]). However, in an actual multi-tier network, the RSS will not be the only aspect considered for association with a given network tier. Instead, the application that the user is using and the required QoS may also play a role in the association decision.

REFERENCES

1. H. ElSawy, E. Hossain, and S. Camorlinga, "Traffic offloading techniques in two-tier femtocell networks," in Proceedings of *IEEE International Conference on Communications (ICC'13)*, Budapest, Hungary, 9–13 June 2013.
2. H. Jo, Y. Sang, P. Xia, and J. Andrews, "Outage probability for heterogeneous cellular networks with biased cell association," in *Proceedings of IEEE Global Communications Conference (Globecom 2011)*, 5–9 December, Houston, TX, USA, 2011.
3. 3GPP TR 36.942, Radio Frequency (RF) system scenarios (Release 10), December 2010.
4. 3GPP TSG RAN WG1, R1-101506, Importance of serving cell selection in heterogeneous networks, Qualcomm Incorporated, San Francisco, February 2010.
5. H. ElSawy and E. Hossain, "Channel assignment and opportunistic spectrum access in two-tier cellular networks with cognitive small cells," in *Proceedings of IEEE Global Communications Conference (Globecom'13)*, Atlanta, USA, 9–13 December 2013.
6. H. ElSawy and E. Hossain, "On cognitive small cells in two-tier heterogeneous networks," in *Proc. of Workshop on Spatial Stochastic Models for Wireless Networks (SpaSWiN)*, in conjunction with *11th Intl. Symposium on Modeling and Optimization in Mobile, Ad Hoc, and Wireless Networks (WiOpt 2013)*, Tsukuba Science City, Japan, May 13–17, 2013.
7. H. ElSawy, E. Hossain, and D. I. Kim, "HetNets with cognitive small cells: User offloading and Distributed channel allocation techniques," *IEEE Communications Magazine*, Special Issue on "Heterogeneous and Small Cell Networks (HetSNets)", vol. 51, no. 6, June 2013.
8. H. ElSawy, E. Hossain, and M. Haenggi, "Stochastic geometry for modeling, analysis, and design of multi-tier and cognitive cellular wireless networks: A survey," *IEEE Communications Surveys and Tutorials*, vol. 15, pp. 996–1019, July 2013.
9. H. Inaltekin, S. B. Wicker, M. Chiang, and H. V. Poor, "On unbounded path-loss models: Effects of singularity on wireless network performance," *IEEE Journal on Selected Areas in Communications*, pp. 1078–1092, 2009.

10. H. Nguyen, F. Baccelli, and D. Kofman, "A stochastic geometry analysis of dense IEEE 802.11 networks," in *Proceedings of 26th IEEE International Conference on Computer Communications (INFOCOM'07)*, May 2007, pp. 1199–1207.

11. M. Haenggi, J. G. Andrews, F. Baccelli, O. Dousse, and M. Franceschetti, "Stochastic geometry and random graphs for the analysis and design of wireless networks," *IEEE Journal on Selected Areas of Communications*, vol. 27, no. 7, pp. 1029–1046, September 2009.

12. M. Haenggi, *Stochastic Geometry for Wireless Networks*. Cambridge University Press, 2012.

13. C.-H. Lee and M. Haenggi, "Interference and outage in Poisson cognitive networks," *IEEE Transactions on Wireless Communications*, vol. 11, pp. 1392–1401, April 2012.

14. J. Andrews, F. Baccelli, and R. Ganti, "A tractable approach to coverage and rate in cellular networks," *IEEE Transactions on Communications*, vol. 59, no. 11, pp. 3122–3134, November 2011.

15. H. Dhillon, R. Ganti, F. Baccelli, and J. Andrews, "Modeling and analysis of K-tier downlink heterogeneous cellular networks," *IEEE Journal on Selected Areas in Communications*, vol. 30, no. 3, pp. 550–560, April 2012.

16. D. Stoyan, W. Kendall, and J. Mecke, *Stochastic Geometry and Its Applications*. John Wiley & Sons, 1995.

17. M. Haenggi and R. Ganti, "Interference in large wireless networks," in Foundations and Trends in Networking. NOW Publishers, 2008, vol. 3, no. 2, pp. 127–248. Available: http://www.nd.edu mhaenggi/pubs/now.pdf

18. H. ElSawy and E. Hossain, "Modeling random CSMA wireless networks in general fading environments," in *Proceedings of IEEE International Conference on Communications (ICC 2012)*, Ottawa, Canada, 10–15 June 2012.

19. H. ElSawy, E. Hossain, and S. Camorlinga, "Characterizing random CSMA wireless networks: A stochastic geometry approach," in *Proceedings of IEEE International Conference on Communications (ICC 2012)*, Ottawa, Canada, 10–15 June 2012.

20. Y. Kim, F. Baccelli, and G. de Veciana, "Spatial reuse and fairness of mobile ad-hoc networks with channel-aware CSMA protocols," in *Proceedings of 17th Workshop on Spatial Stochastic Models for Wireless Networks*, May 2011.

21. M. Kaynia, N. Jindal, and G. Oien, "Improving the performance of wireless ad hoc networks through MAC layer design," *IEEE Transactions on Wireless Communications*, vol. 10, no. 1, pp. 240–252, January 2011.

22. M. Haenggi, "Mean interference in hard-core wireless networks," *IEEE Communications Letters*, vol. 15, pp. 792–794, August 2011.

23. H. ElSawy and E. Hossain, "Two-tier HetNets with cognitive femtocells: Downlink performance modeling and analysis in a multi-channel environment," *IEEE Transactions on Mobile Computing*, to appear.

24. H. Liang, S. Cho, and X. Li, "How to support local IP access from the femtocell," in *IEEE International Conference on Communications Technology and Applications (ICCTA)*, Beijing, October 2009.

INDEX

Radio Resource Management in Multi-Tier Cellular Wireless Networks, First Edition.
Ekram Hossain, Long Bao Le, and Dusit Niyato
© 2014 John Wiley & Sons, Inc. Published 2014 by John Wiley & Sons, Inc.

Wiley Series on
Adaptive and Cognitive Dynamic Systems

Editor: Simon Haykin